AF178126

Guide to Computational Geometry Processing

Jakob Andreas Bærentzen · Jens Gravesen ·
François Anton · Henrik Aanæs

Guide to Computational Geometry Processing

Foundations, Algorithms, and Methods

 Springer

Jakob Andreas Bærentzen
Department of Informatics and
    Mathematical Modelling
Technical University of Denmark
Kongens Lyngby, Denmark

Jens Gravesen
Department of Mathematics
Technical University of Denmark
Kongens Lyngby, Denmark

François Anton
Department of Informatics and
    Mathematical Modelling
Technical University of Denmark
Kongens Lyngby, Denmark

Henrik Aanæs
Department of Informatics and
    Mathematical Modelling
Technical University of Denmark
Kongens Lyngby, Denmark

ISBN 978-1-4471-4074-0   ISBN 978-1-4471-4075-7 (eBook)
DOI 10.1007/978-1-4471-4075-7
Springer London Heidelberg New York Dordrecht

Library of Congress Control Number: 2012940245

© Springer-Verlag London 2012
This work is subject to copyright. All rights are reserved by the Publisher, whether the whole or part of
the material is concerned, specifically the rights of translation, reprinting, reuse of illustrations, recitation,
broadcasting, reproduction on microfilms or in any other physical way, and transmission or information
storage and retrieval, electronic adaptation, computer software, or by similar or dissimilar methodology
now known or hereafter developed. Exempted from this legal reservation are brief excerpts in connection
with reviews or scholarly analysis or material supplied specifically for the purpose of being entered
and executed on a computer system, for exclusive use by the purchaser of the work. Duplication of
this publication or parts thereof is permitted only under the provisions of the Copyright Law of the
Publisher's location, in its current version, and permission for use must always be obtained from Springer.
Permissions for use may be obtained through RightsLink at the Copyright Clearance Center. Violations
are liable to prosecution under the respective Copyright Law.
The use of general descriptive names, registered names, trademarks, service marks, etc. in this publication
does not imply, even in the absence of a specific statement, that such names are exempt from the relevant
protective laws and regulations and therefore free for general use.
While the advice and information in this book are believed to be true and accurate at the date of publication, neither the authors nor the editors nor the publisher can accept any legal responsibility for any
errors or omissions that may be made. The publisher makes no warranty, express or implied, with respect
to the material contained herein.

Printed on acid-free paper

Springer is part of Springer Science+Business Media (www.springer.com)

Preface

This book grew out of a conversation between two of the authors. We were discussing the fact that many of our students needed a set of competencies, which they could not really learn in any course that we offered at the Technical University of Denmark. The specific competencies were at the junction of computer vision and computer graphics, and they all had something to do with "how to deal with" discrete 3D shapes (often simply denoted "geometry").

The tiresome fact was that many of our students at graduate level had to pick up things like registration of surfaces, smoothing of surfaces, reconstruction from point clouds, implicit surface polygonization, etc. on their own. Somehow these topics did not quite fit in a graphics course or a computer vision course. In fact, just a few years before our conversation, topics such as these had begun to crystallize out of computer graphics and vision forming the field of geometry processing. Consequently, we created a course in computational geometry processing and started writing a set of course notes, which have been improved over the course of a few years, and now, after some additional polishing and editing, form the present book.

Of course, the question remains: why was the course an important missing piece in our curriculum, and, by extension, why should anyone bother about this book?

The answer is that optical scanning is becoming ubiquitous. In principle, any technically minded person can create a laser scanner using just a laser pointer, a web cam, and a computer together with a few other paraphernalia. Such a device would not be at the 20 micron precision which an industrial laser scanner touts these days, but it goes to show that the principles are fairly simple. The result is that a number of organizations now have easy access to optical acquisition devices. In fact, many individuals have too—since the Microsoft Kinect contains a depth sensing camera. Geometry also comes from other sources. For instance, medical CT, MR and 3D ultrasound scanners provide us with huge volumetric images from which we can extract surfaces.

However, often we cannot directly use this acquired geometry for its intended purpose. Any measurement is fraught with error, so we need to be able to filter the geometry to reduce noise, and usually acquired geometry is also very verbose and simplification is called for. Often we need to convert between various representations, or we need to put together several partial models into one big model. In other words, raw acquired geometry needs to be processed before it is useful for some envisioned purpose, and this book is precisely about algorithms for such processing of geometry as is needed in order to make geometric data useful.

Overview and Goals

Geometry processing can loosely be defined as the field which is concerned with how geometric objects (points, lines, polygons, polyhedra, smooth curves, or smooth surfaces) are worked upon by a computer. Thus, we are mostly concerned with algorithms that work on a (large) set of data. Often, but not necessarily, we have data that have been acquired by scanning some real object. Dealing with laser scanned data is a good example of what this book is about, but it is by no means the only example.

We could have approached the topic by surveying the literature within the topics covered by the book. That would have led to a book giving an overview of the topics, and it would have allowed us to cover more methods than we actually do. Instead, since we believe that we have a relatively broad practical experience in the areas, we have chosen to focus on methods we actually use, cf. Chap. 1. Therefore, with very few exceptions, the methods covered in this book have been implemented by one or more of the authors. This strategy has allowed us to put emphasis on what we believe to be the core tools of the subject, allowing the reader to gain a deeper understanding of these, and, hopefully, made the text more accessible. We believe that our strategy makes this book very suitable for teaching, because students are able to implement much of the material in this book without needing to consult other texts.

We had a few other concerns too. One is that we had no desire to write a book which was tied to a specific programming library or even a specific programming language, since that tends to make some of the information in a book less general. On the other hand, in our geometry processing course, we use C++ for the exercises in conjunction with a library called GEL[1] which contains many algorithms and functions for geometry processing. In this book, we rarely mention GEL except in the exercises, where we sometimes make a note that some particular problem can be solved in a particular way using the GEL library.

In many ways this is a practical book, but we aim to show the connections to the mathematical underpinnings: Most of the methods rely on theory which it is our desire to explain in as much detail as it takes for a graduate student to not only implement a given method but also to understand the ideas behind it, its limitations and its advantages.

Organization and Features

A problem confronting any author is how to delimit the subject. In this book, we cover a range of topics that almost anyone intending to do work in geometry processing will need to be familiar with. However, we choose not to go into concrete

---

[1]C++ library developed by some of the authors of this book and freely available. URL provided at the end of this preface.

applications of geometry processing. For instance, we do not discuss animation, deformation, 3D printing of prototypes, or topics pertaining to (computer graphics) rendering of geometric data. In the following, we give a brief overview of the contents of the book.

Chapter 1 contains a brief overview of techniques for acquisition of 3D geometry and applications of 3D geometry.

Chapters 2–4 are about mathematical theory which is used throughout the rest of the book. Specifically, these chapters cover vector spaces, metric space, affine spaces, differential geometry, and finite difference methods for computing derivatives and solving differential equations. For many readers these chapters will not be necessary on a first reading, but they may serve as useful points of reference when something in a later chapter is hard to understand.

Chapters 5–7 are about geometry representations. Specifically, these chapters cover polygonal meshes, splines, and subdivision surfaces.

Chapter 8 is about computing curvature from polygonal meshes. This is something often needed either for analysis or for the processing algorithms described in later chapters.

Chapters 9–11 describe algorithms for mesh smoothing, mesh parametrization, and mesh optimization and simplification—operations very often needed in order to be able to use acquired geometry for the intended purpose.

Chapters 12–13 cover point location databases and convex hulls of point sets. Point databases (in particular kD trees) are essential to many geometry processing algorithms, for instance registration. Convex hulls are also needed in numerous contexts such as collision detection.

Chapters 14–18 are about a variety of topics that pertain to the reconstruction of triangle meshes from point clouds: Delaunay triangulation, registration of point clouds (or meshes), surface reconstruction using scattered data interpolation (with radial basis functions), volumetric methods for surface reconstruction and the level set method, and finally isosurface extraction. Together, these chapters should provide a fairly complete overview of the algorithms needed to go from a raw set of scanned points to a final mesh. For further processing of the mesh, the algorithms in Chaps. 9–11 are likely to be useful.

Target Audience

The intended reader of this book is a professional or a graduate student who is familiar with (and able to apply) the main results of linear algebra, calculus, and differential equations. It is an advantage to be familiar with a number of more advanced subjects, especially differential geometry, vector spaces, and finite difference methods for partial differential equations. However, since many graduate students tend to need a brush up on these topics, the initial chapters cover the mathematical preliminaries just mentioned.

The ability to program in a standard imperative programming language such as C++, C, C#, Java or similar will be a distinct advantage if the reader intends to put the material in this book to actual use. Provided the reader is familiar with such a programming language, he or she should be able to implement many of the methods presented in this book. The implementation will, however, be much easier if a library of basic data structures and algorithms for dealing with linear algebra and geometric data is available.

Supplemental Resources

At the web page of this book, http://www.springer.com/978-1-4471-4074-0, we provide three types of supplementary material.
1. Data for exercises. This comprises point sets and polygonal meshes suitable for solving some of the exercise problems which are listed at the end of each chapter.
2. The GEL library. GEL is an abbreviation for *Geometry and Linear algebra Library*—a collection of C++ classes and functions distributed as source code. GEL is useful for geometry processing and visualization tasks in particular and most of the algorithms in this book have been implemented on top of GEL.
3. Example C++ programs. Readers interested in implementing the material in this book using GEL will probably find it very convenient to use our example programs. These programs build on GEL and should make it easy and convenient to get started. The example programs are fairly generic, but for all programming one of the examples should serve as a convenient starting point.

Notes to the Instructor

As mentioned above, the first three chapters in this book are considered to be prerequisite material, and would typically not be part of a course syllabus. For instance, we expect students who follow our geometry processing course to have passed a course in differential geometry, but experience has taught us that not all come with the prerequisites. Therefore, we have provided the four initial chapters to give the students a chance to catch up on some of the basics.

In general, it might be a good idea to consider the grouping of chapters given in the overview above as the "atomic units". We do have references from one chapter to another, but the chapters can be read independently. The exception is that Chap. 5 introduces many notions pertaining to polygonal meshes without which it is hard to understand many of the later chapters, so we recommend that this chapter is not skipped in a course based on this book.

GEL is just one library amongst many others, but it is the one we used in the exercises from the aforementioned course. Since we strongly desire that the book should not be too closely tied to GEL and that it should be possible to use this book with other packages, no reference is made to GEL in the main description of each exercise, but in some of the exercises you will find paragraphs headed by

   [GEL Users]
   These paragraphs contain notes on material that can be used by GEL users.

Acknowledgements

A number of 3D models and images have been provided by courtesy of people or organizations outside the circle of authors.

- The Stanford bunny, provided courtesy of The Stanford Computer Graphics Laboratory, has been used in Chaps. 9, 11, 16, and 17. In most places Greg Turk's reconstruction (Turk and Levoy, Computer Graphics Proceedings, pp. 311–318, 1994) has been used, but in Chap. 17, the geometry of the bunny is reconstructed from the original range scans.
- The 3D scans of the Egea bust and the Venus statue both used in Chap. 11 are provided by the AIM@SHAPE Shape Repository.
- Stine Bærentzen provided the terrain data model used in Chaps. 9 and 11.
- Rasmus Reinhold Paulsen provided the 3D model of his own head (generated from a structured light scan), which was used in Chap. 11.
- In Fig. 10.3 we have used two pictures taken from Wikipedia. The Mercator projection by Peter Mercator, http://en.wikipedia.org/wiki/File:MercNormSph.png and Lambert azimuthal equal-area projection by Strebe, http://en.wikipedia.org/wiki/File:Lambert_azimuthal_equal-area_projection_SW.jpg.
- In Fig. 12.4, we have used the 3D tree picture taken from Wikipedia, http://en.wikipedia.org/wiki/File:3dtree.png.
- In Fig. 12.10, we have used the octree picture taken from Wikipedia, http://fr.wikipedia.org/wiki/Fichier:Octreend.png.
- In Fig. 12.11, we have used the r-tree picture taken from Wikipedia, http://en.wikipedia.org/wiki/File:R-tree.svg.
- In Fig. 12.12, we have used the 3D r-tree picture taken from Wikipedia, http://en.wikipedia.org/wiki/File:RTree-Visualization-3D.svg.

We would like to acknowledge our students, who, through their feedback, have helped us improve the course and the material which grew into this book. We would also like to thank the company 3Shape. Every year, we have taken our class to 3Shape for a brief visit to show them applications of the things they learn. That has been very helpful in motivating the course and, thereby, also the material in this book.

Research and university teaching is becoming more and more of a team sport, and as such we would also like to thank our colleagues at the Technical University of Denmark for their help and support in the many projects where we gained experience that has been distilled into this book.

Last but not least, we would like to thank our families for their help and support.

Kongens Lyngby, Denmark   Jakob Andreas Bærentzen
                                                        Jens Gravesen
                                                        François Anton
                                                        Henrik Aanæs

Contents

List of Notations

Natural numbers	$\mathbb{N}$
Integers	$\mathbb{Z}$
Rational numbers	$\mathbb{Q}$
Real numbers	$\mathbb{R}$
Spaces	$X, Y, \ldots$
Points	$\mathbf{p}, \mathbf{q}, \mathbf{r}, \mathbf{x}, \mathbf{y}, \mathbf{z}$
Vector spaces	T, U, V
Vectors	$\mathbf{e}, \mathbf{f}, \mathbf{u}, \mathbf{v}, \mathbf{w}$
Coordinates of vectors	$\underline{\mathbf{e}}, \underline{\mathbf{f}}, \underline{\mathbf{u}}, \underline{\mathbf{v}}, \underline{\mathbf{w}}$
Linear map	L
Matrices	$\mathbf{A}, \mathbf{B}, \mathbf{C}, \ldots$
Orthogonal matrices	$\mathbf{U}, \mathbf{V}$
Diagonal matrices	$\boldsymbol{\Lambda}$
Functions	f, g, h
Polynomials	p, q
Radial basis function	ψ
Function interpolating scattered data points	s
B-spline	N_ℓ^n
Inner product	$\langle \cdot, \cdot \rangle$
Norm (2-norm unless otherwise stated)	$\| \cdot \|$
Absolute value	$| \cdot |$
Curve	C
Curve parameter	t
Curve parameterization	$\mathbf{r}(t)$
Tangent	$\mathbf{t}$
Curvature vector	$\boldsymbol{\kappa}$
Curvature	κ
Surface	S
Surface parameters	(u, v)
Surface parameterization	$\mathbf{x}(u, v)$
Tangent space	$T_{\mathbf{x}} S$
Surface normal	$\mathbf{n}$
Weingarten map	W
Matrix representation of Weingarten map	$\mathbf{W}$

Shape operator	$\mathbf{S} = -\mathbf{W}$
First fundamental form	I
Matrix representation of first fundamental form	$\mathbf{I} = \begin{bmatrix} g_{11} & g_{12} \\ g_{21} & g_{22} \end{bmatrix}$
Second fundamental form	$\mathbb{II}$
Matrix representation of second fundamental form	$\mathbf{II} = \begin{bmatrix} b_{11} & b_{12} \\ b_{21} & b_{22} \end{bmatrix}$
Normal curvature	κ_n
Geodesic curvature	κ_g
Principal curvatures	κ_1, κ_2
Gaussian curvature	K
Mean curvature	H
Mean curvature normal	$\mathbf{H}$
Signed distance function	d
Scalar field	Φ
Differential	d
Energy	E
Gradient	∇
Laplacian	$\triangle$
Forward difference operator	D^+
Backward difference operator	D^-
Second order central difference operator	D^2
Set of Faces	$\mathcal{F}$
Set of Edges	$\mathcal{E}$
Set of Vertices	$\mathcal{V}$
Set of vertices that are neighbors to vertex i	$\mathcal{N}_i$
Manifold	M
Partition	P

Introduction

Invoking Moore's law and the long term exponential growth of computing power as the underlying reasons for why a particular research field has emerged is perhaps a bit of a cliché. Nevertheless, we cannot get around it here. Computational geometry processing is about practical algorithms that operate on geometric data sets, and these data sets tend to be rather large if they are to be useful. Processing a big polygonal mesh, say a triangulated terrain, an isosurface from a medical volume, or a laser scanned object, would generally not be feasible given a PC from the early 1980s with its limited computational power and a hard disk of around 10 MB. Even in the late 1990s, large geometric data sets might require numerous hard disks. For instance, the raw scans of Michaelangelo's David as carried out by Marc Levoy and his students during the Digital Michelangelo project [1] required 32 GB of space—more than a typical hard disk at the time. However, since then the world has seen not only a sharp decrease in the price of computation and storage but also a proliferation of equipment for acquiring digital models of 3D shapes, and in 2003 also the Symposium on Geometry Processing which was founded by Leif Kobbelt.

Due to its practical nature, geometry processing is a research field which has strong ties to numerous other fields. First of all, computer graphics and computer vision are probably the fields that have contributed most to geometry processing. However, many researchers and practitioner in other fields confront problems of a geometric nature and have at their disposal apparatus which can measure 3D geometric data. The first task is to convert these data to a useable form. This is often (but not always) a triangle mesh. Next, since any type of measurement is fraught with error, we need algorithms for removing noise from the acquired objects. Typically, acquired 3D models also contain a great deal of redundancy, and algorithms for geometry compression or simplification are also important topics in geometry processing. Moreover, we need tools for transmission, editing, synthesis, visualization, and parametrization; and, of course, this is clearly not an exhaustive list. Painting with rather broad strokes, we see the goal of geometry processing as to provide the tools needed in order to analyze geometric data in order to answer questions about

J.A. Bærentzen et al., *Guide to Computational Geometry Processing*, DOI 10.1007/978-1-4471-4075-7_1, © Springer-Verlag London 2012

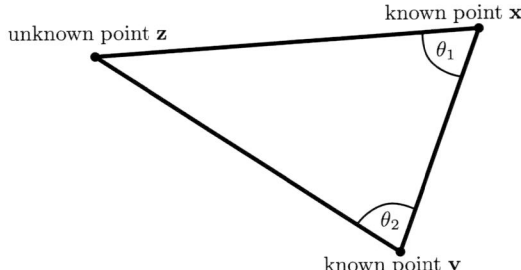

Fig. 1.1 An illustration of basic optical triangulation. Assume that unknown point **z** is observed from points **x** and **y**, of which we know the positions. If we also know the angles θ_1 and θ_2, then point **z**'s position can be determined by simple trigonometry. The observations of point **z** in points **x** and **y** can e.g. be made by cameras or photogrammetrists on hilltops

the real world or to transform it into a form where it can be used as a digital prototype or as digital content for, e.g., geographical information systems, virtual or augmented reality, or entertainment purposes.

The goal of this chapter is to present a selection of the domains in which geometry processing is used. During this overview, we will also discuss methods for acquiring the geometric data and refer to the chapters where we discuss the topics in detail.

1.1   From Optical Scanning to 3D Model

Acquisition of 3D data can be done in a wide variety of ways. For instance, a *touch probe* allows a user to interactively touch an object at various locations. The probe is connected to a stationary base via an articulated arm. Knowing the lengths and angles of this arm, we can compute the position of the tip and hence a point on the object. Clearly, this is a laborious process. There are also automated mechanical procedures for acquisition, but due to the speed and relative ease with which it can be implemented, optical scanning has emerged as the most used method for creating digital 3D models from physical objects.

Almost all optical scanning procedures are based on *optical triangulation*, cf. Fig. 1.1, which is the core of classical photogrammetry, and also of much of navigation. Today, optical triangulation is often done via camera technology. If we observe the same 3D point in the images from two different cameras with known positions and optics, we have a situation like that illustrated in Fig. 1.1: From the relative positions and orientations of the two cameras combined with the positions in the images of the observed points, we can compute the two angles θ_1 and θ_2 and, consequently, the position of the unknown point **z**. Expressed differently: given a camera of known orientation and position, a point in the image produced by the camera corresponds to a line in space. Thus, if we observe the same 3D point in two cameras, we can

Fig. 1.2 To help establish correspondences between images, laser light can e.g. be shone on the object, as illustrated here

find the location of that 3D point as the intersection of two lines. Namely, the lines that correspond to the images of the 3D point in each of the two cameras.

Unfortunately, it is not easy to find points in two images that we can say with certainty to correspond to the same 3D point. This is why *active scanners* are often used. In an active scanner a light source is used in place of one of the cameras. For instance, we can use a laser beam. Clearly, a laser beam also corresponds to a line in space, and it is very easy to detect a laser dot in the image produced by a camera thus obtaining the intersecting line. In actual practice, one generally uses a laser, which emits a planar sheet of light. Since line plane intersection is also unique, this is not much of a constraint. In fact, it allows us to trace a curve on the scanned object in one go. A laser plane shone onto a surface is illustrated in Fig. 1.2. Finally, by projecting a structured pattern of light onto an object with a projector, it can be made much easier to find correspondences. This is known as *structured light* scanning.

Another optical technology for 3D optical acquisition is time of flight (ToF), cf. Fig. 1.3. Here a light pulse (or an array of light pulses) is emitted, and the time it takes for these light pulses to return is measured. Typically, this is done by measuring the difference in phase between the outgoing and the returning light. Note that this modality directly provides a depth value per pixel.

What these scanners have in common is that they produce a, fairly irregular, point cloud. Usually, several geometry processing tasks are required to use the points acquired. First of all, a single optical scanning generally does not capture the entire object: typically we need to perform several scans. In some cases, we do not know the precise orientation of the scanner relative to the object for each scan, and in these cases, we need to register the scans to each other. This is the topic of Chap. 15. Secondly, the density of the acquired point cloud depends on the distance and angle between the object and the scanner. Moreover, many objects have concavities into which the cameras (and the laser) do not reach. Thus, we cannot make many assumptions regarding the point clouds which the scanner produces.

In most cases, we wish to put together a triangle mesh from these points. Triangle meshes, and, more generally, polygonal meshes along with basic operations for editing such meshes are the topic of Chap. 5.

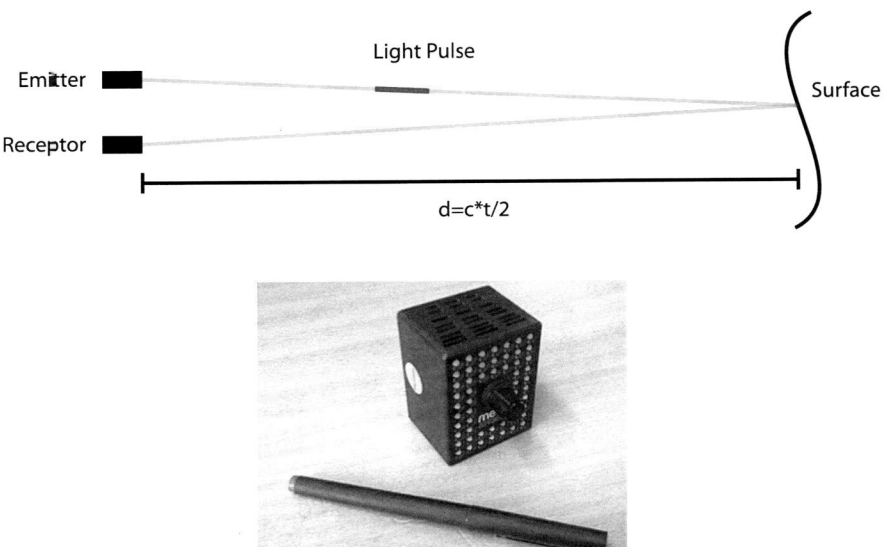

Fig. 1.3 Time of Flight (ToF). *Top*: An illustration of the time of flight principal. A light pulse is emitted from the emitter, bounces of the surface to be measured, and recorded by the receiver. Since the time difference, t, between emission and receival can be measured, the distance to the surface, d, can be calculated, because the speed of light c is known. *Bottom*: An image of a time of flight camera. The lens in the *middle* is in front of a camera, where each pixel is a receiver. As such, a 3D image is recorded at each frame

Generating a mesh from the individual scans is often done by Delaunay triangulation, which we describe in Chap. 14, but when the scans have been registered, one often employs a volumetric method, cf. Chap. 17, for the reconstruction since this class of methods generally performs well when it comes to reconstructing closed, watertight surfaces—even from moderately noisy data. Volumetric methods produce an implicit surface representation in the form of a voxel grid as the immediate output.

For point clouds of just a few thousand points, *radial basis functions* (RBFs) provide an alternative methods for reconstruction as discussed in Chap. 16. With this class of methods, we solve a linear system in order to find coefficients for a set of radial basis functions such that their sum weighted by the coefficients is equal to the data values. This is a very effective and general tool that allows us not only to reconstruct 3D models from 3D point clouds but also to solve many other interpolation problems. However, the simple variations of this method employ dense linear systems. Thus, for optical scans we would generally resort to the voxel-based methods from Chap. 17.

To obtain a triangle mesh from an implicit representation, we need to apply an isosurface polygonization method. Such methods are discussed in Chap. 18. A triangle mesh produced as the final surface reconstruction from a point cloud obtained via structured light scanning is shown in Fig. 1.4.

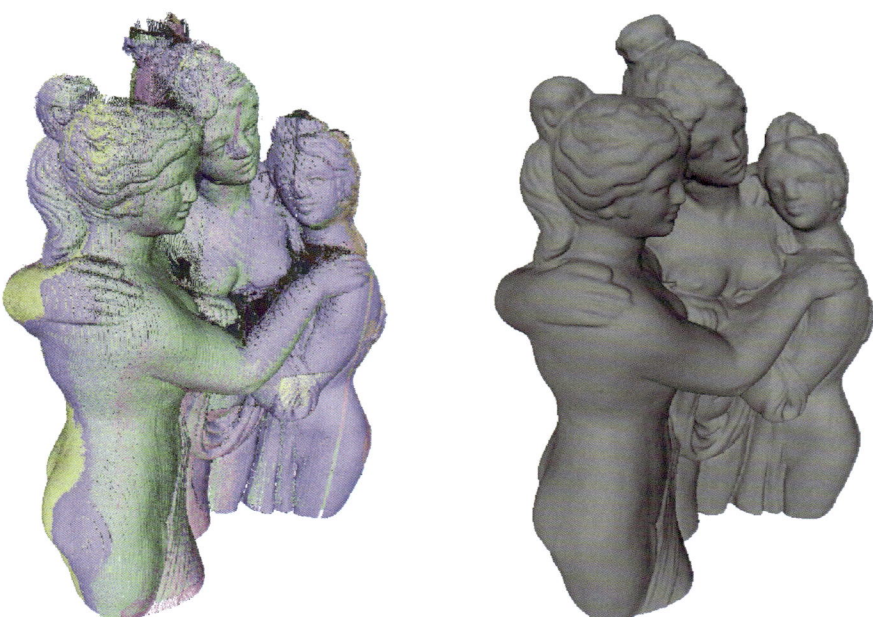

Fig. 1.4 A sculpture scanned with a structured light setup. The eight resulting scans were coarsely aligned manually and then registered using ICP. The registered points are shown on the *left* with different colors for each subscan. From the registered point cloud a mesh was produced (as shown on the *right*) using a volumetric method

Regardless of what method is used for reconstruction, the meshes produced from an optical scanning tend to be both noisy and to contain too many triangles. For this reason, we often employ algorithms for smoothing, mesh reduction and mesh optimization on our acquired meshes. Such algorithms are discussed in Chaps. 9 and 11. Techniques for both optimization and smoothing often rely on measures for curvature in triangle meshes, which are covered in Chap. 8.

1.2   Medical Reverse Engineering

There are many ways in which 3D surfaces acquired by optical scanning can be of great help by allowing us to create cost-efficient, individually customized solutions where we previously had to either use costly human labor or be content with a one-size-fits-all solution.

A striking example is the case of hearing aids. Many modern hearing aids are of the "in-the-ear" type where the apparatus is inserted into the ear canal. The variation in human anatomy makes it impossible to create a hearing aid that will fit a wide range of people, and a tight fit is quite important to avoid issues with sound feedback. The modern way of creating an in-the-ear hearing aid is to first make an

Fig. 1.5 A schematic drawing of the CT-scanning process. In the CT-scanning process several X-ray images are taken at different angles; typically many more than the two images or scans depicted here. These X-rays, which are projections of the volume, can then be combined into a 3D model

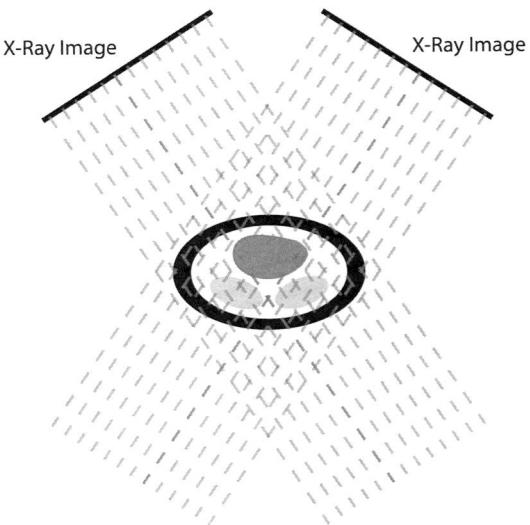

impression of the ear canal and then laser scan this impression. From the scan a shell for the hearing aid can be constructed, and a technician can quickly place the components and thus finish the design. Of course, it would be costly to create a mould for a customized hearing aid, but, fortunately, recent advances in 3D printing mean that it is possible to manufacture just a single shell at reasonable cost.

Dental work is another area where custom solutions are desired. For instance, when a tooth has to be filed down and a crown placed on the stub, it is important that the crown fits the stub snugly and also that it fits the other teeth. Clearly, this is much easier to do if we have a digital model of the teeth in the vicinity of the stub and the stub itself. Again, optical scanning can be used to produce a digital model of the shape. In fact, several companies are developing scanners which can be used directly in the mouth of the patient.

Digital models of 3D geometry do not necessarily come just from optical scanning of surfaces. Another important source of 3D models (in particular) of human anatomy is computed tomography. In X-ray CT, X-ray images of the subject are taken from a wide range of directions. Each of these images show the integral tissue density. Because we know from what directions the images have been taken, we can create a 3D image where we choose densities for all 3D pixels (generally called voxels) such that if we were to create synthetic X-ray images using our 3D image, we would get close to the real X-ray images, see Fig. 1.5. Thus, we are solving a large inverse problem to obtain an anatomical 3D image. Such 3D images can also be created based on ultrasound or MRI (magnetic resonance imagery). In either case, we end up with a 3D image, which can be used for direct visualization or we can extract surfaces by finding isosurfaces—i.e. surfaces where the intensity in the 3D image has constant value. As already mentioned, the topic of isosurface polygonization is covered in Chap. 18.

1.3   **Computer Aided Design**

Polygonal meshes are not the only useful representation for geometry. In fact, a quite different way of representing surfaces is used in more or less all modern CAD (*computer aided design*) systems. CAD models are used for much more than visualization purposes. They can form the basis for both critical numerical simulations and for manufacturing to a very high precision. That means it is necessary to have an accurate and genuine smooth representation of the geometry, and the faceted world of meshes does not always suffice. Instead, the surface is divided into patches and each patch is parametrized using piecewise polynomials or rationals as coordinate functions. Such functions are called splines and can be expressed as a linear combination of B-spline or NURBS basis functions.

This representation has many desirable properties. The B-spline basis has minimal support which implies that the surfaces can be edited and controlled locally. B-splines are also positive and form a partition of unity, which implies that the surface is in the convex hull of its control points. Finally, NURBS patches can exactly represent sections of spheres, cones, or cylinders. In practice this is a very important feature when designing objects that are to be manufactured. Parametric surfaces in general and NURBS in particular are discussed in Chap. 6.

When it comes to computer animated movies or computer games, splines are used less often in spite of the fact that content creation for both games and animated movies could be seen as computer aided design. In computer animated movies, the issue is largely that it is difficult to ensure that a collection of NURBS patches continue to be joined in a smooth fashion as the object, say a character, is animated [2]. Conversely, we can consider a very complex polygonal mesh to be a single subdivision surface, and when subdivision is applied, the limit surface will be smooth. In Chap. 7 we discuss the most widely used types of subdivision surface and develop some of the theory used for their analysis.

For computer games, we would often like the scene to render at 60 frames per second which translates into around 16 ms for doing all the computation needed for a frame. For this reason, characters in games are generally just simple polygonal models with texture. Often, displacement maps are used to add geometric detail while relatively few polygons are actually being rendered. In fact, the ability to map color and other attributes onto a triangle mesh is one of the characterizing features of modern graphics hardware. To apply these maps, we need techniques for parametrization of polygonal meshes, and that is the topic of Chap. 10.

It should also be noted that, recently, modern graphics cards have gained the ability to tessellate smooth surfaces, and that means that we may increasingly see subdivision surfaces or even NURBS used in interactive graphics. This can be seen as a consequence of the increased computational power of the GPU: we can now use it to synthesize content in the form of triangles and textures where, initially, it was simply a machine for drawing textured triangles.

1.4   Geographical Information Systems

Geographic Information Systems (GIS) represent an important area of application of geometry processing since it is assessed that at least 80 % of the data processed by local governments is spatial (see [3]). Geographic Information Systems are computerized systems, which deal with the automated or semi-automated acquisition of spatial data, its storage in a spatial database, management system, the optimization of its retrieval through spatial access methods or spatial indexing (covered in Chap. 12), its rendering in 2D or 3D, and spatial analyses and their rendering.

Geometric data structures like the Voronoi diagram and its dual graph, the Delaunay graph (covered in Chap. 14), play a central role in GIS, due to the fact they solve the proximity problem in GIS (i.e. finding the nearest neighbor from a given location), but Voronoi diagrams and Delaunany graphs are also a way to index spatial data, which allows for several levels of details and a hierarchical decomposition of the space into cells used for indexing. Recent GIS research by Boguslawski and Gold [4] uses recently developed geometric data structures like the Dual Half-Edge (a generalization of the Quad-Edge data structure to 3D).

Geographic data can be acquired using different techniques with varying accuracies corresponding to different scale ranges. The best accuracy is obtained using geodesy (on the order of the millimeters for first order geodetic points), surveying techniques, and Global Navigation Satellite Systems (of the order of centimeters). Aerial photogrammetry allows one to generate 3D models from pairs of aerial photographs with accuracies in the order of tenths of centimeters. With the same order of accuracy, LIDAR (LIght Detection And Ranging) allows one to generate 3D models, but requires more processing to clean the datasets. Note that LIDAR is the same technology in principle as Time of Flight, which was discussed previously. Satellite photogrammetry allows one to generate maps with varying accuracies (from the order of the meter to tenths of meters typically).

In the following, we discuss various GIS related uses of geometric data.

Digital Terrain Models (DTMs) in GIS typically use the TIN (*Triangular Irregular Network*) data structure, which is equivalent to a triangular mesh in the domain of computer graphics. A TIN or triangulation is created with vertices at each point where the planar coordinates and the elevation above a reference mean sea level are known. Then the DTM is constructed by calculating the elevations at the other locations, which are typically interpolated using linear interpolation. However, ordinary and Line Voronoi diagrams and their dual graphs and constrained Delaunay graphs are also used to generate Digital Terrain Models. The locations at which elevations are known are used as generators of the Voronoi diagram. The DTM is constructed by computing the elevations at the other locations by the so-called "natural neighbor" interpolation. Line Voronoi diagrams or constrained Delaunay triangulations are used to generate DTMs, when there are linear discontinuities (see Anton [5]). Voronoi diagrams of points and open oriented line segments allow one to model vertical faults. A fault is a curve along which the elevation changes discontinuously as a point crosses from one side of the fault over to the other side.

3D City Models can be constructed on top of DTMs by decomposing buildings in 3D stereo models arising from Aerial Photogrammetry into parallelepipeds and more generally convex polytopes (i.e. convex polyhedra). Then the appearance of the facets of the buildings can be constructed from aerial photographs. This is an area where geometry processing is heavily used, and where photogrammetry and computer graphics meet very closely.

Convex hulls and decompositions of 3D solids into convex polytopes is central to the constrained data model, which is one of the main spatial data models used in GIS. Moreover, convex hulls and decompositions of 3D solids into convex polytopes is central to spatial indexing, and in particular the Binary Space Partitioning Tree.

Finally, augmented reality is used more and more in GIS in order to give a photo-realistic rendering of existing and projected facilities. Augmented reality is another area where GIS, computer graphics and geometry processing are closely intertwined.

1.5   Summary and Advice

Clearly, we have only lightly touched the topic of acquisition of geometric data and sparsely sampled the multifarious applications of geometry processing. Hopefully, this quick tour has given a sense of the topic. As the reader will have noticed, this chapter has also given us an opportunity to give an overview of the book.

In fact, all but the coming three chapters from Part I were discussed above. These three chapters are not about geometry processing but about some of its mathematical underpinnings: Chapter 2 is about vector spaces, affine spaces, and metric spaces. It contains important elements of linear algebra. Notions are needed in many places of the book. Chapter 3 is about *smooth* differential geometry. This provides background for the discrete approximations needed for instance to compute curvature in triangles meshes. Finally, Chap. 4 gives a very brief introduction to finite difference methods for partial differential equations.

The material in the following three chapters may be a daunting place to start on a first reading. We suggest that readers who do indeed feel daunted jump right into the material of Part II. Much of the material can be understood without recourse to Part I, but the reader might find it useful to peruse these chapters later, to gain a deeper understanding of the underpinnings.

References

1. Levoy, M., Rusinkiewcz, S., Ginzton, M., Ginsberg, J., Pulli, K., Koller, D., Anderson, S., Shade, J., Curless, B., Pereira, L., Davis, J., Fulk, D.: The digital Michelangelo project: 3D scanning of large statues. In: Conference Proceedings on Computer Graphics Proceedings. Annual Conference Series 2000. SIGGRAPH 2000, pp. 131–144 (2000)
2. DeRose, T., Kass, M., Truong, T.: Subdivision surfaces in character animation. In: Proceedings of the 25th Annual Conference on Computer Graphics and Interactive Techniques. SIGGRAPH '98, pp. 85–94. ACM, New York (1998)

3. Franklin, C., Hane, P.: An introduction to gis: linking maps to databases. Database **15**(2), 17–22 (1992)
4. Boguslawski, P., Gold, C.: Construction operators for modelling 3D objects and dual navigation structures. In: Lee, J., Zlatanova, S. (eds.) 3D Geo-Information Sciences. Lecture Notes in Geoinformation and Cartography, pp. 47–59. Springer, Berlin (2009)
5. Anton, F.: Line Voronoi diagram based interpolation and application to digital terrain modelling. In: In XXth ISPRS Congress (2004)

Part I
Mathematical Preliminaries

Vector Spaces, Affine Spaces, and Metric Spaces

This chapter is only meant to give a short overview of the most important concepts in linear algebra, affine spaces, and metric spaces and is not intended as a course; for that we refer to the vast literature, e.g., [1] for linear algebra and [2] for metric spaces. We will in particular skip most proofs.

In Sect. 2.1 on vector spaces we present the basic concepts of linear algebra: vector space, subspace, basis, dimension, linear map, matrix, determinant, eigenvalue, eigenvector, and inner product. This should all be familiar concepts from a first course on linear algebra. What might be less familiar is the abstract view where the basic concepts are vector spaces and linear maps, while coordinates and matrices become derived concepts. In Sect. 2.1.5 we state the singular value decomposition which is used for mesh simplification and in the ICP algorithm for registration.

In Sect. 2.2 on affine spaces we only give the basic definitions: affine space, affine combination, convex combination, and convex hull. The latter concept is used in Delauney triangulation.

Finally in Sect. 2.3 we introduce metric spaces which makes the concepts of open sets, neighborhoods, and continuity precise.

2.1 Vector Spaces and Linear Algebra

A vector space consists of elements, called vectors, that we can add together and multiply with scalars (real numbers), such that the normal rules hold. That is,

Definition 2.1 A real vector space is a set V together with two binary operations $V \times V \to V : (\mathbf{u}, \mathbf{v}) \mapsto \mathbf{u} + \mathbf{v}$ and $\mathbb{R} \times V \to V : (\lambda, \mathbf{v}) \mapsto \lambda \mathbf{v}$, such that:
1. For all $\mathbf{u}, \mathbf{v}, \mathbf{w} \in V$, $(\mathbf{u} + \mathbf{v}) + \mathbf{w} = \mathbf{u} + (\mathbf{v} + \mathbf{w})$.
2. For all $\mathbf{u}, \mathbf{v} \in V$, $\mathbf{u} + \mathbf{v} = \mathbf{v} + \mathbf{u}$.
3. There exists a zero vector $\mathbf{0} \in V$, i.e., for any $\mathbf{u} \in V$, $\mathbf{u} + \mathbf{0} = \mathbf{u}$.
4. All $\mathbf{u} \in V$ has a negative element, i.e., there exists $-\mathbf{u} \in V$ such that $\mathbf{u} + (-\mathbf{u}) = \mathbf{0}$.
5. For all $\alpha, \beta \in \mathbb{R}$ and $\mathbf{u} \in V$, $\alpha(\beta\mathbf{u}) = (\alpha\beta)\mathbf{u}$.

J.A. Bærentzen et al., *Guide to Computational Geometry Processing*, 13
DOI 10.1007/978-1-4471-4075-7_2, © Springer-Verlag London 2012

6. For all $\alpha, \beta \in \mathbb{R}$ and $\mathbf{u} \in V$, $(\alpha + \beta)\mathbf{u} = \alpha\mathbf{u} + \beta\mathbf{u}$.
7. For all $\alpha \in \mathbb{R}$ and $\mathbf{u}, \mathbf{v} \in V$, $\alpha(\mathbf{u} + \mathbf{v}) = \alpha\mathbf{u} + \alpha\mathbf{v}$.
8. Multiplication by $1 \in \mathbb{R}$ is the identity, i.e., for all $\mathbf{u} \in V$, $1\mathbf{u} = \mathbf{u}$.

Remark 2.1 In the definition above the set $\mathbb{R}$ of real numbers can be replaced with the set $\mathbb{C}$ of complex numbers and then we obtain the definition of a complex vector space. We can in fact replace $\mathbb{R}$ with any field, e.g., the set $\mathbb{Q}$ of rational numbers, the set of rational functions, or with finite fields such as $\mathbb{Z}_2 = \{0, 1\}$.

Remark 2.2 We often write the sum $\mathbf{u} + (-\mathbf{v})$ as $\mathbf{u} - \mathbf{v}$.

We leave the proof of the following proposition as an exercise.

Proposition 2.1 *Let V be a vector space and let $\mathbf{u} \in V$ be a vector.*
1. *The zero vector is unique, i.e., if $\mathbf{0}', \mathbf{u} \in V$ are vectors such that $\mathbf{0}' + \mathbf{u} = \mathbf{u}$, then $\mathbf{0}' = \mathbf{0}$.*
2. *If $\mathbf{v}, \mathbf{w} \in V$ are negative elements to $\mathbf{u}$, i.e., if $\mathbf{u} + \mathbf{v} = \mathbf{u} + \mathbf{w} = \mathbf{0}$, then $\mathbf{v} = \mathbf{w}$.*
3. *Multiplication with zero gives the zero vector, i.e., $0\mathbf{u} = \mathbf{0}$.*
4. *Multiplication with -1 gives the negative vector, i.e., $(-1)\mathbf{u} = -\mathbf{u}$.*

Example 2.1 The set of vectors in the plane or in space is a real vector space.

Example 2.2 The set $\mathbb{R}^n = \{(x_1, \ldots, x_n) \mid x_i \in \mathbb{R}, i = 1, \ldots, n\}$ is a real vector space, with addition and multiplication defined as

$$(x_1, \ldots, x_n) + (y_1, \ldots, y_n) = (x_1 + y_1, \ldots, x_n + y_n), \tag{2.1}$$

$$\alpha(x_1, \ldots, x_n) = (\alpha x_1, \ldots, \alpha x_n). \tag{2.2}$$

Example 2.3 The complex numbers $\mathbb{C}$ with usual definition of addition and multiplication is a real vector space.

Example 2.4 The set $\mathbb{C}^n$ with addition and multiplication defined by (2.1) and (2.2) is a real vector space.

Example 2.5 Let Ω be a domain in $\mathbb{R}^n$. A real function $f : \Omega \to \mathbb{R}$ is called a C^n function if all partial derivatives up to order n exist and are continuous, the set of these functions is denoted $C^n(\Omega)$, and it is a real vector space with addition and multiplication defined as

$$(f + g)(x) = f(x) + g(x),$$

$$(\alpha f)(x) = \alpha f(x).$$

Example 2.6 Let Ω be a domain in $\mathbb{R}^n$. A map $f : \Omega \to \mathbb{R}^k$ is called a C^n map if each coordinate function is a C^n function. The set of these functions is denoted $C^n(\Omega, \mathbb{R}^k)$ and it is a real vector space, with addition and multiplication defined as

$$(f + g)(x) = f(x) + g(x),$$

$$(\alpha f)(x) = \alpha f(x).$$

Example 2.7 The set of real polynomials is a real vector space.

Example 2.8 The set of solutions to a system of homogeneous linear equations is a vector space.

Example 2.9 The set of solutions to a system of homogeneous linear ordinary differential equations is a vector space.

Example 2.10 If U and V are real vector spaces, then $U \times V$ is a real vector space too, with addition and multiplication defined as

$$(\mathbf{u}_1, \mathbf{v}_1) + (\mathbf{u}_2, \mathbf{v}_2) = (\mathbf{u}_1 + \mathbf{u}_2, \mathbf{v}_1 + \mathbf{v}_2),$$

$$\alpha(\mathbf{u}, \mathbf{v}) = (\alpha\mathbf{u}, \alpha\mathbf{v}).$$

Example 2.11 Let $a = t_0 < t_1 < \cdots < t_k = b$ be real numbers and let $n, m \in \mathbb{Z}_0$ be non zero integers. The space

$$\left\{ f \in C^n([a, b]) \mid f|_{[t_{\ell-1}, t_\ell]} \text{ is a polynomial of degree at most } m, \ell = 1, \ldots, k \right\}$$

is a real vector space.

2.1.1 Subspaces, Bases, and Dimension

A subset $U \subseteq V$ of a vector space is called a subspace if it is a vector space itself. As it is contained in a vector space we do not need to check all the conditions in Definition 2.1. In fact, we only need to check that it is *stable* with respect to the operations. That is,

Definition 2.2 A subset $U \subseteq V$ of a vector space V is a subspace if
1. For all $\mathbf{u}, \mathbf{v} \in U$, $\mathbf{u} + \mathbf{v} \in U$.
2. For all $\alpha \in \mathbb{R}$ and $\mathbf{u} \in U$, $\alpha\mathbf{u} \in U$.

Example 2.12 The subset $\{(x, y, 0) \in \mathbb{R}^3 \mid (x, y) \in \mathbb{R}^3\}$ is a subspace of $\mathbb{R}^3$.

Example 2.13 The subsets $\{\mathbf{0}\}$, $V \subseteq V$ are subspaces of V called the *trivial* subspaces.

Example 2.14 If $U, V \subseteq W$ are subspaces of W the $U \cap V$ is a subspace too.

Example 2.15 If U and V are vector spaces, then $U \times \{\mathbf{0}\}$ and $\{\mathbf{0}\} \times V$ are subspaces of $U \times V$.

Example 2.16 The subsets $\mathbb{R}, i\mathbb{R} \subseteq \mathbb{C}$ of real and purely imaginary numbers, respectively, are subspaces of $\mathbb{C}$.

Example 2.17 The set of solutions to k real homogeneous linear equations in n unknowns is a subspace of $\mathbb{R}^n$.

Example 2.18 If $m \leq n$ then $C^n([a, b])$ is a subspace of $C^m([a, b])$.

Example 2.19 The polynomial of degree at most n is a subspace of the space of all polynomials.

Definition 2.3 Let $X \subseteq V$ be a non empty subset of a vector space. The subspace spanned by X is the smallest subspace of V that contains X. It is not hard to see that it is the set consisting of all linear combinations of elements from X,

$$\text{span } X = \{\alpha_1 \mathbf{v}_1 + \cdots + \alpha_n \mathbf{v}_n \mid \alpha_i \in \mathbb{R}, \mathbf{v}_1, \ldots, \mathbf{v}_n \in X, n \in \mathbb{N}\}. \qquad (2.3)$$

If $\text{span } X = V$ then we say that X spans V and X is called a spanning set.

Example 2.20 A non zero vector in space spans all vectors on a line.

Example 2.21 Two non zero vectors in space that are not parallel span all vectors in a plane.

Example 2.22 The complex numbers 1 and i span the set of real and purely imaginary numbers, respectively, i.e., $\text{span}\{1\} = \mathbb{R} \subseteq \mathbb{C}$ and $\text{span}\{i\} = i\mathbb{R} \subseteq \mathbb{C}$.

Definition 2.4 The *sum of two subspaces* $U, V \subseteq W$ is the subspace

$$U + V = \text{span}(U \cup V) = \{\mathbf{u} + \mathbf{v} \in W \mid \mathbf{u} \in U \wedge \mathbf{v} \in V\}. \qquad (2.4)$$

If $U \cap V = \{\mathbf{0}\}$ then the sum is called the *direct sum* and is written as $U \oplus V$.

Example 2.23 The complex numbers are the direct sum of the real and purely imaginary numbers, i.e., $\mathbb{C} = \mathbb{R} \oplus i\mathbb{R}$.

Definition 2.5 A finite subset $X = \{\mathbf{v}_1, \ldots, \mathbf{v}_n\} \subseteq V$ is called linearly independent if the only solution to the equation

$$\alpha_1 \mathbf{v}_1 + \cdots + \alpha_n \mathbf{v}_n = \mathbf{0}$$

is the trivial one, $\alpha_1 = \cdots = \alpha_n = 0$. That is, the only linear combination that gives the zero vector is the trivial one. Otherwise, the set is called linearly dependent.

An important property of vector spaces is the existence of a *basis*. This is secured by the following theorem, which we shall not prove.

Theorem 2.1 *For a finite subset $\{v_1, \ldots, v_n\} \subseteq V$ of a vector space the following three statements are equivalent.*
1. $\{v_1, \ldots, v_n\}$ *is a minimal spanning set.*
2. $\{v_1, \ldots, v_n\}$ *is a maximal linearly independent set.*
3. *Each vector $v \in V$ can be written as a unique linear combination*

$$v = \alpha_1 v_1 + \cdots + \alpha_n v_n.$$

If $\{u_1, \ldots, u_m\}$ and $\{v_1, \ldots, v_n\}$ both satisfy these conditions then $m = n$.

Definition 2.6 A finite set $\{v_1, \ldots, v_n\} \subseteq V$ of a vector space is called a basis if it satisfies one, and hence all, of the conditions in Theorem 2.1. The unique number of elements in a basis is called the dimension of the vector space and is denoted $\dim V = n$.

Theorem 2.2 *Let V be a finite dimensional vector space and let $X \subseteq V$ be a subset. Then the following holds:*
1. *If X is linearly independent then we can find a set of vectors $Y \subseteq V$ such that $X \cup Y$ is a basis.*
2. *If X is a spanning set then we can find a basis $Y \subseteq X$.*

The theorem says that we always can supplement a linearly independent set to a basis and that we always can extract a basis from a spanning set.

Corollary 2.1 *If $U, V \subseteq W$ are finite dimensional subspaces of W then*

$$\dim(U) + \dim(V) = \dim(U + V) + \dim(U \cap V). \qquad (2.5)$$

Example 2.24 Two vectors not on the same line are a basis for all vectors in the plane.

Example 2.25 Three vectors not in the same plane are a basis for all vectors in space.

Example 2.26 The vectors

$$e_k = (\underbrace{0, \ldots, 0}_{k-1}, 1, \underbrace{0, \ldots, 0}_{n-k}) \in \mathbb{R}^n, \quad k = 1, \ldots, n, \qquad (2.6)$$

are a basis for $\mathbb{R}^n$ called the *standard basis*, so $\dim(\mathbb{R}^n) = n$.

Example 2.27 The complex numbers 1 and i are a basis for $\mathbb{C}$.

Example 2.28 If $U \cap V = \{\mathbf{0}\}$ are subspaces of a vector space and $\{\mathbf{u}_1, \dots, \mathbf{u}_k\}$ and $\{\mathbf{v}_1, \dots, \mathbf{v}_\ell\}$ are bases for U and V, respectively, then $\{\mathbf{u}_1, \dots, \mathbf{u}_k, \mathbf{v}_1, \dots, \mathbf{v}_\ell\}$ is a basis for $U \oplus V$.

Example 2.29 The monomials $1, x, \dots, x^n$ are a basis for the polynomials of degree at most n.

Example 2.30 The Bernstein polynomials $B_k^n(x) = \binom{n}{k}(1-x)^{n-k}x^k$, $k = 0, \dots, n$ are a basis for the polynomials of degree at most n.

2.1.2 Linear Maps, Matrices, and Determinants

A map between vector spaces is linear if it preserves addition and multiplication with scalars. That is,

Definition 2.7 Let U and V be vector spaces. A map $L : U \to V$ is linear if:
1. For all $\mathbf{u}, \mathbf{v} \in U$, $L(\mathbf{u} + \mathbf{v}) = L(\mathbf{u}) + L(\mathbf{v})$.
2. For all $\alpha \in \mathbb{R}$ and $\mathbf{u} \in U$, $L(\alpha \mathbf{u}) = \alpha L(\mathbf{u})$.

Example 2.31 If V is a vector space and $\alpha \in \mathbb{R}$ is a real number then multiplication by α: $V \to V : \mathbf{v} \mapsto \alpha \mathbf{v}$ is a linear map.

Example 2.32 The map $\mathbb{R} \to \mathbb{R} : x \mapsto ax + b$ with $b \neq 0$ is *not* linear, cf., Exercise 2.7.

Example 2.33 Differentiation $C^n([a, b]) \to C^{n-1}([a, b]) : f \mapsto \frac{\mathrm{d}f}{\mathrm{d}x}$ is a linear map.

Example 2.34 If $L_1, L_2 : U \to V$ are two linear maps, then the sum $L_1 + L_2 : U \to V : \mathbf{u} \mapsto L_1(\mathbf{u}) + L_2(\mathbf{u})$ is a linear map too.

Example 2.35 If $\alpha \in \mathbb{R}$ and $L : U \to V$ is a linear map, then the scalar product $\alpha L : U \to V : \mathbf{u} \mapsto \alpha L(\mathbf{u})$ is a linear map too.

Example 2.36 If $L_1 : U \to V$ and $L_2 : V \to W$ are linear maps, then the composition $L_2 \circ L_1 : U \to W$ is a linear map too.

Example 2.37 If $L : U \to V$ is linear and bijective, then the inverse map $L^{-1} : V \to U$ is linear too.

Examples 2.34 and 2.35 show that the space of linear maps between two vector spaces is a vector space.

Recall the definition of an *injective, surjective,* and *bijective* map.

Definition 2.8 A map $f : A \to B$ between two sets is

- injective if for all $x, y \in A$ we have $f(x) = f(y) \implies x = y$;
- surjective if there for all $y \in B$ exists $x \in A$ such that $f(x) = y$;
- bijective if it is both injective and surjective.

A map is *invertible* if and only if it is bijective.

Definition 2.9 Let $L : U \to V$ be a linear map. The *kernel* of L is the set

$$\ker L = L^{-1}(0) = \{\mathbf{u} \in U \mid L(\mathbf{u}) = \mathbf{0}\}, \tag{2.7}$$

and the *image* of L is the set

$$L(U) = \{f(\mathbf{u}) \in V \mid \mathbf{u} \in U\}. \tag{2.8}$$

We have the following.

Theorem 2.3 *Let $L : U \to V$ be a linear map between two vector spaces. Then the kernel $\ker L$ is a subspace of U and the image $L(U)$ is a subspace of V. If U and V are finite dimensional then*
1. $\dim U = \dim \ker L + \dim L(U)$;
2. L *is injective if and only if* $\ker(L) = \{\mathbf{0}\}$;
3. *if L is injective then* $\dim U \le \dim V$;
4. *if L is surjective then* $\dim U \ge \dim V$;
5. *if* $\dim U = \dim V$ *then L is surjective if and only if L is injective.*

If $L : U \to V$ is linear and $\mathbf{u}_1, \ldots, \mathbf{u}_m$ is a basis for U and $\mathbf{v}_1, \ldots, \mathbf{v}_m$ is a basis for V, then we can write the image of a basis vector $\mathbf{u}_j$ as $L(\mathbf{u}_j) = \sum_{i=1}^{n} a_{ij} \mathbf{v}_i$. Then the image of an arbitrary vector $\mathbf{u} = \sum_{j=1}^{m} x_j \mathbf{u}_j \in U$ is

$$L\left(\sum_{j=1}^{m} x_j \mathbf{u}_j\right) = \sum_{j=1}^{m} x_j L(\mathbf{u}_j) = \sum_{j=1}^{m} x_j \sum_{i=1}^{n} a_{ij} \mathbf{v}_i$$

$$= \sum_{i=1}^{n} \left(\sum_{j=1}^{m} a_{ij} x_j\right) \mathbf{v}_i = \sum_{i=1}^{n} y_i \mathbf{v}_i. \tag{2.9}$$

We see that the coordinates y_i of the image vector $L(\mathbf{u})$ is given by the coordinates x_j of $\mathbf{u}$ by the following matrix equation:

$$\begin{pmatrix} y_1 \\ \vdots \\ y_n \end{pmatrix} = \begin{pmatrix} a_{11} & \cdots & a_{1m} \\ \vdots & \ddots & \vdots \\ a_{n1} & \cdots & a_{nm} \end{pmatrix} \begin{pmatrix} x_1 \\ \vdots \\ x_m \end{pmatrix}. \tag{2.10}$$

The matrix with entries a_{ij} is called the *matrix for L with respect to the bases* $\mathbf{u}_1, \ldots, \mathbf{u}_m$ *and* $\mathbf{v}_1, \ldots, \mathbf{v}_m$. Observe that the columns consist of the coordinates of the image of the basis vectors. Also observe that the first index i in a_{ij} gives the row number while the second index j gives the column number.

We denote the ith row in $\mathbf{A}$ by $\mathbf{A}_{i\_}$ and the jth column by $\mathbf{A}_{|j}$. That is,

$$\mathbf{A} = \begin{pmatrix} \mathbf{A}_{1\_} \\ \vdots \\ \mathbf{A}_{n\_} \end{pmatrix} = \begin{pmatrix} \mathbf{A}_{|1} \cdots \mathbf{A}_{|m} \end{pmatrix}. \tag{2.11}$$

Addition of linear maps now corresponds to addition of matrices,

$$\begin{pmatrix} a_{11} & \cdots & a_{1m} \\ \vdots & \ddots & \vdots \\ a_{n1} & \cdots & a_{nm} \end{pmatrix} + \begin{pmatrix} b_{11} & \cdots & b_{1m} \\ \vdots & \ddots & \vdots \\ b_{n1} & \cdots & b_{nm} \end{pmatrix}$$
$$= \begin{pmatrix} a_{11} + b_{11} & \cdots & a_{1m} + b_{1m} \\ \vdots & \ddots & \vdots \\ a_{n1} + b_{n1} & \cdots & a_{nm} + b_{nm} \end{pmatrix} \tag{2.12}$$

and scalar multiplication of linear maps corresponds to multiplication of a matrix with a scalar

$$\alpha \begin{pmatrix} a_{11} & \cdots & a_{1m} \\ \vdots & \ddots & \vdots \\ a_{n1} & \cdots & a_{nm} \end{pmatrix} = \begin{pmatrix} \alpha a_{11} & \cdots & \alpha a_{1m} \\ \vdots & \ddots & \vdots \\ \alpha a_{n1} & \cdots & \alpha a_{nm} \end{pmatrix}. \tag{2.13}$$

Composition of linear maps corresponds to matrix multiplication, which is defined as follows. If $\mathbf{A}$ is a $k \times m$ matrix with entries a_{ij} and $\mathbf{B}$ is an $m \times n$ matrix with entries b_{ij} then the product is an $k \times n$ matrix $\mathbf{C} = \mathbf{AB}$ where the element c_{ij} is the sum of the products of the elements in the ith row from $\mathbf{A}$ and the jth column from $\mathbf{B}$, i.e.,

$$c_{ij} = \mathbf{A}_{i\_}\mathbf{B}_{|j} = \sum_{k=1}^{m} a_{ik} b_{kj}. \tag{2.14}$$

The identity matrix is the $n \times n$ matrix with ones in the diagonal and zeros elsewhere,

$$\mathbf{I} = \begin{pmatrix} 1 & 0 & \cdots & 0 \\ 0 & 1 & \ddots & \vdots \\ \vdots & \ddots & \ddots & 0 \\ 0 & \cdots & 0 & 1 \end{pmatrix}. \tag{2.15}$$

If $\mathbf{A}$ is an $n \times m$ matrix and $\mathbf{B}$ is an $m \times n$ matrix then

$$\mathbf{IA} = \mathbf{A} \quad \text{and} \quad \mathbf{BI} = \mathbf{B}. \tag{2.16}$$

Definition 2.10 We say an $n \times n$ matrix $\mathbf{A}$ is invertible if there exists a matrix $\mathbf{A}^{-1}$ such that

$$\mathbf{AA}^{-1} = \mathbf{A}^{-1}\mathbf{A} = \mathbf{I}. \tag{2.17}$$

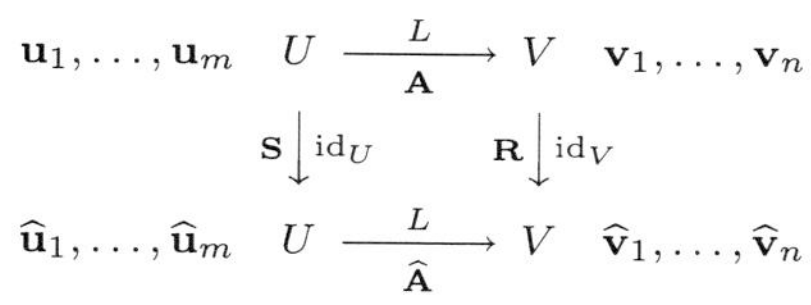

Fig. 2.1 The matrix for a linear map with respect to different bases

$$
\begin{array}{ccccc}
\mathbf{u}_1,\ldots,\mathbf{u}_m & U & \xrightarrow[\;\mathbf{A}\;]{L} & V & \mathbf{v}_1,\ldots,\mathbf{v}_n \\[4pt]
& \mathbf{S}\Big\downarrow \mathrm{id}_U & & \mathbf{R}\Big\downarrow \mathrm{id}_V & \\[4pt]
\widehat{\mathbf{u}}_1,\ldots,\widehat{\mathbf{u}}_m & U & \xrightarrow[\;\widehat{\mathbf{A}}\;]{L} & V & \widehat{\mathbf{v}}_1,\ldots,\widehat{\mathbf{v}}_n
\end{array}
$$

The matrix $\mathbf{A}^{-1}$ is then called the inverse of $\mathbf{A}$.

Theorem 2.4 *Let $\mathbf{A}$ be the matrix for a linear map $L : U \to V$ with respect to the bases $\mathbf{u}_1,\ldots,\mathbf{u}_m$ and $\mathbf{v}_1,\ldots,\mathbf{v}_m$ for U and V, respectively. Then $\mathbf{A}$ is invertible if and only if L is bijective. In that case $\mathbf{A}^{-1}$ is the matrix for L^{-1} with respect to the bases $\mathbf{v}_1,\ldots,\mathbf{v}_m$ and $\mathbf{u}_1,\ldots,\mathbf{u}_m$.*

An in some sense trivial, but still important special case is when $U = V$ and the map is the identity map $\mathrm{id} : \mathbf{u} \mapsto \mathbf{u}$. Let $\mathbf{S}$ be the matrix of id with respect to the bases $\mathbf{u}_1,\ldots,\mathbf{u}_m$ and $\widehat{\mathbf{u}}_1,\ldots,\widehat{\mathbf{u}}_m$. The jth column of $\mathbf{S}$ consists of the coordinates of $\mathrm{id}(\mathbf{u}_j) = \mathbf{u}_j$ with respect to the basis $\widehat{\mathbf{u}}_1,\ldots,\widehat{\mathbf{u}}_m$. Equation (2.10) now reads

$$
\widehat{\underline{\mathbf{u}}} = \mathbf{S}\underline{\mathbf{u}}, \tag{2.18}
$$

and gives us the relation between the coordinates $\underline{\mathbf{u}}$ and $\widehat{\underline{\mathbf{u}}}$ of the same vector $\mathbf{u}$ with respect to the bases $\mathbf{u}_1,\ldots,\mathbf{u}_m$ and $\widehat{\mathbf{u}}_1,\ldots,\widehat{\mathbf{u}}_m$, respectively.

Now suppose we have a linear map $L : U \to V$ between two vector spaces, and two pairs of different bases, $\mathbf{u}_1,\ldots,\mathbf{u}_m$ and $\widehat{\mathbf{u}}_1,\ldots,\widehat{\mathbf{u}}_m$ for U and $\mathbf{v}_1,\ldots,\mathbf{v}_n$ and $\widehat{\mathbf{v}}_1,\ldots,\widehat{\mathbf{v}}_n$ for V. Let $\mathbf{A}$ be the matrix for L with respect to the bases $\mathbf{u}_1,\ldots,\mathbf{u}_m$ and $\mathbf{v}_1,\ldots,\mathbf{v}_n$ and let $\widehat{\mathbf{A}}$ be the matrix for L with respect to the bases $\widehat{\mathbf{u}}_1,\ldots,\widehat{\mathbf{u}}_m$ and $\widehat{\mathbf{v}}_1,\ldots,\widehat{\mathbf{v}}_n$. Let furthermore S be the matrix for the identity $U \to U$ with respect to the bases $\mathbf{u}_1,\ldots,\mathbf{u}_m$ and $\widehat{\mathbf{u}}_1,\ldots,\widehat{\mathbf{u}}_m$ and let R be the matrix for the identity $V \to V$ with respect to the bases $\mathbf{v}_1,\ldots,\mathbf{v}_n$ and $\widehat{\mathbf{v}}_1,\ldots,\widehat{\mathbf{v}}_n$; then

$$
\widehat{\mathbf{A}} = \mathbf{R}\mathbf{A}\mathbf{S}^{-1}, \tag{2.19}
$$

see Fig. 2.1. A special case is when $U = V$, $\mathbf{v}_i = \mathbf{u}_i$, and $\widehat{\mathbf{v}}_i = \widehat{\mathbf{u}}_i$. Then we have $\widehat{\mathbf{A}} = \mathbf{S}\mathbf{A}\mathbf{S}^{-1}$.

Definition 2.11 The transpose of a matrix $\mathbf{A}$ is the matrix $\mathbf{A}^T$ which is obtained by interchanging the rows and columns. That is, if $\mathbf{A}$ has entries a_{ij}, then $\mathbf{A}^T$ has entries α_{ij}, where $\alpha_{ij} = a_{ji}$.

Definition 2.12 An $n \times n$ matrix $\mathbf{A}$ is called *symmetric* if $\mathbf{A}^T = \mathbf{A}$.

Definition 2.13 An $n \times n$ matrix $\mathbf{U}$ is called *orthogonal* if $\mathbf{U}^T\mathbf{U} = \mathbf{I}$, i.e., if $\mathbf{U}^{-1} = \mathbf{U}^T$.

Definition 2.14 An $n \times n$ matrix $\mathbf{A}$ is called *positive definite* if $\mathbf{x}^T\mathbf{A}\mathbf{x} \geq 0$ for all non zero column vectors $\mathbf{x}$.

Before we can define the determinant of a matrix we need the notion of permutations.

Definition 2.15 A permutation is a bijective map $\sigma : \{1, \ldots, n\} \to \{1, \ldots, n\}$. If $i \neq j$, then σ_{ij} denotes the transposition that interchanges i and j, i.e., the permutation defined by

$$\sigma_{ij}(i) = j, \qquad \sigma_{ij}(j) = i, \qquad \sigma_{ij}(k) = k, \quad \text{if } k \neq i, j. \tag{2.20}$$

It is not hard to see that any permutation can be written as the composition of a number of transpositions $\sigma = \sigma_{i_k j_k} \circ \cdots \circ \sigma_{i_2 j_2} \circ \sigma_{i_1 j_1}$. This description is far from unique, but the number k of transpositions needed for a given permutation σ is either always even or always odd. If the number is even σ is called an *even permutation*, otherwise it is called an *odd permutation*. The sign of a sigma is now defined as

$$\text{sign}\,\sigma = \begin{cases} 1 & \text{if } \sigma \text{ is even,} \\ -1 & \text{if } \sigma \text{ is odd.} \end{cases} \tag{2.21}$$

Definition 2.16 The determinant of an $n \times n$ matrix $\mathbf{A}$ is the completely anti symmetric multilinear function of the columns of $\mathbf{A}$ that is 1 on the identity matrix. That is,

$$\det(\mathbf{A}_{|\sigma(1)}, \ldots, \mathbf{A}_{|\sigma(n)}) = \text{sign}(\sigma)\det(\mathbf{A}_{|1}, \mathbf{A}_{|2}, \ldots, \mathbf{A}_{|n}) \tag{2.22}$$

$$\det(\mathbf{A}'_{|1} + \mathbf{A}''_{|1}, \mathbf{A}_{|2}, \ldots, \mathbf{A}_{|n}) = \det(\mathbf{A}'_{|1}, \mathbf{A}_{|2}, \ldots, \mathbf{A}_{|n})$$
$$+ \det(\mathbf{A}''_{|1}, \mathbf{A}_{|2}, \ldots, \mathbf{A}_{|n}), \tag{2.23}$$

$$\det(\alpha\mathbf{A}_{|1}, \mathbf{A}_{|2}, \ldots, \mathbf{A}_{|n}) = \alpha\det(\mathbf{A}_{|1}, \mathbf{A}_{|2}, \ldots, \mathbf{A}_{|n}), \tag{2.24}$$

$$\det(\mathbf{I}) = 1, \tag{2.25}$$

where σ is a permutation. The determinant of $\mathbf{A}$ can be written

$$\det \mathbf{A} = \sum_{\sigma} \text{sign}\,\sigma \prod_{i=1}^{n} a_{i\sigma(i)}, \tag{2.26}$$

where the sum is over all permutations σ of $\{1, \ldots, n\}$.

The definition is not very practical, except in the case of 2×2 and 3×3 matrices. Here we have

$$\det \begin{pmatrix} a_{11} & a_{12} \\ a_{21} & a_{22} \end{pmatrix} = \begin{vmatrix} a_{11} & a_{12} \\ a_{21} & a_{22} \end{vmatrix} = a_{11}a_{22} - a_{12}a_{21}, \tag{2.27}$$

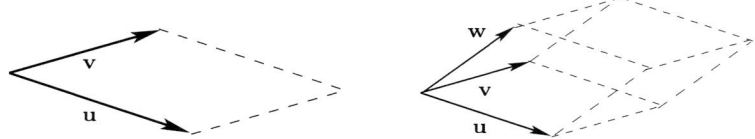

Fig. 2.2 The area and volume can be calculated as determinants: area $= \det(\mathbf{u}, \mathbf{v})$ and volume $= \det(\mathbf{u}, \mathbf{v}, \mathbf{w})$

The determinant of a 2×2 matrix $\mathbf{A}$ can be interpreted as the *signed area* of the parallelogram in $\mathbb{R}^2$ spanned by the vectors $\mathbf{A}_{1\_}$ and $\mathbf{A}_{2\_}$, see Fig. 2.2.

$$
\det \begin{pmatrix} a_{11} & a_{12} & a_{13} \\ a_{21} & a_{22} & a_{23} \\ a_{31} & a_{32} & a_{33} \end{pmatrix} = \begin{vmatrix} a_{11} & a_{12} & a_{13} \\ a_{21} & a_{22} & a_{23} \\ a_{31} & a_{32} & a_{33} \end{vmatrix} = a_{11}a_{22}a_{33} + a_{12}a_{23}a_{31} + a_{13}a_{21}a_{32}
$$

$$
- a_{11}a_{23}a_{32} - a_{12}a_{21}a_{33} - a_{13}a_{22}a_{31}. \tag{2.28}
$$

The determinant of a 3×3 matrix $\mathbf{A}$ can be interpreted as the *signed volume* of the parallelepiped spanned by the vectors $\mathbf{A}_{1\_}$, $\mathbf{A}_{2\_}$, and $\mathbf{A}_{3\_}$, see Fig. 2.2. The same is true in higher dimensions. The determinant of a $n \times n$ matrix A is the signed n-dimensional volume of the n-dimensional parallelepiped spanned by the columns of A.

For practical calculations one makes use of the following properties of the determinant.

Theorem 2.5 *Let $\mathbf{A}$ be an $n \times n$ matrix, then*

$$
\det \mathbf{A}^T = \det \mathbf{A}. \tag{2.29}
$$

The determinant changes sign if two rows or columns are interchanged, in particular

$$
\det \mathbf{A} = 0, \quad \text{if two rows or columns in } \mathbf{A} \text{ are equal}, \tag{2.30}
$$

$$
\det \mathbf{A} = \sum_{i=1}^{n} (-1)^{i+j} a_{ij} \det \mathbf{A}_{ij}, \quad \text{for } i = 1, \dots, n, \tag{2.31}
$$

where $\mathbf{A}_{ij}$ is the matrix obtained from $\mathbf{A}$ by deleting the ith row and jth column, i.e., the row and column where a_{ij} appears. If $\mathbf{B}$ is another $n \times n$ matrix then

$$
\det(\mathbf{AB}) = \det(\mathbf{A}) \det(\mathbf{B}). \tag{2.32}
$$

The matrix $\mathbf{A}$ is invertible if and only if $\det \mathbf{A} \neq 0$, and in that case

$$
\det(\mathbf{A}^{-1}) = \frac{1}{\det \mathbf{A}}. \tag{2.33}
$$

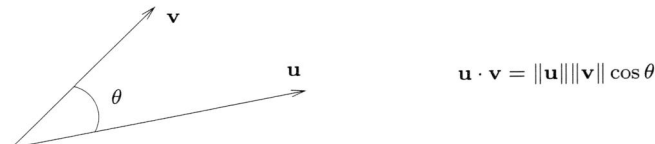

Fig. 2.3 The inner product between two vectors in the plane (or in space)

If $\mathbf{A}$ *is invertible then* $\mathbf{A}^{-1}$ *has entries* α_{ij}, *where*

$$\alpha_{ij} = \frac{(-1)^{i+j} \det \mathbf{A}_{ji}}{\det \mathbf{A}}. \tag{2.34}$$

Suppose $\mathbf{A}$ and $\widehat{\mathbf{A}}$ are matrices for a linear map $L : V \to V$ with respect to two different bases. Then we have $\widehat{\mathbf{A}} = \mathbf{S}\mathbf{A}\mathbf{S}^{-1}$ where $\mathbf{S}$ is an invertible matrix. We now have $\det \widehat{\mathbf{A}} = \det(\mathbf{S}\mathbf{A}\mathbf{S}^{-1}) = \det \mathbf{S} \det \mathbf{A} \det \mathbf{S}^{-1} = \det \mathbf{A}$. Thus, we can define the determinant of L as the determinant of any matrix representation and we clearly see that L is injective if and only if $\det L \neq 0$.

2.1.3 Euclidean Vector Spaces and Symmetric Maps

For vectors in the plane, or in space, we have the concepts of length and angles. This then leads to the definition of the inner product, see Fig. 2.3. For two vectors $\mathbf{u}$ and $\mathbf{v}$ it is given by

$$\langle \mathbf{u}, \mathbf{v} \rangle = \mathbf{u} \cdot \mathbf{v} = \|\mathbf{u}\| \|\mathbf{v}\| \cos \theta, \tag{2.35}$$

where $\|\mathbf{u}\|$ and $\|\mathbf{v}\|$ is the length of a $\mathbf{u}$ and $\mathbf{v}$, respectively, and θ is the angle between $\mathbf{u}$ and $\mathbf{v}$.

A general vector space V does not have the a priori notions of length and angle and in order to be able to have the concepts of length and angle we introduce an abstract inner product.

Definition 2.17 An *Euclidean vector space* is a real vector space V equipped with a positive definite, symmetric, bilinear mapping $V \times V \to \mathbb{R} : (\mathbf{u}, \mathbf{v}) \mapsto \langle \mathbf{u}, \mathbf{v} \rangle$, called the *inner product*, i.e., we have the following:
1. For all $\mathbf{u}, \mathbf{v} \in V$, $\langle \mathbf{u}, \mathbf{v} \rangle = \langle \mathbf{v}, \mathbf{u} \rangle$.
2. For all $\mathbf{u}, \mathbf{v}, \mathbf{w} \in V$, $\langle \mathbf{u} + \mathbf{v}, \mathbf{w} \rangle = \langle \mathbf{u}, \mathbf{w} \rangle + \langle \mathbf{v}, \mathbf{w} \rangle$.
3. For all $\alpha \in \mathbb{R}$ and $\mathbf{u}, \mathbf{v} \in V$, $\langle \alpha \mathbf{u}, \mathbf{v} \rangle = \alpha \langle \mathbf{u}, \mathbf{v} \rangle$.
4. For all $\mathbf{u} \in V$, $\langle \mathbf{u}, \mathbf{u} \rangle \geq 0$.
5. For all $\mathbf{u} \in V$, $\langle \mathbf{u}, \mathbf{u} \rangle = 0 \iff \mathbf{u} = \mathbf{0}$.

Example 2.38 The set of vectors in the plane or in space equipped with the inner product (2.35) is an Euclidean vector space. The norm (2.41) becomes the usual length and the angle defined by (2.44) is the usual angle.

Example 2.39 The set $\mathbb{R}^n$ equipped with inner product

$$\langle (x_1, \ldots, x_n), (y_1, \ldots, y_n) \rangle = x_1 y_1 + \cdots + x_n y_n, \tag{2.36}$$

is an Euclidean vector space.

Example 2.40 The space $C^n([a, b])$ of n times differentiable functions with continuous nth derivative equipped with the inner product

$$\langle f, g \rangle = \int_a^b f(x)g(x)\, dx, \tag{2.37}$$

is an Euclidean vector space. The corresponding norm is called the L^2-norm.

Example 2.41 If $(V_1, \langle \cdot, \cdot \rangle_1)$ and $(V_2, \langle \cdot, \cdot \rangle_2)$ are Euclidean vector spaces, then $V_1 \times V_2$ equipped with the inner product

$$\langle (u_1, u_2), (v_1, v_2) \rangle = \langle u_1, v_1 \rangle_1 + \langle u_2, v_2 \rangle_2, \tag{2.38}$$

is an Euclidean vector space.

Example 2.42 If $(V, \langle \cdot, \cdot \rangle)$ is an Euclidean vector space and $U \subseteq V$ is a subspace then U equipped with the restriction $\langle \cdot, \cdot \rangle|_{U \times U}$ of $\langle \cdot, \cdot \rangle$ to $U \times U$ is an Euclidean vector space too.

Example 2.43 The space $C_0^\infty([a, b]) = \{ f \in C^\infty([a, b]) \mid f(a) = f(b) = 0 \}$ of infinitely differentiable functions that are zero at the endpoints equipped with the restriction of the inner product (2.37) is an Euclidean vector space.

If $\mathbf{u}_1, \ldots, \mathbf{u}_n$ is a basis for V, $\mathbf{v} = \sum_{k=1}^n v_i \mathbf{u}_i$, and $\mathbf{w} = \sum_{k=1}^n w_i \mathbf{u}_i$ then the inner product of $\mathbf{v}$ and $\mathbf{w}$ can be written

$$\langle \mathbf{v}, \mathbf{w} \rangle = \sum_{k, \ell=1}^n v_k w_\ell \langle \mathbf{u}_k, \mathbf{u}_\ell \rangle = \underline{v}^T \mathbf{G} \underline{w}, \tag{2.39}$$

where $\underline{v}$ and $\underline{w}$ are the coordinates with respect to the basis $\mathbf{u}_1, \ldots, \mathbf{u}_n$ of $\mathbf{v}$ and $\mathbf{w}$, respectively, and $\mathbf{G}$ is the matrix

$$\mathbf{G} = \begin{pmatrix} \langle \mathbf{u}_1, \mathbf{u}_1 \rangle & \cdots & \langle \mathbf{u}_1, \mathbf{u}_n \rangle \\ \vdots & & \vdots \\ \langle \mathbf{u}_n, \mathbf{u}_1 \rangle & \cdots & \langle \mathbf{u}_n, \mathbf{u}_n \rangle \end{pmatrix}. \tag{2.40}$$

It is called the matrix for the inner product with respect to the basis $\mathbf{u}_1, \ldots, \mathbf{u}_n$ and it is a positive definite symmetric matrix. Observe that we have the same kind of matrix representation of a symmetric bilinear map, i.e., a map that satisfies condition

(1), (2), and (3) in Definition 2.17. The matrix G is still symmetric but it need not be positive definite.

Let $\widehat{\mathbf{u}}_1, \ldots, \widehat{\mathbf{u}}_n$ be another basis, let $\widehat{\mathbf{G}}$ be the corresponding matrix for the inner product, and let $\mathbf{S}$ be the matrix for the identity on V with respect to the two bases. Then the coordinates of a vector $\mathbf{u}$ with respect to the bases satisfies (2.18) and we see that $\underline{\widehat{\mathbf{u}}}^T \widehat{\mathbf{G}} \underline{\widehat{\mathbf{v}}} = \underline{\mathbf{u}}^T \mathbf{S}^T \widehat{\mathbf{G}} \mathbf{S} \underline{\mathbf{v}}$. That is, $\mathbf{G} = \mathbf{S}^T \widehat{\mathbf{G}} \mathbf{S}$.

Definition 2.18 The norm of a vector $\mathbf{u} \in V$ in an Euclidean vector space $(V, \langle \cdot, \cdot \rangle)$ is defined as

$$\|\mathbf{u}\| = \sqrt{\langle \mathbf{u}, \mathbf{u} \rangle}. \tag{2.41}$$

A very important property of an arbitrary inner product is the Cauchy–Schwartz inequality.

Theorem 2.6 *If $(V, \langle \cdot, \cdot \rangle)$ is an Euclidean vector space then the inner product satisfies the Cauchy–Schwartz inequality*

$$\left| \langle \mathbf{u}, \mathbf{v} \rangle \right| \leq \|\mathbf{u}\| \|\mathbf{v}\|, \tag{2.42}$$

with equality if and only if one of the vectors is a positive multiple of the other.

Corollary 2.2 *The norm satisfies the following conditions*:
1. *For all $\alpha \in \mathbb{R}$ and $\mathbf{u} \in V$, $\|\alpha \mathbf{u}\| = |\alpha| \|\mathbf{u}\|$.*
2. *For all $\mathbf{u}, \mathbf{v} \in V$, $\|\mathbf{u} + \mathbf{v}\| \leq \|\mathbf{u}\| + \|\mathbf{v}\|$.*
3. *For all $\mathbf{u} \in V$, $\|\mathbf{u}\| \geq 0$.*
4. *For all $\mathbf{u} \in V$, $\|\mathbf{u}\| = 0 \iff \mathbf{u} = \mathbf{0}$.*

This is the conditions for an abstract norm on a vector space and not all norms are induced by an inner product. But if a norm is induced by an inner product then this inner product is unique. Indeed, if $\mathbf{u}, \mathbf{v} \in V$ then symmetry and bilinearity imply that

$$\langle \mathbf{u} + \mathbf{v}, \mathbf{u} + \mathbf{v} \rangle = \langle \mathbf{u}, \mathbf{u} \rangle + 2 \langle \mathbf{u}, \mathbf{v} \rangle + \langle \mathbf{v}, \mathbf{v} \rangle.$$

That is, the inner product of two vectors $\mathbf{u}, \mathbf{v} \in V$ can be written as

$$\langle \mathbf{u}, \mathbf{v} \rangle = \frac{1}{2} \left(\|\mathbf{u} + \mathbf{v}\|^2 - \|\mathbf{u}\|^2 - \|\mathbf{v}\|^2 \right). \tag{2.43}$$

The angle θ between two vectors $\mathbf{u}, \mathbf{v} \in V$ in an Euclidean vector space $(V, \langle \cdot, \cdot \rangle)$ can now be defined by the equation

$$\cos \theta = \frac{\langle \mathbf{u}, \mathbf{v} \rangle}{\|\mathbf{u}\| \|\mathbf{v}\|}. \tag{2.44}$$

Two vectors $\mathbf{u}, \mathbf{v} \in V$ are called orthogonal if the angle between them is $\frac{\pi}{2}$, i.e., if $\langle \mathbf{u}, \mathbf{v} \rangle = 0$.

Example 2.44 If $(V, \langle \cdot, \cdot \rangle)$ is an Euclidean vector space and $U \subseteq V$ is a subspace then the *orthogonal complement*

$$U^{\perp} = \left\{ \mathbf{v} \in V \mid \langle \mathbf{u}, \mathbf{v} \rangle = 0 \text{ for all } \mathbf{u} \in U \right\} \tag{2.45}$$

is a subspace of V, and $V = U \oplus U^{\perp}$.

Definition 2.19 A basis $\mathbf{e}_1, \ldots, \mathbf{e}_n$ for an Euclidean vector space is called orthonormal if

$$\langle \mathbf{e}_i, \mathbf{e}_j \rangle = \delta_{ij} = \begin{cases} 1 & \text{if } i = j, \\ 0 & \text{if } i \neq j. \end{cases} \tag{2.46}$$

That is, the elements of the basis are pairwise orthogonal and have norm 1.

If $\mathbf{u}_1, \ldots, \mathbf{u}_n$ is a basis for an Euclidean vector space V then we can construct an orthonormal basis $\mathbf{e}_1, \ldots, \mathbf{e}_n$ by Gram–Schmidt orthonormalization. The elements of that particular orthonormal basis is defined as follows:

$$\mathbf{v}_{\ell} = \mathbf{u}_{\ell} - \sum_{k=1}^{\ell-1} \langle \mathbf{u}_{\ell}, \mathbf{e}_k \rangle \mathbf{e}_k, \qquad \mathbf{e}_{\ell} = \frac{\mathbf{v}_{\ell}}{\|\mathbf{v}_{\ell}\|}, \qquad \ell = 1, \ldots, n. \tag{2.47}$$

Definition 2.20 A linear map $L : U \to V$ between two Euclidean vector spaces is called an *isometry* if it is bijective and $\langle L(\mathbf{u}), L(\mathbf{v}) \rangle_V = \langle \mathbf{u}, \mathbf{v} \rangle_U$ for all $\mathbf{u}, \mathbf{v} \in U$.

So an isometry preserves the inner product. As the inner product is determined by the norm it is enough to check that the map preserves the norm, i.e., if $\|L(\mathbf{u})\|_V = \|\mathbf{u}\|_U$ for all $\mathbf{u} \in U$ then L is an isometry.

Example 2.45 A rotation in the plane or in space is an isometry.

Example 2.46 A symmetry in space around the origin $\mathbf{0}$ or around a line through $\mathbf{0}$ is an isometry.

Theorem 2.7 *Let $L : U \to V$ be a linear map between two Euclidean vector spaces. Let $\mathbf{u}_1, \ldots, \mathbf{u}_m$ and $\mathbf{v}_1, \ldots, \mathbf{v}_m$ be bases for U and V, respectively, and let $\mathbf{A}$ be the matrix for L with respect to these bases. Let furthermore G_U and G_V be the matrices for the inner product on U and V, respectively. Then L is an isometry if and only if*

$$\mathbf{A}^T \mathbf{G}_V \mathbf{A} = \mathbf{G}_U. \tag{2.48}$$

If $\mathbf{u}_1, \ldots, \mathbf{u}_m$ and $\mathbf{v}_1, \ldots, \mathbf{v}_m$ both are orthonormal then $\mathbf{G}_U = \mathbf{G}_V = \mathbf{I}$ and the equation reads

$$\mathbf{A}^T \mathbf{A} = \mathbf{I}, \tag{2.49}$$

i.e., $\mathbf{A}$ is orthogonal.

On a similar note, if $\mathbf{u}_1, \ldots, \mathbf{u}_m$ and $\widehat{\mathbf{u}}_1, \ldots, \widehat{\mathbf{u}}_m$ are bases for an Euclidean vector space U and $\mathbf{u} \in U$ then the coordinates $\underline{\mathbf{u}}$ and $\widehat{\underline{\mathbf{u}}}$ for $\mathbf{u}$ with respect to the two bases are related by the equation $\widehat{\underline{\mathbf{u}}} = \mathbf{S}\underline{\mathbf{u}}$, cf. (2.18). If $\mathbf{G}$ and $\widehat{\mathbf{G}}$ are the matrices for the inner product with respect to the bases then we have

$$\underline{\mathbf{u}}^T \mathbf{G}\underline{\mathbf{u}} = \langle \mathbf{u}, \mathbf{u} \rangle = \widehat{\underline{\mathbf{u}}}^T \widehat{\mathbf{G}}\widehat{\underline{\mathbf{u}}} = (\mathbf{S}\underline{\mathbf{u}})^T \widehat{\mathbf{G}}\mathbf{S}\underline{\mathbf{u}} = \underline{\mathbf{u}}^T \mathbf{S}^T \widehat{\mathbf{G}}\mathbf{S}\underline{\mathbf{u}},$$

i.e., we have

$$\mathbf{G} = \mathbf{S}^T \widehat{\mathbf{G}}\mathbf{S}. \tag{2.50}$$

If the bases both are orthonormal then $\mathbf{G} = \widehat{\mathbf{G}} = \mathbf{I}$ and we see that $\mathbf{S}$ is orthogonal.

Definition 2.21 A linear map $L : V \to V$ from an Euclidean vector space to itself is called symmetric if

$$\langle L(\mathbf{u}), \mathbf{v} \rangle = \langle \mathbf{u}, L(\mathbf{v}) \rangle, \quad \text{for all } \mathbf{u}, \mathbf{v} \in V. \tag{2.51}$$

Example 2.47 The map $f \mapsto f''$ is a symmetric map of the space $(C_0^\infty([a, b]), \langle \cdot, \cdot \rangle)$ to itself, where the inner product $\langle \cdot, \cdot \rangle$ is given by (2.37).

If $\mathbf{A}$ is the matrix for a linear map L with respect to some basis and $\mathbf{G}$ is the matrix for the inner product then L is symmetric if and only if $\mathbf{A}^T \mathbf{G} = \mathbf{G}\mathbf{A}$. If the basis is orthonormal then $\mathbf{G} = \mathbf{I}$ and the condition reads $\mathbf{A}^T = \mathbf{A}$, i.e., $\mathbf{A}$ is a symmetric matrix.

2.1.4 Eigenvalues, Eigenvectors, and Diagonalization

Definition 2.22 Let $L : V \to V$ be a linear map. If there exist a non zero vector $\mathbf{v} \in V$ and a scalar $\lambda \in \mathbb{R}$ such that $L(\mathbf{v}) = \lambda\mathbf{v}$ then $\mathbf{v}$ is called an *eigenvector* with *eigenvalue* λ. If λ is an eigenvalue then the space

$$E_\lambda = \left\{ \mathbf{v} \in V \mid L(\mathbf{v}) = \lambda\mathbf{v} \right\} \tag{2.52}$$

is a subspace of V called the *eigenspace* of λ. The dimension of E_λ is called the *geometric multiplicity* of λ.

If $\mathbf{u}_1, \ldots, \mathbf{u}_m$ is a basis for V, $\mathbf{A}$ is the matrix for L in this basis and a vector $\mathbf{v} \in V$ has coordinates $\underline{\mathbf{v}}$ with respect to this basis then

$$L(v) = \lambda\mathbf{v} \iff \mathbf{A}\underline{\mathbf{v}} = \lambda\underline{\mathbf{v}} \tag{2.53}$$

We say that $\underline{\mathbf{v}}$ is an eigenvector for the matrix $\mathbf{A}$ with eigenvalue λ.

Example 2.48 Consider the matrix $\left(\begin{smallmatrix} 1 & 3 \\ 3 & 1 \end{smallmatrix}\right)$. The vector $\left(\begin{smallmatrix} 1 \\ 1 \end{smallmatrix}\right)$ is an eigenvector with eigenvalue 4 and $\left(\begin{smallmatrix} 1 \\ -1 \end{smallmatrix}\right)$ is an eigenvector with eigenvalue -2.

Example 2.49 The exponential map exp is an eigenvector with eigenvalue 1 for the linear map $C^\infty(\mathbb{R}) \to C^\infty(\mathbb{R}) : f \mapsto f'$.

Example 2.50 The trigonometric functions cos and sin are eigenvectors with eigenvalue -1 for the linear map $C^\infty(\mathbb{R}) \to C^\infty(\mathbb{R}) : f \mapsto f''$.

We see that λ is an eigenvalue for L if and only if the map $L - \lambda \,\mathrm{id}$ is not injective, i.e., if and only if $\det(L - \lambda \,\mathrm{id}) = 0$. In that case $E_\lambda = \ker(L - \lambda \,\mathrm{id})$. If $\mathbf{A}$ is the matrix for L with respect to some basis for V then we see that

$$\det(L - \lambda \,\mathrm{id}) = \det(\mathbf{A} - \lambda \mathbf{I}) = (-\lambda)^n + \mathrm{tr}\,\mathbf{A}(-\lambda)^{n-1} + \cdots + \det \mathbf{A} \qquad (2.54)$$

is a polynomial of degree n in λ. It is called the *characteristic polynomial* of L (or $\mathbf{A}$). The eigenvalues are precisely the roots of the characteristic polynomial and the multiplicity of a root λ in the characteristic polynomial is called the *algebraic multiplicity* of the eigenvalue λ. The relation between the geometric and algebraic multiplicity is given in the following proposition.

Proposition 2.2 *Let* $v_g(\lambda) = \dim(E_\lambda)$ *be the geometric multiplicity of an eigenvalue* λ *and let* $v_a(\lambda)$ *be the algebraic multiplicity of* λ. *Then* $1 \le v_g(\lambda) \le v_a(\lambda)$.

The characteristic polynomial may have complex roots and even though they strictly speaking are not eigenvalues we will still call them *complex eigenvalues*. Once the eigenvalues are determined the eigenvectors belonging to a particular real eigenvalue λ can be found by determining a non zero solution to the linear equation $L(\mathbf{u}) - \lambda \mathbf{u} = \mathbf{0}$ or equivalently a non zero solution to the matrix equation

$$\begin{pmatrix} a_{1,1} - \lambda & a_{1,2} & \cdots & a_{1,n} \\ a_{2,1} & a_{2,2} - \lambda & \ddots & \vdots \\ \vdots & & \ddots & a_{n-1,1} \\ a_{n,1} & \cdots & a_{n,n-1} & a_{n,n} - \lambda \end{pmatrix} \begin{pmatrix} u_1 \\ \vdots \\ u_n \end{pmatrix} = \begin{pmatrix} 0 \\ \vdots \\ 0 \end{pmatrix}. \qquad (2.55)$$

If V has a basis $\mathbf{u}_1, \ldots, \mathbf{u}_n$ consisting of eigenvectors for L, i.e., $L(\mathbf{u}_k) = \lambda_k \mathbf{u}_k$ then the corresponding matrix is diagonal

$$\Lambda = \begin{pmatrix} \lambda_1 & 0 & \cdots & 0 \\ 0 & \lambda_2 & \ddots & \vdots \\ \vdots & \ddots & \ddots & 0 \\ 0 & \cdots & 0 & \lambda_n \end{pmatrix}, \qquad (2.56)$$

and we say that L is *diagonalizable*. Not all linear maps (or matrices) can be diagonalized. The condition is that there is a basis consisting of eigenvectors and this is the same as demanding that $V = \bigoplus_\lambda E_\lambda$ or that all eigenvalues are real and the sum of the geometric multiplicities is the dimension of V. If there is a complex eigenvalue then this is impossible. The same is the case if $v_g(\lambda) < v_a(\lambda)$ for some real eigenvalue λ.

Example 2.51 The matrix $\begin{pmatrix} 0 & -1 \\ 1 & 0 \end{pmatrix}$ has no real eigenvalues.

Example 2.52 The matrix $\begin{pmatrix} \sqrt{2} & 1 \\ 0 & \sqrt{2} \end{pmatrix}$ has the eigenvalue $\sqrt{2}$ which has algebraic multiplicity 2 and geometric multiplicity 1.

In case of a symmetric map the situation is much nicer. Indeed, we have the following theorem, which we shall not prove.

Theorem 2.8 *Let $(V, \langle \cdot, \cdot \rangle)$ be an Euclidean vector space and let $L : V \to V$ be a symmetric linear map. Then all eigenvalues are real and V has an orthonormal basis consisting of eigenvectors for L.*

By choosing an orthonormal basis for V we obtain the following theorem for symmetric matrices.

Theorem 2.9 *A symmetric matrix $\mathbf{A}$ can be decomposed as $\mathbf{A} = \mathbf{U}^T \Lambda \mathbf{U}$, where Λ is diagonal and $\mathbf{U}$ is orthogonal.*

Let $(V, \langle \cdot, \cdot, \rangle)$ be an Euclidean vector space and let $h : V \times V \to \mathbb{R}$ be a symmetric bilinear map, i.e., it satisfies condition (1), (2), and (3) in Definition 2.17. Then there exists a unique symmetric linear map $L : V \to V$ such that $h(\mathbf{u}, \mathbf{v}) = \langle L(\mathbf{u}), \mathbf{v} \rangle$. Theorem 2.8 tells us that V has an orthonormal basis consisting of eigenvectors for L, and with respect to this basis the matrix representation for h is diagonal with the eigenvalues of L in the diagonal. Now suppose we have an arbitrary basis for V and let $\mathbf{G}$ and $\mathbf{H}$ be the matrices for the inner product $\langle \cdot, \cdot \rangle$ and the bilinear map h, respectively. Let furthermore $\mathbf{A}$ be the matrix for L. Then we have $\mathbf{H} = \mathbf{A}^T \mathbf{G}$, or as both $\mathbf{G}$ and $\mathbf{H}$ are symmetric $\mathbf{H} = \mathbf{G}\mathbf{A}$. That is, $\mathbf{A} = \mathbf{G}^{-1}\mathbf{H}$ and the eigenvalue problem $\mathbf{A}\mathbf{v} = \lambda \mathbf{v}$ is equivalent to the *generalized eigenvalue problem* $\mathbf{H}\mathbf{v} = \lambda \mathbf{G}\mathbf{v}$. This gives us the following generalization of Theorem 2.9.

Theorem 2.10 *Let $\mathbf{G}, \mathbf{H}$ be symmetric $n \times n$ matrices with $\mathbf{G}$ positive definite. Then we can decompose $\mathbf{H}$ as $\mathbf{H} = \mathbf{S}^{-1}\Lambda\mathbf{S}$, where Λ is diagonal and $\mathbf{S}$ is orthogonal with respect to $\mathbf{G}$, i.e., $\mathbf{S}^T\mathbf{G}\mathbf{S} = \mathbf{G}$.*

2.1.5 Singular Value Decomposition

Due to its numerically stability the singular value decomposition (SVD) is extensively used for practical calculations such as solving over- and under-determined systems and eigenvalue calculations. We will use it for mesh simplification and in the ICP algorithm for registration. The singular value decomposition can be formulated as

Theorem 2.11 *Let $L : V \to U$ be a linear map between two Euclidean vector spaces of dimension n and m, respectively, and let $k = \min\{m, n\}$. Then there exist an orthonormal basis $\mathbf{e}_1, \ldots, \mathbf{e}_n$ for V, an orthonormal basis $\mathbf{f}_1, \ldots, \mathbf{f}_m$ for U, and non negative numbers $\sigma_1 \geq \sigma_1 \geq \cdots \geq \sigma_k \geq 0$, called the* singular values, *such that $L(\mathbf{u}_\ell) = \sigma_\ell \mathbf{v}_\ell$ for $\ell = 1, \ldots, k$ and $L(\mathbf{u}_\ell) = \mathbf{0}$ for $\ell = k + 1, \ldots, n$.*

We see that $\sigma_1 = \max\{\|L(\mathbf{e})\| \mid \|\mathbf{e}\| = 1\}$ and that $\mathbf{e}_1$ realizes the maximum. We have in general that $\sigma_\ell = \max\{\|L(\mathbf{e})\| \mid \mathbf{e} \in \operatorname{span}\{\mathbf{e}_1, \ldots, \mathbf{e}_{\ell-1}\}^\perp \wedge \|\mathbf{e}\| = 1\}$ and that $\mathbf{e}_\ell$ realizes the maximum. The basis for V is simply given as $\mathbf{f}_\ell = \frac{L(\mathbf{e}_\ell)}{\|L(\mathbf{e}_\ell)\|}$ when $L(\mathbf{e}_\ell) \neq \mathbf{0}$. If this gives $\mathbf{f}_1, \ldots, \mathbf{f}_{k'}$ then the rest of the basis vectors are chosen as an orthonormal basis for $\operatorname{span}\{\mathbf{f}_1, \ldots, \mathbf{f}_{k'}\}^\perp$. In terms of matrices it has the following formulation.

Theorem 2.12 *Let $\mathbf{A}$ be an $m \times n$ matrix and let $k = \min\{m, n\}$. Then $\mathbf{A}$ can be decomposed as $\mathbf{A} = \mathbf{U}\boldsymbol{\Sigma}\mathbf{V}^T$, where $\mathbf{U}$ is an orthogonal $m \times m$ matrix, $\mathbf{V}$ is an orthogonal $n \times n$ matrix, and $\boldsymbol{\Sigma}$ is a diagonal matrix with non zero elements $\sigma_1 \geq \sigma_1 \geq \cdots \geq \sigma_k \geq 0$ in the diagonal.*

The singular values are the square root of the eigenvalues of $\mathbf{A}^T\mathbf{A}$, which is a positive semi definite symmetric matrix. The columns of $\mathbf{V}$, and hence the rows of $\mathbf{V}^T$, are the eigenvectors for $\mathbf{A}^T\mathbf{A}$.

Example 2.53

$$\begin{pmatrix} 0 & -1 \\ 1 & 0 \end{pmatrix} = \begin{pmatrix} 0 & -1 \\ 1 & 0 \end{pmatrix} \begin{pmatrix} 1 & 0 \\ 0 & 1 \end{pmatrix} \begin{pmatrix} 1 & 0 \\ 0 & 1 \end{pmatrix}.$$

Example 2.54

$$\begin{pmatrix} \sqrt{2} & 1 \\ 0 & \sqrt{2} \end{pmatrix} = \begin{pmatrix} \frac{\sqrt{6}}{3} & -\frac{\sqrt{3}}{3} \\ \frac{\sqrt{3}}{3} & \frac{\sqrt{6}}{3} \end{pmatrix} \begin{pmatrix} 2 & 0 \\ 0 & 1 \end{pmatrix} \begin{pmatrix} \frac{\sqrt{3}}{3} & \frac{\sqrt{6}}{3} \\ -\frac{\sqrt{6}}{3} & \frac{\sqrt{3}}{3} \end{pmatrix}.$$

Example 2.55

$$\begin{pmatrix} 1 & 1 & 1 \\ 1 & 1 & -1 \end{pmatrix} = \begin{pmatrix} \frac{\sqrt{2}}{2} & -\frac{\sqrt{2}}{2} \\ \frac{\sqrt{2}}{2} & \frac{\sqrt{2}}{2} \end{pmatrix} \begin{pmatrix} 2 & 0 & 0 \\ 0 & \sqrt{2} & 0 \end{pmatrix} \begin{pmatrix} \frac{\sqrt{2}}{2} & \frac{\sqrt{2}}{2} & 0 \\ 0 & 0 & 1 \\ \frac{\sqrt{2}}{2} & -\frac{\sqrt{2}}{2} & 0 \end{pmatrix}.$$

Definition 2.23 *The Moore–Penrose pseudo inverse of a matrix $\mathbf{A}$ is the matrix $\mathbf{A}^+ = \mathbf{V}\boldsymbol{\Sigma}^+\mathbf{U}^T$ where $\mathbf{A} = \mathbf{U}\boldsymbol{\Sigma}\mathbf{V}^T$ is the singular value decomposition of $\mathbf{A}$ and $\boldsymbol{\Sigma}^+$ is a diagonal matrix with $\frac{1}{\sigma_1}, \ldots, \frac{1}{\sigma_k}$ in the diagonal. So $\mathbf{A}\mathbf{A}^+$ is a diagonal $m \times m$ matrix with $\underbrace{1, \ldots, 1}_{k}, \underbrace{0, \ldots, 0}_{m-k}$ in the diagonal and $\mathbf{A}^+\mathbf{A}$ is a diagonal $n \times n$ matrix with $\underbrace{1, \ldots, 1}_{k}, \underbrace{0, \ldots, 0}_{n-k}$ in the diagonal.*

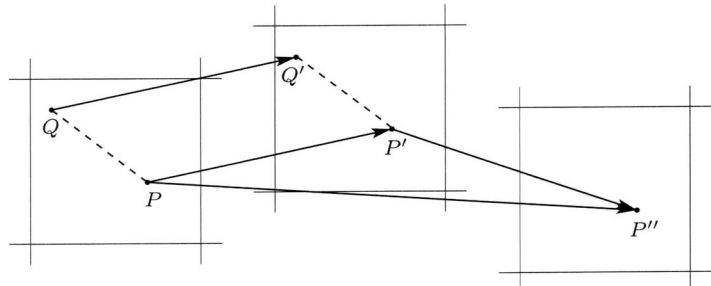

Fig. 2.4 There is a unique translation that maps a point P to another point P'. If it also maps Q to Q' then $\overrightarrow{PP'} = \overrightarrow{QQ'}$. Composition of translations corresponds to addition of vectors, $\overrightarrow{PP''} = \overrightarrow{PP'} + \overrightarrow{P'P''}$

Observe that the pseudo inverse of $\boldsymbol{\Sigma}$ is $\boldsymbol{\Sigma}^+$.

Example 2.56 If we have the equation $\mathbf{Ax} = \mathbf{b}$ and $\mathbf{A} = \mathbf{U}\boldsymbol{\Sigma}\mathbf{V}^T$ is the singular value decomposition of $\mathbf{A}$, then $\mathbf{U}$ and $\mathbf{V}^T$ are invertible, with inverse $\mathbf{U}^{-1} = \mathbf{U}^T$ and $\mathbf{V}^{T-1} = \mathbf{V}$, respectively. We now have $\boldsymbol{\Sigma}\mathbf{V}^T\mathbf{x} = \mathbf{U}^T\mathbf{b}$ and the best we can do is to let $\mathbf{V}^T\mathbf{x} = \boldsymbol{\Sigma}^+\mathbf{U}^T\mathbf{b}$ and hence $\mathbf{x} = \mathbf{V}\boldsymbol{\Sigma}^+\mathbf{U}^T\mathbf{b} = \mathbf{A}^+\mathbf{b}$. If we have an overdetermined system we obtain the least square solution, i.e., the solution to the problem

$$\min_{\mathbf{x}} \|\mathbf{Ax} - \mathbf{b}\|^2. \tag{2.57}$$

If we have an underdetermined system we obtain the least norm solution, i.e., the solution to the problem

$$\min_{\mathbf{x}} \|\mathbf{x}\|^2, \quad \text{such that } \mathbf{Ax} = \mathbf{b}. \tag{2.58}$$

2.2   Affine Spaces

We all know, at least intuitively, two affine spaces, namely the set of points in a plane and the set of points in space. If P and P' are two points in a plane then there is a *unique* translation of the plane that maps P to P', see Fig. 2.4. If the point Q is mapped to Q' then the vector from Q to Q' is the same as the vector from P to P', see Fig. 2.4. That is, we can identify the space of translation in the plane with the set of vectors in the plane. Under this identification addition of vectors corresponds to composition of translations, see Fig. 2.4. Even though we often identify our surrounding space with $\mathbb{R}^3$ and we can add elements of $\mathbb{R}^3$ it does obviously not make sense to add two points in space. The identification with $\mathbb{R}^3$ depends on the choice of coordinate system, and the result of adding the coordinates of two points depends on the choice of coordinate system, see Fig. 2.5.

Fig. 2.5 If we add the coordinates of points in an affine space then the result depends on the choice of origin

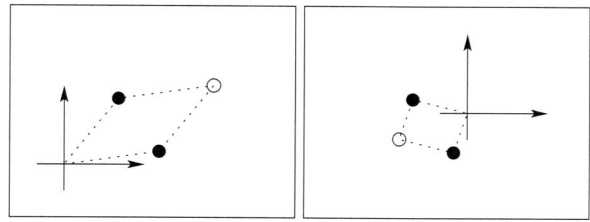

What does make sense in the usual two dimensional plane and three dimensional space is the notion of translation along a vector $\mathbf{v}$. It is often written as adding a vector to a point, $\mathbf{x} \mapsto \mathbf{x} + \mathbf{v}$. An abstract affine space is a space where the notation of translation is defined and where this set of translations forms a vector space. Formally it can be defined as follows.

Definition 2.24 An affine space is a set X that admits a free transitive action of a vector space V. That is, there is a map $X \times V \to X : (\mathbf{x}, \mathbf{v}) \mapsto \mathbf{x} + \mathbf{v}$, called *translation* by the vector $\mathbf{v}$, such that
1. Addition of vectors corresponds to composition of translations, i.e., for all $\mathbf{x} \in X$ and $\mathbf{u}, \mathbf{v} \in V$, $\mathbf{x} + (\mathbf{u} + \mathbf{v}) = (\mathbf{x} + \mathbf{u}) + \mathbf{v}$.
2. The zero vector acts as the identity, i.e., for all $\mathbf{x} \in X$, $\mathbf{x} + \mathbf{0} = \mathbf{x}$.
3. The action is free, i.e., if there for a given vector $\mathbf{v} \in V$ exists a point $\mathbf{x} \in X$ such that $\mathbf{x} + \mathbf{v} = \mathbf{x}$ then $\mathbf{v} = \mathbf{0}$.
4. The action is transitive, i.e., for all points $\mathbf{x}, \mathbf{y} \in X$ exists a vector $\mathbf{v} \in V$ such that $\mathbf{y} = \mathbf{x} + \mathbf{v}$.

The dimension of X is the dimension of the vector space of translations, V.

The vector $\mathbf{v}$ in Condition 4 that translates the point $\mathbf{x}$ to the point $\mathbf{y}$ is by Condition 3 unique, and is often written as $\mathbf{v} = \overrightarrow{\mathbf{x}\mathbf{y}}$ or as $\mathbf{v} = \mathbf{y} - \mathbf{x}$. We have in fact a unique map $X \times X \to V : (\mathbf{x}, \mathbf{y}) \mapsto \mathbf{y} - \mathbf{x}$ such that $\mathbf{y} = \mathbf{x} + (\mathbf{y} - \mathbf{x})$ for all $\mathbf{x}, \mathbf{y} \in X$. It furthermore satisfies
1. For all $\mathbf{x}, \mathbf{y}, \mathbf{z} \in X$, $\mathbf{z} - \mathbf{x} = (\mathbf{z} - \mathbf{y}) + (\mathbf{y} - \mathbf{x})$.
2. For all $\mathbf{x}, \mathbf{y} \in X$ and $\mathbf{u}, \mathbf{v} \in V$, $(\mathbf{y} + \mathbf{v}) - (\mathbf{x} + \mathbf{u}) = (\mathbf{y} - \mathbf{x}) + \mathbf{v} - \mathbf{u}$.
3. For all $\mathbf{x} \in X$, $\mathbf{x} - \mathbf{x} = \mathbf{0}$.
4. For all $\mathbf{x}, \mathbf{y} \in X$, $\mathbf{y} - \mathbf{x} = -(\mathbf{x} - \mathbf{y})$.

Example 2.57 The usual two dimensional plane and three dimensional space are affine spaces and the vector space of translations is the space of vectors in the plane or in space.

Example 2.58 If the set of solutions to k real inhomogeneous linear equations in n unknowns is non empty then it is an affine space and the vector space of translations is the space of solutions to the corresponding set of homogeneous equations.

Example 2.59 If (X, U) and (Y, V) are affine spaces then $(X \times Y, U \times V)$ is an affine space with translation defined by $(\mathbf{x}, \mathbf{y}) + (\mathbf{u}, \mathbf{v}) = (\mathbf{x} + \mathbf{u}, \mathbf{y} + \mathbf{v})$.

A *coordinate system* in an affine space (X, V) consists of a point $O \in X$, called the origin, and a basis $\mathbf{v}_1, \ldots, \mathbf{v}_n$ for V. Any point $\mathbf{x} \in X$ can now be written as

$$\mathbf{x} = O + (\mathbf{x} - O) = O + \sum_{k=1}^{n} x_k \mathbf{v}_k, \tag{2.59}$$

where the numbers $x_1, \ldots, x_n$ are the coordinates for the vector $\mathbf{x} - O$ with respect to the basis $\mathbf{v}_1, \ldots, \mathbf{v}_n$, they are now also called the coordinates for $\mathbf{x}$ with respect to the coordinate system $O, \mathbf{v}_1, \ldots, \mathbf{v}_n$.

2.2.1   Affine and Convex Combinations

We have already noticed that it does not make sense to add points in an affine space, or more generally to take linear combination of points, see Fig. 2.5. So when a coordinate system is chosen it is important to be careful. It is of course possible to add the coordinates of two points and regard the result as the coordinates for a third point. But it is not meaningful. In fact, by changing the origin we can obtain any point by such a calculation.

But even though linear combination does not make sense, affine combination does.

Definition 2.25 A formal sum $\sum_{\ell=1}^{k} \alpha_\ell \mathbf{x}_\ell$ of k points $\mathbf{x}_1, \ldots, \mathbf{x}_k$ is called an *affine combination* if the coefficients sum to 1, i.e., if $\sum_{\ell=1}^{k} \alpha_\ell = 1$. Then we have

$$\sum_{\ell=1}^{k} \alpha_\ell \mathbf{x}_\ell = O + \sum_{\ell=1}^{k} \alpha_\ell (\mathbf{x}_\ell - O), \tag{2.60}$$

where $O \in X$ is an arbitrary chosen point.

Observe that in the last sum we have a linear combination of vectors so the expression makes sense. If we choose an other point O' then the vector between the two results are

$$\left(O + \sum_{\ell=1}^{k} \alpha_\ell (\mathbf{x}_\ell - O) \right) - \left(O' + \sum_{\ell=1}^{k} \alpha_\ell (\mathbf{x}_\ell - O') \right)$$

$$= (O - O') + \sum_{\ell=1}^{k} \alpha_\ell \left((\mathbf{x}_\ell - O) - (\mathbf{x}_\ell - O') \right)$$

$$= (O - O') + \sum_{\ell=1}^{k} \alpha_\ell \left((\mathbf{x}_\ell - \mathbf{x}_\ell) + (O - O') \right)$$

Fig. 2.6 The plane spanned
by three points

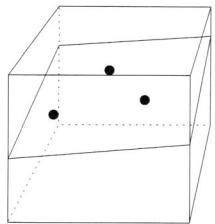

$$= (O - O') - \left(\sum_{\ell=1}^{k} \alpha_\ell\right)(O - O')$$

$$= (O - O') - (O - O') = \mathbf{0}. \tag{2.61}$$

That is, the result does not depend on the auxiliary point O.

Example 2.60 The line spanned by two different points $\mathbf{x}$ and $\mathbf{y}$ in an affine space consists of affine combinations of the two points, that is, the points $(1 - t)\mathbf{x} + t\mathbf{y} = \mathbf{x} + t(\mathbf{y} - \mathbf{x})$, $t \in \mathbb{R}$.

Example 2.61 The plane spanned by three points in space (not on the same line) consists of all affine combinations of the three points, see Fig. 2.6.

Unless the vector space of translations is equipped with an inner product there is no notion of lengths in an affine space. But for points on a line the *ratio of lengths* makes sense. Let $\mathbf{x}_1, \mathbf{x}_2, \mathbf{y}_1, \mathbf{y}_2$ be four points on a line and choose a non zero vector $\mathbf{v}$ on the line, e.g., the difference between two of the given points. Then there exist numbers $t_1, t_2 \in \mathbb{R}$ such that we have $\mathbf{y}_1 - \mathbf{x}_1 = t_1\mathbf{v}$ and $\mathbf{y}_2 - \mathbf{x}_2 = t_2\mathbf{v}$. The ratio between the line segments $\mathbf{x}_1\mathbf{y}_2$ and $\mathbf{x}_2\mathbf{y}_2$ is now defined as $\frac{t_1}{t_2}$. If we had chosen another vector $\mathbf{w}$ then $\mathbf{v} = \alpha\mathbf{w}$ and $\mathbf{y}_k - \mathbf{x}_k = t_k\alpha\mathbf{w}$ and the ratio $\frac{\alpha t_1}{\alpha t_2} = \frac{t_1}{t_2}$ is the same. Observe that we even have a well defined signed ratio.

Definition 2.26 A *convex combination* of points $\mathbf{x}_1, \ldots, \mathbf{x}_k$ is an affine combination $\sum_{\ell=1}^{k} \alpha_\ell \mathbf{x}_\ell$ where all the coefficients are non negative, i.e., $\alpha_\ell \geq 0$ for all $\ell = 1, \ldots, k$.

Example 2.62 The line segment between two points consists of all convex combination of the two points.

Let X be an affine space of dimension n and let $\mathbf{x}_0, \ldots, \mathbf{x}_n$ be $n + 1$ points that are affinely independent, i.e., none of the points can be written as an affine combination of the others. This is equivalent to the vectors $\mathbf{x}_1 - \mathbf{x}_0, \ldots, \mathbf{x}_n - \mathbf{x}_0$ being linearly independent. Then any point $\mathbf{y}$ in X can be written uniquely as an affine combination of the given points, $\mathbf{y} = \sum_{k=0}^{n} \alpha_k \mathbf{x}_k$. The numbers $\alpha_0, \ldots, \alpha_n$ are called *barycentric coordinates* for $\mathbf{y}$ with respect to the points $\mathbf{x}_0, \ldots, \mathbf{x}_n$. The case $n = 2$ is illustrated

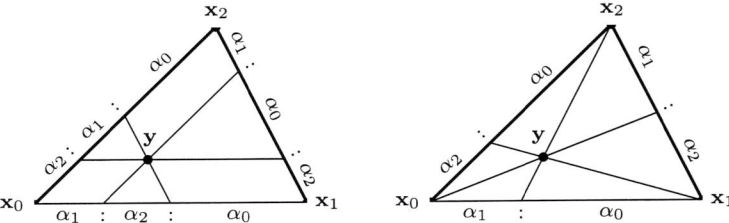

Fig. 2.7 The barycentric coordinates of a point $\mathbf{y} = \alpha_0\mathbf{x}_0 + \alpha_1\mathbf{x}_1 + \alpha_2\mathbf{x}_2$ can be given in terms of the ratio between different line segments

in Fig. 2.7 where two different geometric interpretations of the barycentric coordinates are shown.

2.2.2 Affine Maps

Definition 2.27 An affine map between two affine spaces X and Y is a map $f : X \to Y$ that preserves affine combinations, i.e.,

$$f\left(\sum_{\ell=1}^{k}\alpha_k\mathbf{x}_k\right) = \sum_{\ell=1}^{k}\alpha_k f(\mathbf{x}_k). \tag{2.62}$$

There is a close connection between affine maps between X and Y and linear maps between their vector spaces of translations U and V. More precisely we have the following proposition.

Proposition 2.3 *Let f be an affine map between two affine spaces (X, U) and (Y, V). Then there is a unique linear map $L : U \to V$ such that $f(\mathbf{x} + \mathbf{v}) = f(\mathbf{x}) + L(\mathbf{v})$ for all $\mathbf{x} \in X$ and $\mathbf{u} \in U$.*

We see that $L(\mathbf{v}) = f(\mathbf{x} + \mathbf{v}) - f(\mathbf{x})$ and it turns out that this expression does not depend on $\mathbf{x}$ and that L is linear. If we now choose an origin $O \in X$ and $O' \in Y$ then

$$f(O + \mathbf{v}) = O' + \big(f(O) - O'\big) + L(\mathbf{v}) = O' + \mathbf{y} + L(\mathbf{v}). \tag{2.63}$$

and we see that f is the sum of the linear map L and the translation defined by $\mathbf{y} = f(O) - O' \in V$.

Definition 2.28 A hyperplane in an affine space X is a subset, H, of the form

$$H = f^{-1} = \big\{x \in X \mid f(x) = c\big\}, \tag{2.64}$$

where $f : X \to \mathbb{R}$ is affine and $c \in \mathbb{R}$.

Fig. 2.8 To the left a convex set and to the right a non convex set

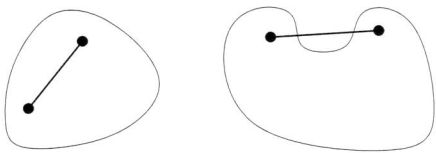

If $x_1, \ldots, x_n$ are coordinates for points in X with respect to some coordinate system, then a hyperplane is given by an equation of the form

$$a_1 x_1 + \cdots + a_n x_n = c. \tag{2.65}$$

A half space is defined in a similar manner.

Definition 2.29 A half space in an affine space X is a subset, H, of the form

$$H = f^{-1} = \{x \in X \mid f(x) \geq c\}, \tag{2.66}$$

where $f : X \to \mathbb{R}$ is affine and $c \in \mathbb{R}$.

If $x_1, \ldots, x_n$ are coordinates for points in X with respect to some coordinate system, then a half space is given by an equation of the form

$$a_1 x_1 + \cdots + a_n x_n \geq c. \tag{2.67}$$

2.2.3 Convex Sets

Definition 2.30 A subset $C \subseteq X$ of an affine space is called *convex* if for each pair of points in C the line segment between the points are in C, see Fig. 2.8.

Definition 2.31 Let $A \subseteq X$ be an arbitrary subset of an affine space X. The *convex hull* of A is the smallest convex set containing A and is denoted $CH(A)$.

There are alternative, equivalent, definitions of the convex hull.
1. The convex hull is the intersection of all convex sets containing A:

$$CH(A) = \bigcap_{\substack{A \subseteq C \\ C \text{ is convex}}} C. \tag{2.68}$$

2. The convex hull is the intersection of all half spaces containing A:

$$CH(A) = \bigcap_{\substack{A \subseteq H \\ H \text{ is a half space}}} H. \tag{2.69}$$

Fig. 2.9 All convex combinations of two, three, and nine points

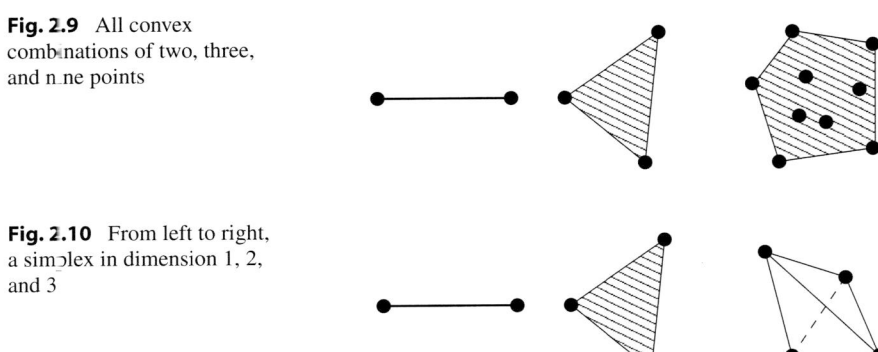

Fig. 2.10 From left to right, a simplex in dimension 1, 2, and 3

3. The convex hull is the set of all convex combinations of points in A:

$$CH(A) = \left\{ \sum_{\ell=1}^{k} \alpha_k \mathbf{x}_k \ \bigg| \ \sum_{\ell=1}^{k} \alpha_k = 1 \wedge \alpha_1, \ldots, \alpha_k \geq 0 \wedge \mathbf{x}_1, \ldots, \mathbf{x}_k \in A \right\}, \quad (2.70)$$

see Fig. 2.9

Definition 2.32 A *simplex* is the convex hull of $n + 1$ affinely independent points in a n dimensional affine space, see Fig. 2.10

Example 2.63 The convex hull of two points is the line segment between the two points, a two simplex.

Example 2.64 The convex hull of three points is a triangle, and its interior a three simplex.

Example 2.65 The convex hull of the unit circle $S^1 = \{(x, y) \in \mathbb{R}^2 \mid x^2 + y^2 = 1\}$ is the closed disk $\{(x, y) \in \mathbb{R}^2 \mid x^2 + y^2 \leq 1\}$.

2.3   Metric Spaces

A metric space is a space where an abstract notion of distance is defined. When we have such a notion we can define continuity of mappings between metric spaces, the notion of convergence and of open and closed sets, and the notion of neighborhoods of a point.

Definition 2.33 A metric space (X, d) is a set X equipped with a map $d : X \times X \to \mathbb{R}$ that satisfies the following three conditions:
1. Symmetry, for all $x, y \in X$: $d(x, y) = d(y, x)$.
2. The triangle inequality, for all $x, y, z \in X$: $d(x, z) \leq d(x, y) + d(y, z)$.

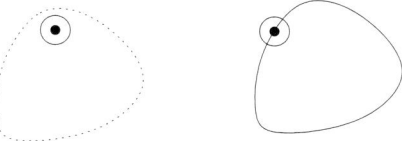

Fig. 2.11 To the *left* an open set, there is room for a ball around each point. To the *right* a non open set, any ball around a point on the boundary is not contained in the set

3. Positivity, for all $x, y \in X$: $d(x, y) \geq 0$ and $d(x, y) = 0 \iff x = y$.

Example 2.66 if $V, \langle \cdot, \cdot \rangle$ is an Euclidean vector space and $\| \cdot \|$ is the corresponding norm, then V equipped with the distance $d(\mathbf{u}, \mathbf{v}) = \|\mathbf{v} - \mathbf{u}\|$ is a metric space.

Example 2.67 If (X, V) is an affine space and V is an Euclidean vector space with norm $\| \cdot \|$, then X equipped with the distance $d(\mathbf{x}, \mathbf{y}) = \|\mathbf{y} - \mathbf{x}\|$ is a metric space.

Example 2.68 If $Y \subseteq X$ is a subset of a metric space (X, d), then Y equipped with the restriction of d to $Y \times Y$ is a metric space.

Example 2.69 If (X_1, d_1) and (X_2, d_2) are metric spaces then the Cartesian product $X_1 \times X_2$ equipped with the distance $d((\mathbf{x}_1, \mathbf{x}_2), (\mathbf{y}_1, \mathbf{y}_2)) = d_1(\mathbf{x}_1, \mathbf{y}_1) + d_2(\mathbf{x}_2, \mathbf{y}_2)$ is a metric space.

Example 2.70 If X is an arbitrary set and we define d by

$$d(x, y) = \begin{cases} 1 & \text{if } x \neq y, \\ 0 & \text{if } x = y, \end{cases} \tag{2.71}$$

then (X, d) is a metric space. This metric is called the *discrete metric*.

Definition 2.34 Let (X, d) be a metric space. The *open ball* with radius $r > 0$ and center $x \in X$ is the set $B(x, r) = \{y \in X \mid d(x, y) < r\}$.

Example 2.71 If X is equipped with the discrete metric and $x \in X$ is an arbitrary point then

$$B(x, r) = \begin{cases} \{x\} & \text{if } r \leq 1, \\ X & \text{if } r > 1. \end{cases}$$

Definition 2.35 Let X be a metric space. A subset $U \subseteq X$ is called an *open set* if there for all points $x \in U$ exists an open ball $B(x, r) \subseteq U$, see Fig. 2.11.

Example 2.72 If X is equipped with the discrete metric then all subsets are open.

Theorem 2.13 *If X is a metric space then the set of open sets has the following three properties:*

1. *The empty set $\emptyset$ and the whole space X are open sets.*
2. *If U_i, $i \in I$ is an arbitrary collection of open sets then their union $\bigcup_{i \in I} U_i$ is an open set.*
3. *If $U_1, \ldots, U_n$ is a finite collection of open sets then their intersection $U_1 \cap \cdots \cap U_n$ is an open set.*

The three properties above are the defining properties of a topological space which is a more general concept. There exist many topological spaces that are not induced by a metric.

Definition 2.36 Let X be a metric space. A subset $F \subseteq X$ is called a *closed set* if the complement $X \setminus F$ is open.

Example 2.73 If X is equipped with the discrete metric then all subsets are closed.

Theorem 2.13 implies that the closed sets have the following three properties:

1. The empty set $\emptyset$ and the whole space X are closed sets.
2. If F_i, $i \in I$ is an arbitrary collection of closed sets then their intersection $\bigcap_{i \in I} F_i$ is a closed set.
3. If $F_1, \ldots, F_n$ is a finite collection of closed sets then their union $F_1 \cup \cdots \cup F_n$ is an closed set.

Definition 2.37 Let $A \subseteq X$ be a subset of a metric space. The *interior* A° of A is the largest open set contained in A. The *closure* $\overline{A}$ is the smallest closed set containing A.

The interior and closure of a subset A of X can equivalently be defined as

$$A^\circ = \left\{ x \in A \mid \exists r > 0 : B(x, r) \subseteq A \right\}, \tag{2.72}$$

$$\overline{A} = \left\{ x \in X \mid \forall r > 0 : B(x, r) \cap A \neq \emptyset \right\}. \tag{2.73}$$

In other words all points in the interior has a surrounding ball contained in A, and all balls centered at points in the closure intersects A.

A subset A of a metric space X is a *neighborhood* of a set $B \subseteq X$ if $B \in A^\circ$, i.e., if and only if there for each $x \in Y$ exists an $r > 0$ such that $B(x, r) \subseteq A$.

Definition 2.38 A sequence $(x_n)_{n \in \mathbb{N}}$ in a metric space (X, d) is called *convergent* with *limit* x if

$$\forall \epsilon > 0 \; \exists n_0 \in \mathbb{N} : n > n_0 \implies d(x, x_n) < \epsilon. \tag{2.74}$$

We write $x_n \to x$ for $n \to \infty$ or $\lim_{n \to \infty} x_n = x$. The formal definition of continuity is as follows.

Definition 2.39 Let (X, d) and (Y, d') be metric spaces. A map $f : X \to Y$ is a *continuous* in a point $x \in X$ if

$$\forall \epsilon > 0 \ \exists \delta > 0 : \forall y \in X : d(x, y) < \delta \implies d'\big(f(x), f(y)\big) < \epsilon.$$

A map $f : X \to Y$ is a *continuous map* if it is continuous at all points of X and it is a *homeomorphism* if it is bijective, continuous and the inverse $f^{-1} : Y \to X$ is continuous too.

There are alternative definitions of continuity.

Theorem 2.14 *Let (X, d) and (Y, d') be metric spaces. A map $f : X \to Y$ is a continuous map if and only if for all convergent sequences $(x_n)_{n \in \mathbb{N}}$ in X, the sequence $(f(x_n))_{n \in \mathbb{N}}$ is convergent in Y and $\lim_{n \to \infty} f(x_n) = f(\lim_{n \to \infty} x_n)$.*

Theorem 2.15 *Let (X, d) and (Y, d') be metric spaces. A map $f : X \to Y$ is a continuous map if and only if for all open set $U \subseteq Y$ the preimage $f^{-1}(U) = \{x \in X \mid f(x) \in U\}$ is an open set in X.*

The last concept we need is compactness.

Definition 2.40 A subset C of a metric space X is called compact if always when C is covered by a collection of open sets, i.e., $C \subseteq \bigcup_{i \in I} U_i$, where U_i is open for all $i \in I$, then there exists a finite number of $U_{i_1}, \ldots, U_{i_n}$ of the given open sets that cover C, i.e., $C \subseteq U_{i_1} \cup \cdots \cup U_{i_n}$.

There is an alternative definition of compact sets.

Theorem 2.16 *A subset C of a metric space X is compact if and only if each sequence $(x_n)_{n \in \mathbb{N}}$ in C has a convergent sub sequence $(x_{n_k})_{k \in \mathbb{N}}$.*

Theorem 2.17 *If C is a compact subset of a metric space X then C is closed and bounded.*

If $X = \mathbb{R}^n$ then the converse is true.

Theorem 2.18 *A subset C of $\mathbb{R}^n$ is compact if and only if C is closed and bounded.*

One of the important properties of compact sets is the following result.

Theorem 2.19 *A continuous function $f : C \to \mathbb{R}$ on a compact set has a minimum and a maximum, i.e., there exist $x_0, x_1 \in C$ such that $f(x_0) \leq f(x) \leq f(x_1)$ for all $x \in C$.*

2.4   Exercises

Exercise 2.1 Prove Proposition 2.1.

Exercise 2.2 Prove that the examples in Examples 2.1–2.11 are vector spaces.

Exercise 2.3 Prove that if V is a vector space and $U \subseteq V$ satisfies the conditions in Definition 2.2 then U is a vector space.

Exercise 2.4 Prove that the examples in Examples 2.13–2.19 are subspaces.

Exercise 2.5 Prove Corollary 2.1.

Exercise 2.6 Show that the monomials as well as the Bernstein polynomials are a basis, cf. Examples 2.29 and 2.30.

Exercise 2.7 Show that the map $\mathbb{R} \to \mathbb{R} : x \mapsto ax + b$ is linear if and only if $b = 0$.

Exercise 2.8 Prove that the maps in Examples 2.31–2.37 are linear.

Exercise 2.9 Prove that the spaces in Examples 2.38–2.43 are Euclidean vector spaces.

Exercise 2.10 Prove the Cauchy–Schwartz inequality, Theorem 2.6. Hint: first note that the theorem is trivial if one of the vectors is the zero vector. Next, use $\langle \mathbf{u} - \alpha \mathbf{v}, \mathbf{u} - \alpha \mathbf{v} \rangle = \|\mathbf{u} - \alpha \mathbf{v}\|^2 \geq 0$ for all $\alpha \in \mathbb{R}$ and find the α that minimize the expression.

Exercise 2.11 Prove the statements in Example 2.44.

Exercise 2.12 Prove that the map in Example 2.47 is symmetric.

Exercise 2.13 Prove the statements in Examples 2.49 and 2.50.

Exercise 2.14 Prove Theorem 2.9. Hint: use Theorem 2.8.

Exercise 2.15 Prove Theorem 2.12. Hint: use Theorem 2.11.

Exercise 2.16 Prove the statements in Example 2.56.

Exercise 2.17 Prove that the spaces in Examples 2.57–2.59 are affine spaces.

Exercise 2.18 Prove Proposition 2.3.

Exercise 2.19 Determine the convex hull of the set

$$A = \big\{(x, y) \in \mathbb{R}^2 \mid x = 0 \wedge y > 0\big\} \cup \big\{(x, y) \in \mathbb{R}^2 \mid x > 0 \wedge xy = 1\big\}.$$

Exercise 2.20 Prove that the spaces in Examples 2.66–2.70 are metric spaces.

Exercise 2.21 Prove the statements in Examples 2.71, 2.72, and 2.73.

Exercise 2.22 Let $\mathbf{A}$ be a real $n \times m$ matrix of rank $m \leq n$, let $\mathbf{x} \in \mathbb{R}^m$, and let $\mathbf{b} \in \mathbb{R}^n$. What is the solution to

$$\min_{\mathbf{x}} f(\mathbf{x})|_{\|\mathbf{x}\|=1},$$

where

$$f(\mathbf{x}) = \mathbf{x}^T \mathbf{A}^T \mathbf{A} \mathbf{x}?$$

Hint: try first the case where $\mathbf{A}$ is a 2×2-matrix, e.g., $\mathbf{A}^T \mathbf{A} = \begin{bmatrix} 5 & -1 \\ -1 & 5 \end{bmatrix}$.

Exercise 2.23 What are the solutions to

$$\max_{\mathbf{x}} f(\mathbf{x}),$$

where

$$f(\mathbf{x}) = \frac{\mathbf{b}^T \mathbf{x}}{\|\mathbf{b}\| \|\mathbf{x}\|}?$$

Exercise 2.24 What geometric object do the points $\mathbf{x} \in \mathbb{R}^3$, fulfilling the equation

$$\mathbf{n}^T \mathbf{x} = \alpha,$$

describe? Here $\mathbf{n} \in \mathbb{R}^3$ and $\alpha \in \mathbb{R}$. Please explain.

References

1. Strang, G.: Linear Algebra and Its Applications, 4th edn. Brooks Cole (2006)
2. Hansen, V.L.: Fundamental Concepts in Modern Analysis. World Scientific, River Edge (1999)

Differential Geometry

This chapter is not a course in differential geometry, for that we refer to the vast literature, e.g., some with a classical approach [1–4], some with a more modern approach [5–7] (closer to the general theory of manifolds). There are also other books [8, 9] and some online notes [10, 11]. This text is only a short overview of the most important concepts in surface theory. There are a few calculations, but for examples or complete proofs the reader is referred to the references.

3.1 First Fundamental Form, Normal, and Area

A surface in space or, after choosing a coordinate system, in $\mathbb{R}^3$, is a set of points $S \subset \mathbb{R}^3$ which is two dimensional in nature. That is, each point $\mathbf{p} \in S$ has a neighborhood which can be parametrized by two coordinates:

$$\mathbf{x} : \mathbb{R}^2 \supseteq U \to \mathbb{R}^3 : (u, v) \mapsto \mathbf{x}(u, v). \tag{3.1}$$

Just as a parametrization of a curve is called regular if the derivative is non vanishing, $\mathbf{x}$ is a *regular parametrization* if the partial derivatives

$$\mathbf{x}_1 = \mathbf{x}_u = \frac{\partial \mathbf{x}}{\partial u}, \qquad \mathbf{x}_2 = \mathbf{x}_v = \frac{\partial \mathbf{x}}{\partial v} \tag{3.2}$$

are linearly independent for each $(u, v) \in U$. In that case they span the tangent space $T_{\mathbf{x}(u,v)}S$, see Fig. 3.1. The *first fundamental form* is the quadratic form on the tangent plane giving the inner product. If ξ, η are coordinates with respect to the basis $\mathbf{x}_u, \mathbf{x}_v$, that is, if $\mathbf{v} = \xi \mathbf{x}_u + \eta \mathbf{x}_v$ then

$$\mathrm{I}(\mathbf{v}) = \mathbf{v} \cdot \mathbf{v} = g_{11}\xi^2 + 2g_{12}\xi\eta + g_{22}\eta^2, \tag{3.3}$$

where the coefficients are given by

$$g_{ij} = \mathbf{x}_i \cdot \mathbf{x}_j, \tag{3.4}$$

J.A. Bærentzen et al., *Guide to Computational Geometry Processing*,
DOI 10.1007/978-1-4471-4075-7_3, © Springer-Verlag London 2012

Fig. 3.1 A tangent vector at
a point $\mathbf{x}$ is the derivative of a
curve on surface through $\mathbf{x}$

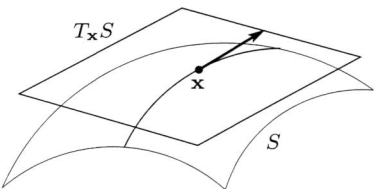

The notation varies in the literature, and often the coefficients g_{ij} are denoted
E, F, G. With respect to the basis $\mathbf{x}_u, \mathbf{x}_v$ the first fundamental form I has the matrix

$$\mathbf{I} = \begin{bmatrix} g_{11} & g_{12} \\ g_{21} & g_{22} \end{bmatrix} = \begin{bmatrix} E & F \\ F & G \end{bmatrix}. \tag{3.5}$$

That is, if a tangent vector $\mathbf{v} = \xi \, \mathbf{x}_u + \eta \, \mathbf{x}_v$ has coordinates (ξ, η) with respect to the
basis $\mathbf{x}_u, \mathbf{x}_v$ then

$$\mathrm{I}(\mathbf{v}) = \begin{pmatrix} \xi & \eta \end{pmatrix} \mathbf{I} \begin{pmatrix} \xi \\ \eta \end{pmatrix}. \tag{3.6}$$

The *normal* is

$$\mathbf{n} = \frac{\mathbf{x}_u \times \mathbf{x}_v}{\|\mathbf{x}_u \times \mathbf{x}_v\|} \quad \text{and} \quad \|\mathbf{x}_u \times \mathbf{x}_v\|^2 = \det \mathbf{I} = g_{11}g_{22} - g_{12}^2. \tag{3.7}$$

By definition $\|\mathbf{x}_u \times \mathbf{x}_v\|$ is the area of the parallelogram spanned by the partial
derivatives $\mathbf{x}_u$ and $\mathbf{x}_v$. So if $A \subset U$ then the *area* of the corresponding subset of the
surface is

$$\mathrm{Area}\big(\mathbf{x}(A)\big) = \int_A \|\mathbf{x}_u \times \mathbf{x}_v\| \, \mathrm{d}u \, \mathrm{d}v = \int_A \sqrt{g_{11}g_{22} - g_{12}^2} \, \mathrm{d}u \, \mathrm{d}v. \tag{3.8}$$

3.2   Mapping of Surfaces and the Differential

First we consider a real function, $f : S \to \mathbb{R}$ defined on surface $S \subseteq \mathbb{R}^3$. It is
called *smooth* if it expressed in the composition with a parametrization (3.1),
$f \circ \mathbf{x} : U \to \mathbb{R}$, is smooth. The *differential* $\mathrm{d}_p f : T_p S \to \mathbb{R}$ at a point $p \in S$ is a
linear map between the tangent space $T_p S$ to S at p to $\mathbb{R}$. The differential is defined
in the following way. If $\mathbf{w} \in T_p S$ is a tangent vector and γ is a smooth curve in S
with $\gamma'(0) = \mathbf{w}$, see Fig. 3.1, then

$$\mathrm{d}_p f \, \mathbf{w} = (f \circ \gamma)'(0). \tag{3.9}$$

A map $f : S \to \mathbb{R}^n$ is given by n coordinate functions, $f = (f_1, \ldots, f_n)$. It is
called *smooth* if all coordinate functions $f_i : S \to \mathbb{R}$ is smooth. The differential of
f at $p \in S$ is the linear map $\mathrm{d}_p f : T_p S \to \mathbb{R}^n$ given by $\mathrm{d}_p f = (\mathrm{d}_p f_1, \ldots, \mathrm{d}_p f_n)$.

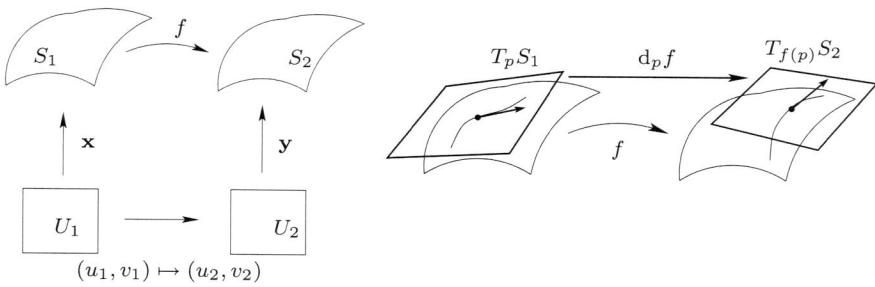

Fig. 3.2 A map between two surfaces and its differential

Finally we consider a map $f : S_1 \to S_2$ between two surfaces, see Fig. 3.2. It *smooth* if it considered as a map into $\mathbb{R}^3$ is smooth. Equivalently, it is smooth if it, expressed in local coordinates, i.e., composed with parametrizations

$$\mathbf{y}^{-1} \circ f \circ \mathbf{x} : U_1 \to U_2 : (u_1, v_1) \mapsto (u_2, v_2), \tag{3.10}$$

is smooth. The image of curve γ is S_1 is a curve in S_2 and its derivative is a tangent vector to S_2. So the differential $\mathrm{d}_p f : T_p S_1 \to \mathbb{R}^3$ maps into $T_{f(p)} S_2$. That is, the *differential* is a linear map $\mathrm{d}_p f : T_p S_1 \to T_{f(p)} S_2$ between the tangent space $T_p S_1$ to S_1 at p to the tangent space $T_{f(p)} S_2$ to S_2 at $f(p)$. If we in particular consider the curve $\gamma(t) = \mathbf{x}(u_0 + tw_1, v_0 + tw_2)$ then we have

$$f \circ \gamma(t) = \mathbf{y} \circ \big(\mathbf{y}^{-1} \circ f \circ \mathbf{x}\big)(u_0 + tw_1, v_0 + tw_2)$$
$$= \mathbf{y}\big(u_2(u_0 + tw_1, v_0 + tw_2), v_2(u_0 + tw_1, v_0 + tw_2)\big),$$

hence

$$\mathbf{w} = \gamma'(0) = w_1 \mathbf{x}_u + w_2 \mathbf{x}_v,$$

$$\mathrm{d}_{\gamma(0)} f \mathbf{w} = (f \circ \gamma)'(0)$$
$$= \left(\frac{\partial u_2}{\partial u_1} w_1 + \frac{\partial u_2}{\partial v_1} w_2\right) \mathbf{y}_u + \left(\frac{\partial v_2}{\partial u_1} w_1 + \frac{\partial v_2}{\partial v_1} w_2\right) \mathbf{y}_v.$$

So we see that with respect to the basis $\mathbf{x}_u, \mathbf{x}_v$ in $T_{\gamma(0)} S_1$ and the basis $\mathbf{y}_u, \mathbf{y}_v$ in $T_{f(\gamma(0))} S_2$ the differential $\mathrm{d}_{\gamma(0)} f$ has the matrix

$$\mathrm{d}f \sim \mathbf{J} = \begin{pmatrix} \frac{\partial u_2}{\partial u_1} & \frac{\partial u_2}{\partial v_1} \\ \frac{\partial v_2}{\partial u_1} & \frac{\partial v_2}{\partial v_1} \end{pmatrix}. \tag{3.11}$$

That is, we have the Jacobian $\mathbf{J}$ of the local expression $(u_1, v_1) \mapsto (u_2, v_2)$.

Definition 3.1 A smooth map $f : S_1 \to S_2$ between two surfaces is
1. an *isometry* if it preserves the length of curves;

2. an *conformal map* if it preserves angles;
3. *equiarea* if it preserves area.

These three properties are used in Chap. 10 and can be expressed in terms of the differential of f.

Theorem 3.1 *Let* I_1 *and* I_2 *be the first fundamental form of the surfaces* S_1 *and* S_2, *respectively. Then we have the following:*

- *The map* f *is an isometry if and only if the differential* $d_p f$ *is an isometry at each point* $p \in S_1$. *This is the case if and only if* $\mathbf{I}_1 = \mathbf{J}^T \mathbf{I}_2 \mathbf{J}$.
- *The map* f *is conformal if and only if the differential* $d_p f$ *is a scaling at each point* $p \in S_1$. *This is the case if and only if* $\mathbf{I}_1 = \lambda \mathbf{J}^T \mathbf{I}_2 \mathbf{J}$, *where* λ *is a function on* S_1.
- *The map* f *is equiarea if and only if the differential* $d_p f$ *is equiarea at each point* $p \in S_1$. *This is the case if and only if* $\det \mathbf{I}_1 = (\det \mathbf{J})^2 \det \mathbf{I}_2$.

We immediately see that if the map is both conformal and equiarea then $\mathbf{I}_1 = \lambda \mathbf{J}^T \mathbf{I}_2 \mathbf{J}$ and $\lambda = 1$ so the map is an isometry.

3.3 Second Fundamental Form, the Gauß Map and the Weingarten Map

Just as the curvature of a curve expresses how fast the tangent varies along the curve, the curvature of a surface expresses how fast the normal varies along the surface. The *Gauß map* takes a point on the surface to the normal, i.e., in local coordinates it is given by $(u, v) \mapsto \mathbf{n}(u, v)$. This is obviously a map from the surface S to the unit sphere S^2. Furthermore, the normal to S^2 at the point $\mathbf{n} \in S^2$ is $\mathbf{n}$ itself so the tangent spaces $T_{\mathbf{n}(u,v)} S^2$ and $T_{\mathbf{x}(u,v)} S$ are parallel. Thus, there is a unique translation in $\mathbb{R}^3$ which maps one onto the other. The *shape operator* or *Weingarten map* is the differential of the Gauß map

$$W = d\mathbf{n} : T_{\mathbf{x}(u,v)} S \to T_{\mathbf{n}(u,v)} S^2 \cong T_{\mathbf{x}(u,v)} S. \tag{3.12}$$

If $\mathbf{v} = \xi \mathbf{x}_u + \eta \mathbf{x}_v$ is a tangent vector then

$$W(\mathbf{v}) = d\mathbf{n}(\mathbf{v}) = \xi \mathbf{n}_u + \eta \mathbf{n}_v \tag{3.13}$$

and

$$\mathbf{v} \cdot W(\mathbf{v}) = (\xi \mathbf{n}_u + \eta \mathbf{n}_v) \cdot (\xi \mathbf{x}_u + \eta \mathbf{x}_v)$$
$$= \xi^2 \mathbf{n}_u \cdot \mathbf{x}_u + \xi \eta (\mathbf{n}_u \cdot \mathbf{x}_v + \mathbf{n}_v \cdot \mathbf{x}_u) + \eta^2 \mathbf{n}_v \cdot \mathbf{x}_v. \tag{3.14}$$

This is a measure of the rate of change of the normal in the direction $\mathbf{v}$. Differentiation of the equations $\mathbf{n} \cdot \mathbf{x}_u = 0$ and $\mathbf{n} \cdot \mathbf{x}_v = 0$ shows that

$$\mathbf{x}_{ij} \cdot \mathbf{n} = -\mathbf{x}_i \cdot \mathbf{n}_j = -\mathbf{x}_j \cdot \mathbf{n}_i. \tag{3.15}$$

The *second fundamental form* is the following quadratic form on the tangent space:

$$\mathbb{II}(\mathbf{v}) = -W(\mathbf{v}) \cdot \mathbf{v} = b_{11}\xi^2 + 2b_{12}\xi\eta + b_{22}\eta^2, \tag{3.16}$$

where the coefficients b_{ij} are given by

$$b_{ij} = \mathbf{x}_{ij} \cdot \mathbf{n} = -\mathbf{x}_i \cdot \mathbf{n}_j = -\mathbf{x}_j \cdot \mathbf{n}_i. \tag{3.17}$$

Again, the notation varies in the literature, the coefficients (b_{ij}) can also be denoted (e, f, g), (L, M, N), or (l, m, n). With respect to the basis $\mathbf{x}_u, \mathbf{x}_v$ the second fundamental form $\mathbb{II}$ has the matrix

$$\mathbb{II} = \begin{bmatrix} b_{11} & b_{12} \\ b_{21} & b_{22} \end{bmatrix} = \begin{bmatrix} L & M \\ M & N \end{bmatrix} = \begin{bmatrix} e & f \\ f & g \end{bmatrix} = \begin{bmatrix} l & m \\ m & n \end{bmatrix} = \cdots. \tag{3.18}$$

That is, if a tangent vector $\mathbf{v} = \xi\,\mathbf{x}_u + \eta\,\mathbf{x}_v$ has coordinates (ξ, η) with respect to the basis $\mathbf{x}_u, \mathbf{x}_v$ then

$$\mathbb{II}(\mathbf{v}) = (\xi \ \eta)\mathbb{II}\begin{pmatrix} \xi \\ \eta \end{pmatrix}. \tag{3.19}$$

If the Weingarten map W has the matrix $\mathbf{W}$ with respect to the basis $\mathbf{x}_u, \mathbf{x}_v$ in the tangent plane $T_{\mathbf{x}(u,v)}S$ then (3.14) can be written $\mathbb{II} = -\mathbf{I}\,\mathbf{W}$. Thus, with respect to the basis $\mathbf{x}_u, \mathbf{x}_v$ the Weingarten map has the matrix

$$\mathbf{W} = -\mathbf{I}^{-1}\mathbb{II} = -\begin{bmatrix} g_{11} & g_{12} \\ g_{21} & g_{22} \end{bmatrix}^{-1}\begin{bmatrix} b_{11} & b_{12} \\ b_{21} & b_{22} \end{bmatrix}$$

$$= \frac{-1}{g_{11}g_{22} - g_{12}^2}\begin{bmatrix} g_{22} & -g_{12} \\ -g_{21} & g_{11} \end{bmatrix}\begin{bmatrix} b_{11} & b_{12} \\ b_{21} & b_{22} \end{bmatrix}. \tag{3.20}$$

The Weingarten map W is a *symmetric map*, i.e., $W(\mathbf{u}) \cdot \mathbf{v} = \mathbf{u} \cdot W(\mathbf{v})$, but $\mathbf{W}$ is a symmetric matrix if and only if $\mathbf{x}_u$ and $\mathbf{x}_v$ are orthogonal and have the same length.

3.4   Smoothness of a Surface

If we compose a C^k-parametrization (see Example 2.6) with a change of a parameter of class C^{k-1} then the resulting new parametrization is of class C^{k-1}, but the surface is of course the same. That is, the smoothness of a surface is not the same as the smoothness of an arbitrary parametrization. We define the *smoothness of a surface* as the *maximal smoothness of a regular parametrization*. It is not hard to see that this is the same as the *smoothness of the projection from the tangent plane to the surface*. The representation of the surface as a graph over its tangent plane, as shown in Fig. 3.3, is interesting from other points of view. Let the surface be C^2, let $\mathbf{v}_1, \mathbf{v}_2$ be a positive basis (not necessarily orthonormal) for the tangent plane, i.e., $\mathbf{n} = $

Fig. 3.3 The surface as a graph over its tangent plane

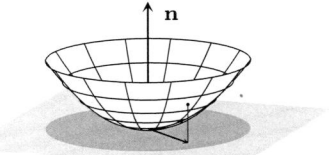

Fig. 3.4 A planar section of a surface. If $\theta = \frac{\pi}{2}$ we have a normal section. In that case $\kappa_n = \pm\kappa$ and $\kappa_g = 0$ at the origin

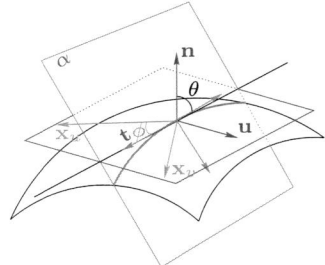

$\mathbf{v}_1 \times \mathbf{v}_2 / \|\mathbf{v}_1 \times \mathbf{v}_2\|$, and let u, v, w be coordinates on $\mathbb{R}^3$ with respect to the basis $\mathbf{v}_1, \mathbf{v}_2, \mathbf{n}$. Then the projection from the tangent plane to the surface has the expansion

$$w = b_{11}u^2 + 2b_{12}uv + b_{22}v^2 + o(u^2 + v^2), \qquad (3.21)$$

i.e., the second order Taylor polynomial of the projection from the tangent plane to the surface is given by the second fundamental form.

3.5   Normal and Geodesic Curvature

Let $\mathbf{r}(t) = \mathbf{x}(u(t), v(t))$ be a curve on the surface S. Denote the curve tangent by $\mathbf{t}$, the curvature vector by $\kappa = d\mathbf{t}/ds$, where s is arc length on the curve. If $\mathbf{u} = \mathbf{n} \times \mathbf{t}$ then the *Darboux frame* is $\mathbf{t}, \mathbf{u}, \mathbf{n}$, see Fig. 3.4. The tangent vector is

$$\mathbf{t} = \frac{d\mathbf{r}}{ds} = \frac{du}{ds}\mathbf{x}_u + \frac{dv}{ds}\mathbf{x}_v, \qquad (3.22)$$

and the curvature vector is

$$\begin{aligned}
\kappa &= \frac{d\mathbf{t}}{ds} = \frac{d^2\mathbf{r}}{ds^2} \\
&= \frac{d^2 u}{ds^2}\mathbf{x}_u + \frac{d^2 v}{ds^2}\mathbf{x}_v + \left(\frac{du}{ds}\right)^2\mathbf{x}_{uu} + 2\frac{du}{ds}\frac{dv}{ds}\mathbf{x}_{uv} + \left(\frac{dv}{ds}\right)^2\mathbf{x}_{vv}.
\end{aligned} \qquad (3.23)$$

The curvature vector is orthogonal to the tangent vector, so it can be written

$$\kappa = \kappa_n \mathbf{n} + \kappa_g \mathbf{u}, \qquad (3.24)$$

where κ_n is called the *normal curvature* and κ_g is called the *geodesic curvature*. The normal curvature is

$$\kappa_n = \boldsymbol{\kappa} \cdot \mathbf{n} = \left(\frac{du}{ds}\right)^2 b_{11} + 2\frac{du}{ds}\frac{dv}{ds}b_{12} + \left(\frac{dv}{ds}\right)^2 b_{22} = \mathbb{I}(\mathbf{t}). \qquad (3.25)$$

Observe that it *depends only on the tangent!* Consequently it only depends on the first derivative. If we use another parameter than arc length s on the curve, say t, then we have

$$\frac{du}{dt} = \frac{ds}{dt}\frac{du}{ds} \quad \text{and} \quad \frac{dv}{dt} = \frac{ds}{dt}\frac{dv}{ds}, \qquad (3.26)$$

so

$$\mathrm{I}\big(\mathbf{r}'(t)\big) = \left(\frac{ds}{dt}\right)^2 \mathrm{I}(\mathbf{t}) = \left(\frac{ds}{dt}\right)^2, \qquad (3.27)$$

and

$$\mathbb{I}\big(\mathbf{r}'(t)\big) = \left(\frac{ds}{dt}\right)^2 \mathbb{I}(\mathbf{t}) = \mathrm{I}\big(\mathbf{r}'(t)\big)\mathbb{I}(\mathbf{t}). \qquad (3.28)$$

Thus

$$\kappa_n = \mathbb{I}(\mathbf{t}) = \frac{\mathbb{I}(\mathbf{r}')}{\mathrm{I}(\mathbf{r}')}. \qquad (3.29)$$

The *geodesic curvature* is the remaining part of the curvature

$$\kappa_g = \boldsymbol{\kappa} \cdot \mathbf{u}. \qquad (3.30)$$

The minimal curvature at a point of a curve with a certain tangent vector $\mathbf{t}$ is the normal curvature $\kappa_n = \mathbb{I}(\mathbf{t})$. It is obtained when the geodesic curvature vanishes. A *geodesic* is a curve with vanishing geodesic curvature, i.e., a curve with minimal curvature and it is not surprising that it is a curve that minimizes length locally.

3.6 Principal Curvatures and Direction

The Weingarten map is a symmetric linear map so it has real eigenvalues and the corresponding eigenvectors are orthogonal. The *principal curvatures* and *principal directions* are the eigenvalues and eigenvectors of minus the Weingarten map, i.e., they are solutions to

$$-W(\mathbf{v}) = \kappa\mathbf{v} \iff \mathbf{I}^{-1}\mathbf{II}\underline{\mathbf{v}} = \kappa\underline{\mathbf{v}} \iff \mathbf{II}\underline{\mathbf{v}} = \kappa\mathbf{I}\underline{\mathbf{v}}, \qquad (3.31)$$

where $\underline{\mathbf{v}}$ is the column vector containing the coordinates of $\mathbf{v}$. If $W\mathbf{v}_i = -\kappa_i\mathbf{v}_i$, where $\mathbf{v}_1, \mathbf{v}_2$ are unit eigenvectors for W, and consequently is an orthonormal basis for the tangent plane, then substituting $\mathbf{t} = \cos\theta\,\mathbf{v}_1 + \sin\theta\,\mathbf{v}_2$ into (3.25) shows *Euler's formula*:

$$\kappa_n = \kappa_1\cos^2\theta + \kappa_2\sin^2\theta, \qquad (3.32)$$

Fig. 3.5 A normal section of
a surface

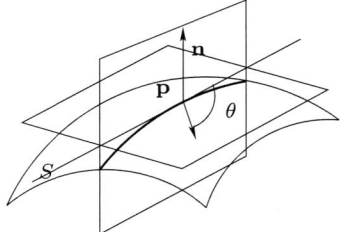

where θ is the angle between the first principal direction and the tangent of the
curve.

Hence, the principal curvatures and principal directions are the maximum and
minimum value of the normal curvature and the directions in which they are obtained.

If we intersect a surface with a plane containing the surface normal **n** at a point
$\mathbf{p} \in S$ we obtain a *normal section* of the surface, see Fig. 3.5. At the point **p** the
signed planar curvature and the normal curvature of the normal section agrees, up
to a sign. So it is a surprising fact that the curvature of a normal section is given by
Euler's formula (3.32). In particular, when the intersecting plane of a normal section
is rotated around the normal then the curvature of intersection is either constant
or has exactly one minimum and one maximum, which are attained in orthogonal
directions.

3.7 The Gaußian and Mean Curvature

The *Gaußian curvature* is the product of the principal curvatures:

$$K = \kappa_1\kappa_2 = \det W = \det \mathbf{W} = \frac{\det \mathbb{I\!I}}{\det \mathbf{I}} = \frac{b_{11}b_{22} - b_{12}^2}{g_{11}g_{22} - g_{12}^2}. \tag{3.33}$$

Points on a surface are classified according to the sign of the Gaußian curvature K.
If $K > 0$ then we have an *elliptic point* and the surface curves the same way in
all directions. If $K < 0$ then we have a *hyperbolic point* and the surface curves
towards the normal in some directions and away from the normal in other directions,
see Fig. 3.6. If $K = 0$ and not both principal curvatures are zero then we have a
parabolic point, and finally if both principal curvatures are zero then we have a
planar point. On a negatively curved surface there are two directions where the
normal curvature is zero and they are called the *asymptotic directions*.

The *mean curvature* is the mean value of the principal curvatures,

$$H = \frac{\kappa_1 + \kappa_2}{2} = \frac{-\operatorname{tr}\mathbf{W}}{2} = \frac{\operatorname{tr}(\mathbf{I}^{-1}\mathbb{I\!I})}{2}$$

$$= \frac{1}{2}\frac{g_{11}b_{22} - 2g_{12}b_{12} + g_{22}b_{11}}{g_{11}g_{22} - g_{12}^2}. \tag{3.34}$$

Fig. 3.6 On the *left* a positively curved surface, on the *right* a negatively curved surface

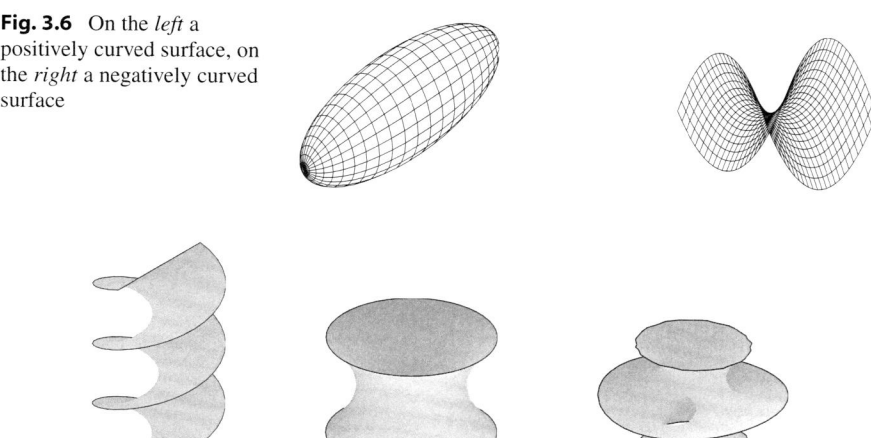

Fig. 3.7 Minimal surfaces. On the *left* a helicoid, in the *middle* a catenoid, and on the *right* Costa's minimal surface

It follows from (3.39) below that if a surface has zero mean curvature then the area of the surface is minimal, at least locally. Consequently, a surface with $H = 0$ is called a *minimal surface*, and it is the mathematical model of a soap film, see Fig. 3.7.

Calculation of curvature requires second order derivatives and it is unstable to estimate derivatives. Integrals are stable to estimate and we will later use integrals to estimate the mean and Gaußian curvature of meshes.

Let $U \subseteq S$ be a connected subset of the surface S and consider the area of the image $\mathbf{n}(U)$ on the unit sphere under the Gauß map. By the mean value theorem we have

$$\text{Area}\big(\mathbf{n}(U)\big) = \int_U \det(\mathrm{d}\mathbf{n})\,\mathrm{d}A = \int_U \det W\,\mathrm{d}A = \int_U K\,\mathrm{d}A$$

$$= K(\mathbf{q}) \int_U \mathrm{d}A = K(\mathbf{q})\text{Area}(U),$$

where $\mathbf{q} \in U$ is a suitably chosen point. Thus, for a point $\mathbf{p}$ on the surface we have the following expression for the Gaußian curvature:

$$K(\mathbf{p}) = \lim_{U \to \mathbf{p}} \frac{\text{Area}(\mathbf{n}(U))}{\text{Area}(U)}, \tag{3.35}$$

where $U \to \mathbf{p}$ means $\mathbf{p} \in U$ and diam $U \to 0$. This, in fact, is Gauß' original definition and we will use it to give an estimate of the Gaußian curvature of a mesh, see (8.7).

Now consider an offset $\widetilde{S}$ of the surface S,

$$\widetilde{\mathbf{x}}(u, v) = \mathbf{x}(u, v) + h(u, v)\,\mathbf{n}(u, v). \tag{3.36}$$

The partial derivatives are $\widetilde{\mathbf{x}}_i = \mathbf{x}_i + h_i \mathbf{n} + h \mathbf{n}_i$. If we calculate the first fundamental form of the offset $\widetilde{S}$ to first order in h then we get

$$\widetilde{g}_{ij} = \widetilde{\mathbf{x}}_i \cdot \widetilde{\mathbf{x}}_j \approx \mathbf{x}_i \cdot \mathbf{x}_j + h\,(\mathbf{x}_i \cdot \mathbf{n}_j + \mathbf{n}_i \cdot \mathbf{x}_j) = g_{ij} - 2h b_{ij}. \tag{3.37}$$

The determinant is

$$\det\widetilde{\mathbf{I}} \approx \det(\mathbf{I} - 2h\mathbb{I}) = \det(\mathbf{I})\det\big(1 - 2h\mathbf{I}^{-1}\mathbb{I}\big)$$

$$\approx \det(\mathbf{I})\big(1 - 2h\operatorname{tr}\big(\mathbf{I}^{-1}\mathbb{I}\big)\big) = \det(\mathbf{I})(1 - 4hH), \tag{3.38}$$

and the square root is $\sqrt{\det\widetilde{\mathbf{I}}} \approx \sqrt{\det(\mathbf{I})}(1 - 2hH)$. So if $U \subseteq S$ and $\widetilde{U} \subseteq \widetilde{S}$ is the offset of U, then $\widetilde{\mathbf{x}}^{-1}(\widetilde{U}) = \mathbf{x}^{-1}(U)$, and

$$\operatorname{Area}(\widetilde{U}) = \int_{\widetilde{\mathbf{x}}^{-1}(\widetilde{U})} \sqrt{\det\widetilde{\mathbf{I}}}\, du\, dv \approx \int_{\mathbf{x}^{-1}(U)} \sqrt{\det(\mathbf{I})}(1 - 2hH)\, du\, dv$$

$$= \big(1 - 2h(Q)H(Q)\big)\int_{\mathbf{x}^{-1}(U)} \sqrt{\det(\mathbf{I})}\, du\, dv$$

$$= \big(1 - 2h(Q)H(Q)\big)\operatorname{Area}(U), \tag{3.39}$$

for a suitably chosen $Q \in U$. We finally get the following expression for the mean curvature:

$$H(\mathbf{p}) = -\lim_{\substack{h\to 0 \\ U\to\mathbf{p}}} \frac{\operatorname{Area}(\widetilde{U}) - \operatorname{Area}(U)}{2h\,\operatorname{Area}(U)} = \lim_{U\to\mathbf{p}} \frac{\nabla_h\operatorname{Area}(U)}{2\operatorname{Area}(U)}. \tag{3.40}$$

We use this to give an estimate of the mean curvature of a mesh, see (8.4).

3.8   The Gauß–Bonnet Theorem

Consider a region $R \subseteq S$ bounded by a piecewise regular curve C with exterior angles ϕ_i at the break points, see Fig. 3.8. The *Gauß–Bonnet theorem* states that

$$\int_R K\, dA + \int_C \kappa_g\, ds + \sum \phi_i = 2\pi\,\chi(R), \tag{3.41}$$

where $\chi(R)$ denotes the *Euler characteristic* of R. It is defined by

$$\chi(R) = \#\text{Faces} - \#\text{Edges} + \#\text{Vertices} \tag{3.42}$$

in an arbitrary triangulation of R. We have in particular for a triangle T: $\chi(T) = 1$ and hence

$$\int_T K\, dA + \sum_{i=1}^{3} \int_{C_i} \kappa_g\, ds + \sum_{i=1}^{3} \phi_i = 2\pi. \tag{3.43}$$

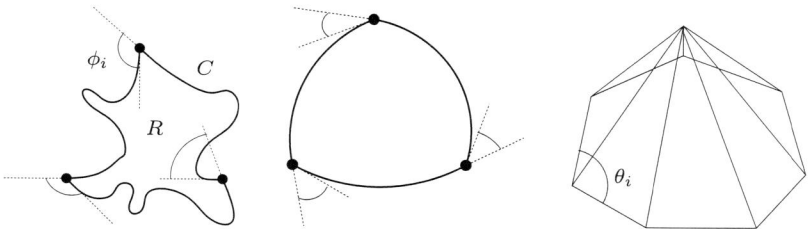

Fig. 3.8 A general triangle, a geodesic triangle, and a geodesic polygon. The exterior angle ϕ_i is the angle between the incoming and outgoing tangent vector

If the triangle is geodesic then the geodesic curvature of the edges vanishes and we have

$$\int_T K \, \mathrm{d}A = 2\pi - \sum_{i=1}^{3} \phi_i = \underbrace{\sum_{i=1}^{3} (\pi - \phi_i)}_{\text{interior angle} \,=\, \theta_i} - \pi. \tag{3.44}$$

For a general geodesic n-gon P we have

$$\int_P K \, \mathrm{d}A = \sum_{i=1}^{n} \theta_i - (n-2)\pi, \tag{3.45}$$

where θ_i denotes the interior angles. We get a new estimate for the Gaußian curvature which also can be used to justify the estimate (8.7).

3.9 Differential Operators on Surfaces

Let $f : S \to \mathbb{R}$ be a real function defined on a surface S. The (*intrinsic*) *gradient*, $\nabla f(p)$, of f at $p \in S$ is defined by the equation

$$\nabla f(p) \cdot \mathbf{v} = \mathrm{d}_p f \, \mathbf{v}, \quad \text{for all } \mathbf{v} \in T_p S. \tag{3.46}$$

Before we can express the gradient in local coordinates we need the inverse of the first fundamental form.

Definition 3.2 The entries of the matrix which is the inverse of matrix, (g_{ij}), of the first fundamental form are denoted g^{ij}. That is,

$$\sum_{k} g^{ik} g_{kj} = \delta^i_j = \begin{cases} 1 & \text{if } i = j, \\ 0 & \text{if } i \neq j. \end{cases} \tag{3.47}$$

We leave the proof of the following theorem as Exercise 3.5.

Theorem 3.2 *If* $\mathbf{x} : U \to S$ *is a parametrization of a surface* S *and* $f : S \to \mathbb{R}$ *is a smooth function, then the gradient of* f *is given by*

$$\nabla f = \sum_{k,\ell} g^{k\ell} \frac{\partial(f \circ \mathbf{x})}{\partial u_\ell} \mathbf{x}_k. \tag{3.48}$$

The collection of all tangent spaces to a surface S is called the tangent bundle, $TS = \bigcup_{p \in S} T_p S$, and a *vector field* is a map $\mathbf{v} : S \to TS$ such that $\mathbf{v}(p) \in T_p S$. If $\mathbf{x} : U \to S$ is a parametrization of S then a vector field can be written as $\mathbf{v} = v^1 \mathbf{x}_1 + v^2 \mathbf{x}_2$, where v^1 and v^2 are real functions. The vector field is called smooth if these functions are smooth.

If $\mathbf{v}$ is a smooth vector field on S then there exists a map $\Phi : \mathbb{R} \times S \to S$, called the *flow* of $\mathbf{v}$, such that

$$\Phi(0, p) = p, \qquad \frac{\partial \Phi}{\partial t}(t, p) = \mathbf{v}\big(\Phi(t, p)\big),$$
$$\Phi(t + s, p) = \Phi\big(t, \Phi(s, p)\big). \tag{3.49}$$

The curves $\gamma : t \mapsto \Phi(t, p)$ are integral curves of $\mathbf{v}$, i.e., $\gamma'(t) = \mathbf{v}(\gamma(t))$. Pictorially the points in S flows along the integral curves.

One way of thinking about the *(intrinsic) divergence*, div $\mathbf{v}$, of a smooth vector field $\mathbf{v}$ is that it measures how volume changes as it flows along the integral curves of $\mathbf{v}$. That is, the divergence is characterized by the equation

$$\int_{\mathbf{x}^{-1}(B)} \text{div}\,\mathbf{v} \, du \, dv = \frac{d}{dt} \int_{\mathbf{x}^{-1}(\Phi(t,B))} \sqrt{\det(g_{ij})} \, du \, dv \bigg|_{t=0}, \tag{3.50}$$

for all subsets $B \subseteq S$, where $\mathbf{x} : U \to S$ is a parametrization, and g_{ij} are the components of the first fundamental form. The divergence theorem also holds for surfaces, i.e., we have the following theorem.

Theorem 3.3 *Let* $\mathbf{v}$ *be smooth vector field on a surface* S *and let* $B \subseteq S$ *be a domain with a piecewise smooth boundary* ∂B. *If* $\mathbf{n}$ *is the outward normal of the domain* B *and* s *is arc-length on* ∂B *then*

$$\int_{\mathbf{x}^{-1}(B)} \text{div}\,\mathbf{v} \, du \, dv = \int_{\partial B} \mathbf{v} \cdot \mathbf{n} \, ds. \tag{3.51}$$

It will take us to far afield to derive the expression for the divergence in local coordinates, so we will just state the result without proof.

Theorem 3.4 *If* $\mathbf{x} : U \to S$ *is a parametrization of a surface* S *and* $\mathbf{v}$ *is a smooth vector field on* S, *then the divergence of* $\mathbf{v} \circ \mathbf{x} = \sum_k v^k \mathbf{x}_k$ *is given by*

$$\text{div}\,\mathbf{v} = \sum_k \frac{1}{\sqrt{\det(g_{ij})}} \frac{\partial}{\partial u_k}\big(\sqrt{\det(g_{ij})}\, v^k\big). \tag{3.52}$$

Just as for the usual Laplacian, the *Laplace–Beltrami operator* is defined as

$$\triangle f = \operatorname{div} \nabla f. \tag{3.53}$$

Knowing Theorems 3.2 and 3.4 it is not hard to derive the expression for Laplace–Beltrami operator in local coordinates. We leave it as Exercise 3.6.

Theorem 3.5 *If $\mathbf{x} : U \to S$ is a parametrization of a surface S and f is a smooth function on S, then the Laplace–Beltrami operator of f is given by*

$$\triangle f = \sum_{k,\ell} \frac{1}{\sqrt{\det(g_{ij})}} \frac{\partial}{\partial u_k} \left(\sqrt{\det(g_{ij})} g^{k\ell} \frac{\partial(f \circ \mathbf{x})}{\partial u_\ell} \right). \tag{3.54}$$

3.10 Implicitly Defined Surfaces

Alternatively a surface can be given as the solution to an equation in three unknowns, $f(x, y, z) = c$. The implicit function theorem [12] ensures that if f is a C^n function and at a point on the surface the gradient ∇f is non vanishing, then the surface can be parametrized locally by a C^n parametrization.

Sometimes this implicit representation has some advantages over the representation as a parametrized surface

$$\mathbf{x} : (u, v) \mapsto \big(x(u, v), y(u, v), z(u, v)\big).$$

E.g., if we want to find the intersection of two surfaces, then we have the following three possibilities:

- Two parametric surfaces. Here we have to solve three equations in four unknowns (u, v, s, t):

$$x_1(u, v) = x_2(s, t), \qquad y_1(u, v) = y_2(s, t), \qquad z_1(u, v) = z_2(s, t).$$

- One parametric surface and one implicit surface. Here we have to solve one equation in two unknowns (u, v):

$$f\big(x(u, v), y(u, v), z(u, v)\big) = c.$$

- Two implicit surfaces. Here we have to solve two equations in three unknowns (x, y, z):

$$f_1(x, y, z) = c_1, \qquad f_2(x, y, z) = c_2.$$

Clearly the second option is the best, so it would be best if we could have a surface represented both as a parametric surface and as an implicit surface. Mathematically that is not a problem. Any surface can be represented in both ways. In practice there are problems and the best one can hope for is to maintain a hybrid representation where the parametric and implicit representations agree up to some given tolerance.

In this note we will not be concerned with that aspect, but we will just demonstrate how to determine the second fundamental form, the principal curvatures and directions, and Gaußian and mean curvature for an implicit surface.

3.10.1 The Signed Distance Function

First we will assume that the surface is given as

$$S = h^{-1}(0) = \{(x, y, z) \in \mathbb{R}^3 \mid h(x, y, z) = 0\}, \tag{3.55}$$

where h is a C^2 function and

$$\left(\frac{\partial h}{\partial x}\right)^2 + \left(\frac{\partial h}{\partial y}\right)^2 + \left(\frac{\partial h}{\partial z}\right)^2 = 1. \tag{3.56}$$

It is not hard to see that (3.56) is equivalent to the condition that h is the *signed distance function* of the surface. The function h is also called the *normal form* of the surface. The signed distance function will in general not be smooth in all of $\mathbb{R}^3$, but we only need the condition (3.56) in a neighborhood of S.

If $\mathbf{r}(t) = (x(t), y(t), z(t))$ is a curve in S, i.e., $h(x(t), y(t), z(t)) = 0$, then differentiation with respect to t yields

$$\frac{\partial h(\mathbf{r}(t))}{\partial x} x'(t) + \frac{\partial h(\mathbf{r}(t))}{\partial y} y'(t) + \frac{\partial h(\mathbf{r}(t))}{\partial z} z'(t) = 0,$$

i.e., the tangent vector $\mathbf{r}'$ is orthogonal to the gradient ∇h. So we immediately find that ∇h is the unit normal of S.

If we differentiate (3.56) with respect to x, y, and z then we obtain three equations:

$$2\frac{\partial^2 h}{\partial x^2}\frac{\partial h}{\partial x} + 2\frac{\partial^2 h}{\partial x \partial y}\frac{\partial h}{\partial y} + 2\frac{\partial^2 h}{\partial x \partial z}\frac{\partial h}{\partial z} = 0,$$

$$2\frac{\partial^2 h}{\partial y \partial x}\frac{\partial h}{\partial x} + 2\frac{\partial^2 h}{\partial y^2}\frac{\partial h}{\partial y} + 2\frac{\partial^2 h}{\partial y \partial z}\frac{\partial h}{\partial z} = 0,$$

$$2\frac{\partial^2 h}{\partial z \partial x}\frac{\partial h}{\partial x} + 2\frac{\partial^2 h}{\partial z \partial y}\frac{\partial h}{\partial y} + 2\frac{\partial^2 h}{\partial z^2}\frac{\partial h}{\partial z} = 0.$$

If we divide by 2 and write the equations in matrix form then we obtain

$$\underbrace{\begin{bmatrix} \frac{\partial^2 h}{\partial x^2} & \frac{\partial^2 h}{\partial x \partial y} & \frac{\partial^2 h}{\partial x \partial z} \\ \frac{\partial^2 h}{\partial y \partial x} & \frac{\partial^2 h}{\partial y^2} & \frac{\partial^2 h}{\partial y \partial z} \\ \frac{\partial^2 h}{\partial z \partial x} & \frac{\partial^2 h}{\partial z \partial y} & \frac{\partial^2 h}{\partial z^2} \end{bmatrix}}_{H(h)} \underbrace{\begin{bmatrix} \frac{\partial h}{\partial x} \\ \frac{\partial h}{\partial y} \\ \frac{\partial h}{\partial z} \end{bmatrix}}_{\nabla h} = \begin{bmatrix} 0 \\ 0 \\ 0 \end{bmatrix}. \tag{3.57}$$

That is, the gradient ∇h of the signed distance function h is an eigenvector for the Hessian $H(h)$ with eigenvalue 0.

We will now see that there is a close connection between the Weingarten map $\mathbf{W}$ for the surface and the $H(h)$ for the signed distance function. Assume that $(x(t), y(t), z(t))$ is a curve on the surface, i.e., $h(x(t), y(t), z(t)) = 0$ for all t. As ∇h is the unit normal, the Gauß map maps this curve to the following curve on the unit sphere:

$$\nabla h = \begin{bmatrix} \frac{\partial h}{\partial x}(x(t), y(t), z(t)) \\ \frac{\partial h}{\partial y}(x(t), y(t), z(t)) \\ \frac{\partial h}{\partial z}(x(t), y(t), z(t)) \end{bmatrix}.$$

If $\mathbf{v} = [x', y', z']^T$ then the action of the Weingarten map on $\mathbf{v}$ is given by

$$\mathbf{W}\mathbf{v} = -\frac{d\nabla h}{dt} = -\begin{bmatrix} \frac{\partial^2 h}{\partial x^2}x' + \frac{\partial^2 h}{\partial x \partial y}y' + \frac{\partial^2 h}{\partial x \partial z}z' \\ \frac{\partial^2 h}{\partial y \partial x}x' + \frac{\partial^2 h}{\partial y^2}y' + \frac{\partial^2 h}{\partial y \partial z}z' \\ \frac{\partial^2 h}{\partial z \partial x}x' + \frac{\partial^2 h}{\partial z \partial y}y' + \frac{\partial^2 h}{\partial z^2}z' \end{bmatrix}$$

$$= -\begin{bmatrix} \frac{\partial^2 h}{\partial x^2} & \frac{\partial^2 h}{\partial x \partial y} & \frac{\partial^2 h}{\partial x \partial z} \\ \frac{\partial^2 h}{\partial y \partial x} & \frac{\partial^2 h}{\partial y^2} & \frac{\partial^2 h}{\partial y \partial z} \\ \frac{\partial^2 h}{\partial z \partial x} & \frac{\partial^2 h}{\partial z \partial y} & \frac{\partial^2 h}{\partial z^2} \end{bmatrix} \begin{bmatrix} x' \\ y' \\ z' \end{bmatrix} = -H(h)\mathbf{v}.$$

In other words, the Weingarten map of the surface $h(x, y, z) = 0$ is minus the Hessian $H(h)$ restricted to the tangent space. The Hessian is a symmetric matrix so we can find an orthonormal basis consisting of eigenvectors, and as ∇h is an eigenvector there exist two eigenvectors that are orthogonal to ∇h, i.e., that are tangent vectors. These two vectors are then eigenvectors for the Weingarten map and hence give the principal directions. Thus we have

Theorem 3.6 *Let h be the signed distance function for a surface S and let $\mathbf{e}_1, \mathbf{e}_2, \mathbf{e}_3$ be pairwise orthogonal eigenvectors for the Hessian $H(h)$ such that $\mathbf{e}_3 = \nabla h$.*

The principal directions for S are then $\mathbf{e}_1$ and $\mathbf{e}_2$ and if $\lambda_1, \lambda_2, \lambda_3$ are the eigenvalues for $H(h)$, then $\lambda_3 = 0$ and the principal curvatures are $\kappa_1 = -\lambda_1$ and $\kappa_2 = -\lambda_2$.

We find in particular that the mean curvature is

$$M = \frac{\kappa_1 + \kappa_2}{2} = -\frac{\lambda_1 + \lambda_2}{2} = -\frac{\operatorname{tr} H(h)}{2} = -\frac{1}{2}\left(\frac{\partial^2 h}{\partial x^2} + \frac{\partial^2 h}{\partial y^2} + \frac{\partial^2 h}{\partial z^2}\right),$$

and the Gaußian curvature is

$$K = \kappa_1 \kappa_2 = \lambda_1 \lambda_2 = \begin{vmatrix} \frac{\partial^2 h}{\partial x^2} & \frac{\partial^2 h}{\partial x \partial y} \\ \frac{\partial^2 h}{\partial y \partial x} & \frac{\partial^2 h}{\partial y^2} \end{vmatrix} + \begin{vmatrix} \frac{\partial^2 h}{\partial x^2} & \frac{\partial^2 h}{\partial x \partial z} \\ \frac{\partial^2 h}{\partial z \partial x} & \frac{\partial^2 h}{\partial z^2} \end{vmatrix} + \begin{vmatrix} \frac{\partial^2 h}{\partial y^2} & \frac{\partial^2 h}{\partial y \partial z} \\ \frac{\partial^2 h}{\partial y \partial z} & \frac{\partial^2 h}{\partial z^2} \end{vmatrix}.$$

Proof The formula for the mean curvature is obvious. So we only need to show the formula for the Gaußian curvature. Here we note that the characteristic polynomial for $H(h)$ on one hand is

$$(\lambda_1 - t)(\lambda_2 - t)(0 - t) = -t^3 + (\lambda_1 + \lambda_2)t^2 - \lambda_1 \lambda_2 t,$$

and on the other hand is

$$\begin{vmatrix} \frac{\partial^2 h}{\partial x^2} - t & \frac{\partial^2 h}{\partial x \partial y} & \frac{\partial^2 h}{\partial x \partial z} \\ \frac{\partial^2 h}{\partial y \partial x} & \frac{\partial^2 h}{\partial y^2} - t & \frac{\partial^2 h}{\partial y \partial z} \\ \frac{\partial^2 h}{\partial z \partial x} & \frac{\partial^2 h}{\partial z \partial y} & \frac{\partial^2 h}{\partial z^2} - t \end{vmatrix}.$$

All that is left is to compare the coefficients to t. $\square$

3.10.2 An Arbitrary Function

We now let f be an arbitrary function with non vanishing gradient. We are interested in the surface given by $f(x, y, z) = 0$. Observe that another level set $f(x, y, z) = c$ can be given as $f(x, y, z) - c = 0$. Let h denote the signed distance function for the surface. Then we can write

$$f(x, y, z) = \lambda(x, y, x)\, h(x, y, z),\tag{3.58}$$

where $\lambda : \mathbb{R}^3 \to \mathbb{R}_+$ is some positive function. Differentiation of (3.58) yields

$$\begin{bmatrix} \frac{\partial f}{\partial x} \\ \frac{\partial f}{\partial y} \\ \frac{\partial f}{\partial z} \end{bmatrix} = h \begin{bmatrix} \frac{\partial \lambda}{\partial x} \\ \frac{\partial \lambda}{\partial y} \\ \frac{\partial \lambda}{\partial z} \end{bmatrix} + \lambda \begin{bmatrix} \frac{\partial h}{\partial x} \\ \frac{\partial h}{\partial y} \\ \frac{\partial h}{\partial z} \end{bmatrix}.\tag{3.59}$$

At the surface we have $h = 0$ so *on the surface* we have $\nabla f = \lambda \nabla h$, i.e.,

$$\lambda = \|\nabla f\|, \qquad \nabla h = \frac{1}{\lambda} \nabla f.\tag{3.60}$$

Differentiation of (3.59) yields

$$
\begin{bmatrix}
\frac{\partial^2 f}{\partial x^2} & \frac{\partial^2 f}{\partial x \partial y} & \frac{\partial^2 f}{\partial x \partial z} \\
\frac{\partial^2 f}{\partial y \partial x} & \frac{\partial^2 f}{\partial y^2} & \frac{\partial^2 f}{\partial y \partial z} \\
\frac{\partial^2 f}{\partial z \partial x} & \frac{\partial^2 f}{\partial z \partial y} & \frac{\partial^2 f}{\partial z^2}
\end{bmatrix}
= h
\begin{bmatrix}
\frac{\partial^2 \lambda}{\partial x^2} & \frac{\partial^2 \lambda}{\partial x \partial y} & \frac{\partial^2 \lambda}{\partial x \partial z} \\
\frac{\partial^2 \lambda}{\partial y \partial x} & \frac{\partial^2 \lambda}{\partial y^2} & \frac{\partial^2 \lambda}{\partial y \partial z} \\
\frac{\partial^2 \lambda}{\partial z \partial x} & \frac{\partial^2 \lambda}{\partial z \partial y} & \frac{\partial^2 \lambda}{\partial z^2}
\end{bmatrix}
+
\begin{bmatrix}
\frac{\partial \lambda}{\partial x} \\
\frac{\partial \lambda}{\partial y} \\
\frac{\partial \lambda}{\partial z}
\end{bmatrix}
\begin{bmatrix}
\frac{\partial h}{\partial x} \\
\frac{\partial h}{\partial y} \\
\frac{\partial h}{\partial z}
\end{bmatrix}^T
$$

$$
+
\begin{bmatrix}
\frac{\partial h}{\partial x} \\
\frac{\partial h}{\partial y} \\
\frac{\partial h}{\partial z}
\end{bmatrix}
\begin{bmatrix}
\frac{\partial \lambda}{\partial x} \\
\frac{\partial \lambda}{\partial y} \\
\frac{\partial \lambda}{\partial z}
\end{bmatrix}^T
+ \lambda
\begin{bmatrix}
\frac{\partial^2 h}{\partial x^2} & \frac{\partial^2 h}{\partial x \partial y} & \frac{\partial^2 h}{\partial x \partial z} \\
\frac{\partial^2 h}{\partial y \partial x} & \frac{\partial^2 h}{\partial y^2} & \frac{\partial^2 h}{\partial y \partial z} \\
\frac{\partial^2 h}{\partial z \partial x} & \frac{\partial^2 h}{\partial z \partial y} & \frac{\partial^2 h}{\partial z^2}
\end{bmatrix}. \qquad (3.61)
$$

Multiplying with ∇h on the right and using that we know that $H(h)\nabla h = 0$ and $h = 0$ on the surface, we obtain

$$
\begin{bmatrix}
\frac{\partial^2 f}{\partial x^2} & \frac{\partial^2 f}{\partial x \partial y} & \frac{\partial^2 f}{\partial x \partial z} \\
\frac{\partial^2 f}{\partial y \partial x} & \frac{\partial^2 f}{\partial y^2} & \frac{\partial^2 f}{\partial y \partial z} \\
\frac{\partial^2 f}{\partial z \partial x} & \frac{\partial^2 f}{\partial z \partial y} & \frac{\partial^2 f}{\partial z^2}
\end{bmatrix}
\begin{bmatrix}
\frac{\partial h}{\partial x} \\
\frac{\partial h}{\partial y} \\
\frac{\partial h}{\partial z}
\end{bmatrix}
$$

$$
=
\begin{bmatrix}
\frac{\partial \lambda}{\partial x} \\
\frac{\partial \lambda}{\partial y} \\
\frac{\partial \lambda}{\partial z}
\end{bmatrix}
\begin{bmatrix}
\frac{\partial h}{\partial x} \\
\frac{\partial h}{\partial y} \\
\frac{\partial h}{\partial z}
\end{bmatrix}^T
\begin{bmatrix}
\frac{\partial h}{\partial x} \\
\frac{\partial h}{\partial y} \\
\frac{\partial h}{\partial z}
\end{bmatrix}
+
\begin{bmatrix}
\frac{\partial h}{\partial x} \\
\frac{\partial h}{\partial y} \\
\frac{\partial h}{\partial z}
\end{bmatrix}
\begin{bmatrix}
\frac{\partial \lambda}{\partial x} \\
\frac{\partial \lambda}{\partial y} \\
\frac{\partial \lambda}{\partial z}
\end{bmatrix}^T
\begin{bmatrix}
\frac{\partial h}{\partial x} \\
\frac{\partial h}{\partial y} \\
\frac{\partial h}{\partial z}
\end{bmatrix}.
$$

That is, *on the surface* we have

$$
H(f)\nabla h = \|\nabla h\|^2 \nabla \lambda + (\nabla \lambda \cdot \nabla h)\nabla h = \nabla \lambda + (\nabla \lambda \cdot \nabla h)\nabla h.
$$

Multiplying with ∇h^T on the left yields

$$
\nabla h^T H(f)\nabla h = \nabla h^T \nabla \lambda + (\nabla \lambda \cdot \nabla h)\nabla h^T \nabla h = 2\nabla \lambda \cdot \nabla h.
$$

So now we see that

$$
\nabla \lambda \cdot \nabla h = \frac{1}{2}\nabla h^T H(f)\nabla h,
$$

$$
\nabla \lambda = H(f)\nabla h - \frac{1}{2}\bigl(\nabla h^T H(f)\nabla h\bigr)\nabla h,
$$

and hence

$$
H(f) = \nabla \lambda \nabla h^T + \nabla h \nabla \lambda^T + \lambda H(h)
$$

$$
= \left(H(f)\nabla h - \frac{1}{2}\bigl(\nabla h^T H(f)\nabla h\bigr)\nabla h \right)\nabla h^T
$$

$$+ \nabla h \left(H(f)\nabla h - \frac{1}{2}(\nabla h^T H(f)\nabla h)\nabla h \right)^T + \lambda H(h)$$

$$= H(f)\nabla h \nabla h^T + \nabla h \nabla h^T H(f) - \left(\nabla h^T H(f)\nabla h\right)\nabla h \nabla h^T + \lambda H(h),$$

where we have used that $H(f)$ is symmetric. We now have

Theorem 3.7 *Let $f : \mathbb{R}^3 \to \mathbb{R}$ be a smooth function with non vanishing gradient and let $h : \mathbb{R}^3 \to \mathbb{R}$ be the signed distance function for the surface*

$$S = f^{-1}(c) = \left\{ (x, y, z) \in \mathbb{R}^3 \mid f(x, y, z) = c \right\}.$$

On the surface we have

$$\nabla h = \frac{\nabla f}{\|\nabla f\|},$$

$$H(a) = \frac{H(f)}{\|\nabla f\|} - \frac{H(f)\nabla f \nabla f^T + \nabla f \nabla f^T H(f)}{\|\nabla f\|^3} + \frac{(\nabla f^T H(f)\nabla f)\nabla f \nabla f^T}{\|\nabla f\|^5}.$$

Proof We have already seen the case $c = 0$. To show the general case we only have to note that the functions $f - c$ and f have the same gradient and Hessian. $\square$

The principal directions, the principal curvatures, and the mean and Gaußian curvature can now be found using Theorem 3.6. For more on the normal form and its applications, see [13].

3.11 Exercises

Exercise 3.1 Consider the surface in $\mathbb{R}^3$ given by the equation $z = xy$.
1. Show that the tangent space at the point $(0, 0, 0)$ is the xy-plane.
2. Explain why you can use (3.21) to find the second fundamental form at the point $(0, 0, 0)$ and do it.
3. Find the principal curvatures and directions at the point $(0, 0, 0)$.
4. Find a parametrization of the surface.
5. Use this parametrization to:
    (a) Find the surface normal at an arbitrary point of the surface.
    (b) Find the first and second fundamental form at an arbitrary point of the surface.
    (c) Find the principal curvatures and directions at the point $(1, 1, 1)$.
    (d) Find the Gaußian and mean curvature at an arbitrary point of the surface.
6. Find an implicit representation of the surface.
7. Use this implicit form to:
    (a) Find the surface normal at an arbitrary point of the surface.
    (b) Find the principal curvatures and directions at the point $(1, 1, 1)$.
    (c) Find the Gaußian and mean curvature at an arbitrary point of the surface.

Exercise 3.2 Consider the circle in the xz-plane with center $(2, 0, 0)$ and radius 1. By rotating the circle around the z-axis we obtain a surface, called a *torus*, in $\mathbb{R}^3$.
1. Find a parametrization of the surface.
2. Find the surface normal at an arbitrary point of the surface.
3. Find the first and second fundamental form at an arbitrary point of the surface.
4. Find the Gaußian and mean curvature at an arbitrary point of the surface.
5. Find the principal curvatures and directions at an arbitrary point of the surface.

Exercise 3.3 Consider the *helicoid* given by the parametrization

$$\mathbf{x}(u, v) = (v \cos u, v \cos u, u), \quad (u, v) \in \mathbb{R}^2.$$

1. Find the surface normal at an arbitrary point of the surface.
2. Find the first and second fundamental form at an arbitrary point of the surface.
3. Find the Gaußian and mean curvature at an arbitrary point of the surface.
4. Find the principal curvatures and directions at an arbitrary point of the surface.

Exercise 3.4 Consider the *catenoid* given by the parametrization

$$\mathbf{x}(u, v) = (\cosh u \cos v, \cosh u \sin v, v)$$

1. Find the surface normal at an arbitrary point of the surface.
2. Find the first and second fundamental form at an arbitrary point of the surface.
3. Find the Gaußian and mean curvature at an arbitrary point of the surface.
4. Find the principal curvatures and directions at an arbitrary point of the surface.

Exercise 3.5 Prove Theorem 3.2.

Exercise 3.6 Prove Theorem 3.5. Hint: use Theorem 3.2 and 3.4.

Exercise 3.7 Show that if $g_{ij} = \delta_{ij}$ then (3.48), (3.52), and (3.54) reduce to the usual expressions for the gradient, divergence, and Laplace operator.

References

1. Graustein, W.C.: Differential Geometry. Dover, New York (1966)
2. Guggenheimer, H.W.: Differential Geometry. McGraw-Hill, New York (1963)
3. Lipschutz, M.: Theory and Problems of Differential Geometry. Schaum's Outline Series. McGraw-Hill, New York (1969)
4. Struik, D.J.: Lectures on Classical Differential Geometry. Addison-Wesley, London (1961)
5. do Carmo, M.P.: Differential Geometry of Curves and Surfaces. Prentice-Hall, Englewood Cliffs (1976)
6. Oprea, J.: The Mathematics of Soap Films: Explorations with Maple®. Student Mathematical Library, vol. 10. Amer. Math. Soc., Providence (2000)
7. Pressley, A.: Elementary Differential Geometry. Springer, London (2001)
8. Gallier, J.: Geometric Methods and Applications. For Computer Science and Engineering. Springer, New York (2001)

9. Gray, A.: Modern Differential Geometry of Curves and Surfaces with Mathematica. CRC Press, Boca Raton (1998)
10. Gravesen, J.: Differential geometry and design of shape and motion. http://www2.mat.dtu.dk/people/J.Gravesen/cagd.pdf (2002)
11. Raussen, M.: Elementary differential geometry: Curves and surfaces (2001). http://www.math.auc.dk/~raussen/MS/01/index.html
12. Hansen, V.L.: Fundamental Concepts in Modern Analysis. World Scientific, River Edge (1999)
13. Hartmann, E.: On the curvature of curves and surfaces defined by normalforms. Comput. Aided Geom. Des. **16**(5), 355–376 (1999)

Finite Difference Methods for Partial Differential Equations

4

When we seek to manipulate geometry, we often arrive at *differential equations*. For instance, a simple height field may be described by a function $f : \mathbb{R}^2 \to \mathbb{R}$. Assume we want the shape to be as smooth as possible; we could try to minimize the so called membrane energy of f,

$$E_M[f] = \frac{1}{2} \int f_x^2 + f_y^2 \, dx \, dy, \tag{4.1}$$

which basically penalizes stretch (cf. Chap. 9). From variational calculus we know that a minimizer of E_M must be a solution to

$$\Delta f = f_{xx} + f_{yy} = \frac{\partial^2 f}{\partial x^2} + \frac{\partial^2 f}{\partial y^2} = 0. \tag{4.2}$$

We would typically solve this equation (known as the *Laplace Equation*) on some bounded domain where we are given specific boundary conditions. To give another example, in the *level set method* (cf. Chap. 17), we need to solve

$$\frac{\partial \Phi}{\partial t} + F \|\nabla \Phi\| = 0, \tag{4.3}$$

where $\Phi(t, \mathbf{x}) : \mathbb{R} \times \mathbb{R}^3 \to \mathbb{R}$ is a scalar function of time and spatial position and $F : \mathbb{R}^3 \to \mathbb{R}$ is known as the speed function. In this case, we would start from a known value of Φ and solve forward in time. One can look at Φ as a function which describes a shape: Φ is positive outside the shape and negative inside, and F is a function which describes how the boundary deforms.

Above, we have seen examples of a *boundary value problem* and an *initial value problem*. In both cases, we do not expect to be able to find closed form solutions but instead we divide both time and space into discrete grids, and look for numerical solutions.

One particular family of methods for solving differential equations on grids is known as *finite difference* methods. The main purpose of this chapter is to briefly review the basics of finite difference methods. This is quite a large field, and we

J.A. Bærentzen et al., *Guide to Computational Geometry Processing*,
DOI 10.1007/978-1-4471-4075-7_4, © Springer-Verlag London 2012

will have to restrict the discussion to simple methods which are directly relevant to the topics covered in this book. For more information about the finite difference method, the interested reader is referred to [1] and [2] which are the main references for the following chapter.

4.1   Discrete Differential Operators

In order to solve partial differential equations on discrete uniform grids, we need discrete analogues of the continuous differential operators. Typically, we use one of the three following discrete operators.

Definition 4.1 The *forward difference* operator, D^+, approximates $\frac{\partial}{\partial x}$ using the current grid point and the next

$$D^+ f = f(i+1) - f(i). \tag{4.4}$$

The *backward difference* operator, D^-, uses the current grid point and the previous

$$D^- f = f(i) - f(i-1). \tag{4.5}$$

The *central differences* operator, D, uses the grid points on either side of the point, we are interested in

$$Df = \frac{f(i+1) - f(i-1)}{2}. \tag{4.6}$$

One way of arriving at these operators is through *Taylor polynomials*. Recall that Taylor's theorem states that if we know the derivatives of all orders of a function f at point i, we can compute f at a point x as a sum

$$f(x) = \sum_n \frac{f^{(n)}(i)}{n!}(x-i)^n$$

$$= f(i) + f'(i)(x-i) + \frac{1}{2}f''(i)(x-i)^2 + \cdots \tag{4.7}$$

Assuming that we know the first-order partial derivative of f with respect to x at a point i, we can write its value at $i+1$

$$f(i+1) = f(i) + f'(i) + \cdots$$

where the dots indicate that we ignore all but the first two terms of the Taylor series. Subtracting $f(i)$ from both sides, we obtain

$$D^+ f = f(i+1) - f(i).$$

To simplify the formulas, it was assumed that we were employing a grid with unit spacing. This need not be the case. Let h be the distance between two grid points; then the Taylor approximation is

$$f(i + h) = f(i) + f'(i)h + \frac{1}{2}f''(i)h^2 + \cdots \tag{4.8}$$

and ignoring all but the first two terms, we can rearrange to obtain the forward difference operator

$$D^+ f(i) = \frac{f(i + h) - f(i)}{h}. \tag{4.9}$$

To give one more example, we will find the second-order derivative in the x direction.

$$f(i + 1) = f(i) + hf'(i) + \frac{1}{2}h^2 f''(i) + \cdots, \tag{4.10}$$

$$f(i - 1) = f(i) - hf'(i) + \frac{1}{2}h^2 f''(i) + \cdots. \tag{4.11}$$

By adding the two formulas and reordering, we obtain

$$D^2 f(i) = \frac{f(i + 1) + f(i - 1) - 2f(i)}{h^2}. \tag{4.12}$$

We could also arrive at higher-order derivatives by repeated application of the difference operators. For instance, applying the forward operator to the result of the backward difference operator produces

$$D^+ D^- f = D^+\big(f(i) - f(i - 1)\big) = \big(f(i + 1) - f(i)\big) - \big(f(i) - f(i - 1)\big)$$
$$= f(i + 1) + f(i - 1) - 2f(i). \tag{4.13}$$

A different result is obtained if we apply the central differences operator twice

$$DDf = D\big(f(i + 1) - f(i - 1)\big) = f(i + 2) - f(i) - \big(f(i) - f(i - 2)\big)$$
$$= f(i + 2) + f(i - 2) - 2f(i) \tag{4.14}$$

but, arguably, it is preferable to start from the polynomials since that gives a clear interpretation. Interestingly, we could also subtract the second formula from the first, and in that case the second-order derivatives would cancel out leading to the central differences formula.

No matter how we look at it, the difference operators simply give us the derivatives of the polynomial which interpolates the data values, and, in fact, the method whereby we obtain the polynomial does not really matter since it is uniquely determined by the data points. We might, for instance, use interpolating polynomials in Lagrange form (cf. Sect. 4.6). The main advantage of this method is that it also works if the points are not equidistant.

For simplicity the examples above dealt with derivatives of a function of one variable, however, partial derivatives are treated in an identical fashion. For instance, let a regular 2D grid be given and let $f : \mathbb{R}^2 \to \mathbb{R}$ be a function of two variables. The *Laplacian* of f at a grid point (i, j) is often computed,

$$\Delta f \approx D_x^2 f(i, j) + D_y^2 f(i, j)$$
$$= f(i + 1, j) + f(i - 1, j) + f(i, j + 1) + f(i, j - 1) - 4f(i, j). \quad (4.15)$$

4.2   Explicit and Implicit Methods

A PDE which often turns up is the so called *heat equation*, which (in 1D) describes how the heat distribution changes over time in a rod. The 1D version of this equation is

$$\frac{\partial f}{\partial t} = \frac{\partial^2 f}{\partial x^2}, \quad (4.16)$$

where f is the heat at a given point x and a given time t. This is an initial value problem which we can solve provided that we have
- an *initial condition*: we know $f(0, x)$;
- *boundary conditions*: we know $f(t, 0)$ and $f(t, 1)$

where both the domain $[0, 1]$ and the fact that we start at time 0 are arbitrary choices.

To solve this equation we might begin by simply replacing the partial derivatives with discrete differences. This leads to

$$\frac{F_j^{i+1} - F_j^i}{h} = \frac{F_{j+1}^i + F_{j-1}^i - 2F_j^i}{k^2}, \quad (4.17)$$

where we have replaced $f(t, x)$ with $F_j^i \approx f(ih, jk)$ to distinguish between the function we find using our discrete method and the true function. Of course, the constants used in the finite difference approximations to the partial derivative, h and k, are the grid spacings in the time and space axes, respectively. Now, let the boundary and initial conditions be given. Following Morton and Mayers [2] we choose a simple scenario which corresponds to a rod that is cooled to precisely zero degrees at either end and warmed to precisely 1 degree at the center:

$$f(0, x) = \begin{cases} 2x & x < 0.5 \\ 2 - 2x & x \geq 0.5 \\ 0 & x < 0 \vee x > 1 \end{cases} \qquad f(t, 0) = f(t, 1) = 0.$$

It has been kept in this state until an equilibrium (the initial condition) has settled which we assume is the case at $t = 0$. At time $t > 0$ we keep it at zero degrees at either end (the boundary condition), but what happens in between is precisely what the PDE models.

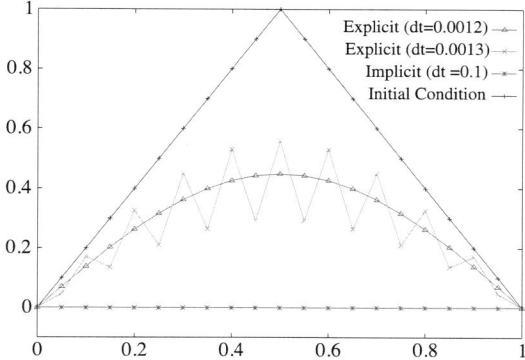

Fig. 4.1 This figure shows the initial condition for the heat equation and the solution after 50 time steps for three different methods. The explicit solution at a time step of 0.0012 is not nearly as good as the implicit solution where we can use a time step of 0.1—almost two orders of magnitude more. Unfortunately, as is plain to see, increasing the time step by the slightest amount (to 0.0013) leads to instability

Now, to evolve F forward in time, we just have to reorganize (4.17) to

$$F_j^{i+1} = F_j^i + \frac{h}{k^2}\big(F_{j+1}^i + F_{j-1}^i - 2F_j^i\big). \tag{4.18}$$

The only thing left is to choose h and k. In the example shown in Fig. 4.1 we have chosen a spatial discretization of $k = 0.05$. There are two examples given; one where $h = 0.0012$ and another where $h = 0.0013$. The former works while the latter does not. To understand why this is the case, observe that we can write (4.18) in matrix notation

$$F^{i+1} = \mathbf{A}F^i, \tag{4.19}$$

where the matrix $\mathbf{A}$ encodes the explicit finite difference scheme from (4.18). Each row has three entries centered around the diagonal consisting of $\frac{h}{k^2}$, $1 - 2\frac{h}{k^2}$, and $\frac{h}{k^2}$. However, the first and last rows have 1 in the diagonal and 0 elsewhere since they encode the boundary condition. If N denotes the number of points in our spatial grid, we can write pseudo code (in Matlab syntax, which is convenient for this purpose) for constructing the matrix $\mathbf{A}$ as shown in Algorithm 4.1.

If we keep the spatial discretization constant and analyze what happens to the eigenvalues of $\mathbf{A}$ for a range of h values, we find that the numerically greatest eigenvalue is 1 until $h = 0.00125$ at which point it increases linearly as seen in Fig. 4.2. Now, if the spectral radius of $\mathbf{A}$, $\rho(\mathbf{A})$, which is precisely the numerically greatest eigenvalue of $\mathbf{A}$ is greater than 1, we cannot expect F^n to be bounded as n tends to infinity. To understand this, consider that if V is an eigenvector of $\mathbf{A}$ corresponding to an eigenvalue $\lambda > 1$, then $\mathbf{A}^n V = \lambda^n V \to \infty$ for $n \to \infty$. Finally, F can be written as a linear combination of eigenvectors, most likely including V. While this is just a vague but credible argument, it is, in fact, true that for a PDE whose solution does not increase with time, the scheme is *stable* if the spectral radius does not

---

Algorithm 4.1 Constructing the matrix A for an explicit solution to the heat equation

```
A = null_matrix(N,N);
float v = h/sqr(k);
for (i = 1 to N-2) {
  A(i,i-1) = v;
  A(i,i) = 1-2*v;
  A(i,i+1) = v;
}
A(0,0) = A(N-1,N-1) = 1;
```

Fig. 4.2 This is a plot of the spectral radius, $\rho(\mathbf{A})$, i.e. the numerically greatest eigenvalue, for the matrix $\mathbf{A}$ as a function of h for $h \in [0, 0.0025]$

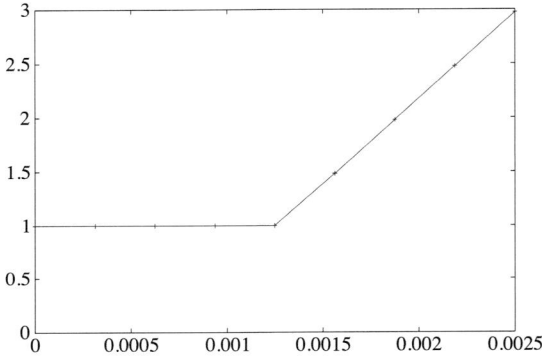

exceed 1 [1]. Stability means that the scheme limits the amplification of all components of the initial condition [1]. Note that this does not necessarily mean that we get the true solution. More precisely, stability does not imply *convergence*. Convergence means that as we refine both the space and time discretizations we approach the true solution. One big piece missing from the picture is *consistency*, which means that, as we refine space and time, the difference equation converges to the true solution to the PDE. In fact, for some combinations of schemes and partial differential equations, consistency and stability imply convergence. We will return to these matters in Sect. 4.5.

It is worth pointing out that the condition $h < 0.00125$ from the example is really a condition on the ratio between the time step and the spatial discretization, namely $\frac{h}{k^2} < 0.5$. In this example, the time step allowed for seems very small. Fortunately, we can take much larger time steps if we use a so called *implicit method*. The difference between an *explicit method* and an implicit method is that, in the latter case, we compute the spatial derivatives at the next time step. This seems contradictory because we do not know the values at the next time step. However, this simply means that we need to solve a linear system rather than just multiply the previous solution onto a matrix.

Algorithm 4.2 Constructing the A matrix for an implicit solution to the heat equation

```
v = dt/sqr(dx);
A = null_matrix(N,N);
for (i = 1 to N-2) {
  A(i,i-1) = -v;
  A(i,i)   = 1+2*v;
  A(i,i+1) = -v;
}
A(0,0) = A(N-1,N-1) = 1;
```

In the case of our example problem, an implicit scheme looks as follows:

$$\frac{F_j^{i+1} - F_j^i}{h} = \frac{F_{j+1}^{i+1} + F_{j-1}^{i+1} - 2F_j^{i+1}}{k^2}, \tag{4.20}$$

which we can reorder

$$F_j^{i+1} + \frac{h}{k^2}\left(F_{j+1}^{i+1} + F_{j-1}^{i+1} - 2F_j^{i+1}\right) = F_j^i, \tag{4.21}$$

and express as the following linear system:

$$\mathbf{A}F^{i+1} = F^i, \tag{4.22}$$

where (4.22) is a tridiagonal system and Gaussian elimination is an appropriate method. Pseudo code for **A** would look as shown in Algorithm 4.2.

The advantage of the implicit method is seen very clearly in Fig. 4.1. For this scheme we are able to use a time step of 0.1 which means that the method converges to the true solution (uniform zero temperature) in the same number of time steps. In fact, the implicit method is unconditionally stable, which means that if we choose a larger time step, we get a bigger error, but the scheme does not "explode" numerically [2].

4.3   Boundary Conditions

We have already mentioned the need to impose conditions on the boundary of a domain when solving a partial differential equation. An obvious possibility is to specify the value the function must take on the boundary, e.g.

$$f(0) = 0. \tag{4.23}$$

This is known as a *Dirichlet* boundary condition [2], which is what we imposed in the preceding section. Another possibility is the use of *von Neumann* boundary conditions, where a value is imposed on the derivative [2]. For instance,

$$\frac{\partial f}{\partial x}\bigg|_{x=0} = f(0). \tag{4.24}$$

There are many further possibilities. We can have different conditions on different parts of the boundary or combine the two types of boundary condition.

To impose von Neumann boundary conditions on a problem, we express the derivative in the condition using finite differences. This gives us a particular equation for the boundary point which we can use when solving the finite difference equations. We continue our example with the heat equation and impose the von Neumann boundary conditions,

$$\frac{\partial f}{\partial x}\bigg|_{x=0} = \frac{\partial f}{\partial x}\bigg|_{x=1} = 0.$$

The continuous conditions translate into the following finite difference equations:

$$\frac{F_{0+1}^i - F_{0-1}^i}{2k} = 0 \tag{4.25}$$

and

$$\frac{F_N^i - F_{N-2}^i}{2k} = 0 \tag{4.26}$$

where we have used central differences and N points. For each of these two equations, we can combine (4.17) with the boundary condition in order to annihilate the grid points outside the grid (-1, and N, respectively). This leads to

$$F_0^{i+1} = F_0^i + 2\frac{h}{k^2}\left(F_1^i - F_0^i\right) \tag{4.27}$$

$$F_{N-1}^{i+1} = F_{N-1}^i + 2\frac{h}{k^2}\left(F_{N-2}^i - F_{N-1}^i\right). \tag{4.28}$$

An example where von Neumann boundary conditions have been used is shown in Fig. 4.3. By imposing a zero derivative, we model a rod that is fully insulated at either end. As expected, the example shows that the heat is simply redistributed along the rod until all points have the average value.

Fig. 4.3 An example of von Neumann boundary conditions which prescribe a zero derivative at either end point but have no effect on the endpoint value. The result is that the heat is redistributed along the rod until all points have the average value

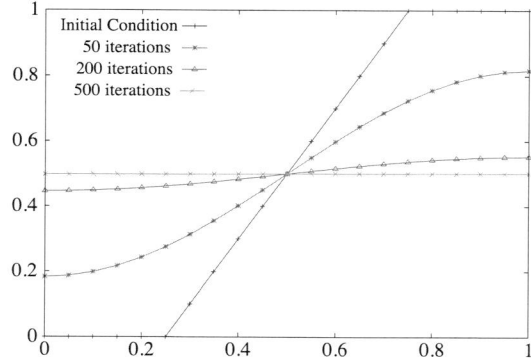

4.4   Hyperbolic, Parabolic, and Elliptic Differential Equations

Partial differential equations are often divided into *hyperbolic*, *parabolic*, and *elliptic* equations. This taxonomy is straightforward in the case of second-order PDEs. Assume that we have a PDE of the form

$$A\frac{\partial^2 f}{\partial x^2} + B\frac{\partial^2 f}{\partial x \partial y} + C\frac{\partial^2 f}{\partial y^2} + \cdots = 0 \qquad (4.29)$$

where the discriminant $B^2 - 4AC$ determines whether the PDE is hyperbolic ($B^2 - 4AC > 0$), parabolic ($B^2 - 4AC = 0$) or elliptic ($B^2 - 4AC < 0$) [1].

The naming is clearly related to conic sections: If we replace the partial derivatives with powers of x and y, the contours are precisely those of a hyperbola if the discriminant is negative and an ellipse if it is positive. Unfortunately, the picture is a little less clear because a PDE can be hyperbolic in some regions and parabolic or elliptic in other regions, since the constants A, B, and C may depend on x, y, f or even on the first-order partial derivatives.

There are significant differences between the types of system that we model with these three different types of partial differential equation and to some extent also on the methods we use to solve them.

4.4.1   Parabolic Differential Equations

Parabolic differential equations are often initial value problems where information is propagated forward in time but also smoothed. The heat equation, which we have already seen, is a good example:

$$\frac{\partial f}{\partial t} = \frac{\partial^2 f}{\partial x^2}. \qquad (4.30)$$

4.4.2 Hyperbolic Differential Equations

Hyperbolic differential equations on the other hand are also usually time dependent, but they tend not to diffuse the initial condition. Discontinuities in the initial condition may be preserved as we solve forward in time. A common example is the wave equation

$$\frac{\partial^2 f}{\partial t^2} = \frac{\partial^2 f}{\partial x^2},$$

(4.31)

but simple linear propagation, i.e.

$$\frac{\partial f}{\partial t} = a \frac{\partial f}{\partial x},$$

(4.32)

and, more generally, conservation laws

$$\frac{\partial f}{\partial t} = \frac{\partial}{\partial x} G(f),$$

(4.33)

where G is the flux function [3], are also considered hyperbolic. This might seem odd since it is a first-order equation which does not fit into the taxonomy based on the discriminant. However, it has an important point in common with the second-order hyperbolic equations, namely that information is propagated along so called characteristics.

A characteristic is a curve along which the PDE reduces to an ODE, i.e. an ordinary differential equation [1]. Given a hyperbolic PDE of the form

$$a \frac{\partial f}{\partial x} + b \frac{\partial f}{\partial y} = c,$$

(4.34)

let us introduce a curve $\Gamma = \{(x(t), y(t)) \in \mathbb{R}^2\}$. By the chain rule, we know that along the curve

$$\frac{df}{dt} = \frac{\partial f}{\partial x} \frac{dx}{dt} + \frac{\partial f}{\partial y} \frac{dy}{dt}.$$

(4.35)

We can get rid of the first partial derivative by combining (4.34) with (4.35):

$$\frac{df}{dt} = \left(\frac{c}{a} - \frac{b}{a} \frac{\partial f}{\partial y} \right) \frac{dx}{dt} + \frac{\partial f}{\partial y} \frac{dy}{dt}.$$

(4.36)

Gathering terms which include the other partial derivative, we get

$$\frac{df}{dt} = \frac{\partial f}{\partial y} \left(\frac{dy}{dt} - \frac{b}{a} \frac{dx}{dt} \right) + \frac{c}{a}.$$

(4.37)

Thus, if we define the curve C by the differential equation

$$\frac{dy}{dt} - \frac{b}{a} \frac{dx}{dt} = 0,$$

(4.38)

we get rid of the other partial derivative in (4.37), and along this curve,

$$\frac{df}{dt} = \frac{c}{a}. \tag{4.39}$$

In other words, if C fulfills (4.38), the differential equation for f becomes the ordinary differential equation (4.39), and we denote by C a characteristic. For a second-order hyperbolic PDE, we have not one but two characteristic directions, given by the roots of the polynomial (in $\frac{dy}{dx}$)

$$D\left(\frac{dy}{dx}\right)^2 + E\frac{dy}{dx} + F = 0. \tag{4.40}$$

Thus, the discriminant mentioned earlier (which tells us whether a second-order PDE is hyperbolic, parabolic, or elliptic) is really the discriminant of this polynomial and it indicates whether the characteristic directions are well defined [1].

4.4.3 Elliptic Differential Equations

Elliptic differential equations tend not to be time dependent but are often modeled by steady state scenarios. For instance, the Poisson equation

$$\frac{\partial^2 f}{\partial x^2} = g, \tag{4.41}$$

and the Laplace equation

$$\frac{\partial^2 f}{\partial x^2} = 0, \tag{4.42}$$

are examples of elliptic equations.

4.5 Consistency, Stability, and Convergence

Of course, the goal is to solve a partial differential equation, and an important question to ask when confronted with a finite difference scheme is whether the solution of the difference equation *converges* to the true PDE as we refine the spatial discretization and the time step.

When examining a finite difference scheme for a partial differential equation, the three important questions to ask are whether the scheme is *stable*, *consistent*, and, as mentioned, convergent [2].

Stability means that the result does not "explode" numerically. In some cases the partial differential equation has a particular structure which allows us to employ a *maximum principle* [2]. For instance, the heat equation, which we have studied so much in this chapter, must attain its maximum on one the three sides of its domain, defined by the initial and boundary equations. If the numerical solution can in fact grow, the scheme is unstable.

Consistency means that the finite difference scheme converges to the true PDE as the resolution of the spatiotemporal discretization is increased (h and k going toward zero). If a scheme is not consistent with the partial differential equation that we are trying to solve, we might get a result which appears reasonable but is really an approximate solution to a different PDE. To check consistency, we look at the *truncation error* of the scheme. The truncation error is the difference between the true solution and one step of the finite difference scheme based on values of the true solution. Based on a Taylor expansion of the true solution, we can analyze what happens to the error as $h, k \rightarrow 0$.

One might think that if a finite difference scheme is both stable and consistent, it must also be convergent. In fact, that is not necessarily true, but for linear partial differential equations and linear finite difference schemes, it is in fact true under some additional assumptions regarding the boundary and initial conditions [1, 2, 4]. This is known as the *Lax Equivalence Theorem*.

Another important result is the *Courant–Friederichs–Lewy Condition* [2, 4]. This condition states that for a finite difference scheme (or really any numerical method) to be stable, then in the limit the numerical domain of dependence must include the mathematical domain of dependence.

Let us consider a 1D time dependent partial differential equation. For instance, we take the hyperbolic equation (4.32)

$$\frac{\partial f}{\partial t} = a \frac{\partial f}{\partial x}.$$

In this case, the mathematical domain of dependence is the subset of the real axis at $t = 0$ upon which the value at a given later point in time depends. In this case, we have a first-order hyperbolic equation, so the dependency is a single point on the x axis.

Now, if we look at a grid point (i, j) where i is the temporal index, and j is the spatial, there is a similar set of grid points at $t = 0$ on which the solution at $t = ih$ depends, which is known as the numerical domain of dependence. If we refine the grid (both space and time) then in the limit this domain must include the mathematical domain.

Perhaps the CFL condition is used mostly in conjunction with hyperbolic partial differential equations since they evolve along characteristics. Thus, checking the CFL condition is tantamount to finding out whether the characteristics intersect the $t = 0$ line within the numerical domain of dependence. The CFL condition can be used to set time step restrictions, and it is a useful "sanity" check for a scheme. Unfortunately, it is only a necessary but not a sufficient condition for stability.

4.6   2D and 3D problems and Irregular Grids

So far, nothing has been said about irregular grids or grids in dimensions above 2D. Most of the concepts carry over directly from 1D, but some special attention may be required.

If the grid is irregular, it is convenient to compute the derivatives using interpolating polynomials in *Lagrange form*. Say, we have three values f_1, f_2, f_3 and three corresponding points t_1, t_2, t_3; then we can define an interpolating polynomial,

$$P(t) = f_1 \frac{t - t_2}{t_1 - t_2} \frac{t - t_3}{t_1 - t_3} + f_2 \frac{t - t_1}{t_2 - t_1} \frac{t - t_3}{t_2 - t_3} + f_3 \frac{t - t_1}{t_3 - t_1} \frac{t - t_2}{t_3 - t_2},$$

which is a second-order polynomial in Lagrange form that interpolates our three data points (as the reader may assure himself) with no restrictions on equal grid spacing. A Lagrange polynomial is a sum of products where each product attains the data value at a particular point and is zero at the other points. Thus, the sum interpolates all data values.

In the case of 2D irregular grids (e.g. triangle meshes) things may be a little more tricky—partly because we do not know how many grid points are in the neighborhood of the point at which we wish to compute the derivatives. However, in practice the differential operators have a very similar form: a, possibly normalized, linear combination of data points. In Chaps. 8 and 9 we will discuss a Laplace operator for triangle meshes.

In Chap. 17, we will discuss techniques for deforming 3D surfaces which are embedded as level-sets of a 3D scalar field stored as samples in a 3D grid of points, and also volumetric methods for reconstruction of surfaces from points which simply boils down to solving a 3D Poisson or Laplace problem using finite differences. In all these examples, we generally assume that the domain is completely regular. However, that need not be the case and we refer the interested reader to [2] for a discussion on how to handle irregular domain boundaries.

4.7   Linear Interpolation

In many cases, we need the value of a sampled function, f, between grid points. Assuming that we only have the samples, $F^i = f(hi)$, how do we find a continuous function, $\hat{f}$, which approximates f? If smoothness is not required, one would, typically, use *linear interpolation*:

$$\hat{f}(t) = \alpha F^{i+1} + (1 - \alpha) F^i, \quad i = \lfloor t/h \rfloor, \ \alpha = t/h - i.$$

The notation $\lfloor \cdot \rfloor$ means "nearest integer below". This can be generalized as follows.

Definition 4.2 We have the linear interpolation operator

$$I_F(t) = \alpha F^{i+1} + (1 - \alpha) F^i, \quad i = \lfloor t \rfloor, \ \alpha = t - i, \tag{4.43}$$

where we make the simplifying assumption that the grid spacing $h = 1$.

With the aid of I, we can define interpolation on a 2D grid,

$$I_F(x, y) = I_{I_F(x,\cdot)}(y)$$
$$= \beta\big(\alpha F^{i+1,j+1} + (1-\alpha)F^{i,j+1}\big)$$
$$+ (1-\beta)\big(\alpha F^{i+1,j} + (1-\alpha)F^{i,j}\big), \tag{4.44}$$

where $i = \lfloor x \rfloor$, $j = \lfloor y \rfloor$, $\alpha = x - i$, and $\beta = y - j$. In other words, *bilinear interpolation* can be written as three linear interpolations. We leave it as an exercise to the reader to show how trilinear interpolation on a 3D grid can be implemented using seven linear interpolations.

In many cases we would like to be able to estimate derivatives of a discrete function at arbitrary points in space. For instance, we need the gradient in some algorithms for *isosurface polygonization* (see Chap. 18). Clearly, we can compute the gradient for each grid point as described and then interpolate it to an arbitrary point using linear interpolation. However, we can also center the gradient filter on the point where we wish to compute the gradient. The needed samples will then no longer be grid points, so we need to interpolate those. For an analysis of the efficiency and accuracy of these and other schemes for computing interpolated gradients, the reader is referred to Möller et al. [5].

Another common scenario is that we would like to interpolate a value defined at the vertices of a triangle to an arbitrary point inside the triangle.

Definition 4.3 Linear interpolation using barycentric coordinates (illustrated in Fig. 2.7). Let a triangle be given by the three points $\mathbf{p}_i \in \mathbb{R}^2$, where $i \in \{0, 1, 2\}$. Linear interpolation of the corresponding values f_i to a point $\mathbf{q}$ is given by

$$\hat{f}(\mathbf{q}) = \alpha f_0 + \beta f_1 + \gamma f_2, \tag{4.45}$$

where

$$\alpha = \frac{A(\mathbf{q}, \mathbf{p}_1, \mathbf{p}_2)}{A(\mathbf{p}_0, \mathbf{p}_1, \mathbf{p}_2)}, \qquad \beta = \frac{A(\mathbf{p}_0, \mathbf{q}, \mathbf{p}_2)}{A(\mathbf{p}_0, \mathbf{p}_1, \mathbf{p}_2)}, \quad \text{and} \quad \gamma = \frac{A(\mathbf{p}_0, \mathbf{p}_1, \mathbf{q})}{A(\mathbf{p}_0, \mathbf{p}_1, \mathbf{p}_2)},$$

where $A(\cdot, \cdot, \cdot)$ is the area of the triangle given by the three arguments.

α, β, and γ are referred to as the *barycentric coordinates* of the point $\mathbf{q}$. They have the important properties that they sum to one, $\alpha + \beta + \gamma = 1$, and for any point $\mathbf{q}$ inside the triangle $\alpha, \beta, \gamma > 0$.

Clearly, there is much more to say about interpolation, but linear interpolation is very frequently the tool we need. For interpolation of data values associated with points in space that do not lie on a regular grid, we refer the reader to Chap. 16, which deals precisely with the issue of *scattered data interpolation*.

4.8   Exercises

Exercise 4.1 Verify that the wave equation is indeed hyperbolic, and the heat equation parabolic.

Exercise 4.2 Using Matlab or some other programming environment, reproduce the example in Fig. 4.1.

Exercise 4.3 Compute the characteristic curve of $\frac{\partial f}{\partial t} + a \frac{\partial f}{\partial x} = 0$ from a point (x_0, t_0) and find where it intersects the $t = 0$ curve.

Exercise 4.4 Compute the analytic solution for all time, $t > 0$, of the PDE above for any given initial condition $f_0(x)$.

References

1. Smith, G.D.: Numerical Solution of Partial Differential Equations: Finite Difference Methods, 3rd edn. Oxford University Press, Oxford (1985)
2. Morton, K., Mayers, D.: Numerical Solution of Partial Differential Equations: An Introduction. Cambridge University Press, Cambridge (2005)
3. Osher, S.J., Fedkiw, R.P.: Level Set Methods and Dynamic Implicit Surfaces, 1st edn. Springer, Berlin (2002)
4. Leveque, R.J.: Numerical Methods for Conservation Laws. Birkhäuser, Boston (1992)
5. Moller, T., Machiraju, R., Mueller, K., Yagel, R.: A comparison of normal estimation schemes. In: Proceedings. Visualization'97 (Cat. No. 97CB36155), pp. 19–26, 525 (1997)

Part II
Computational Geometry Processing

Polygonal Meshes

The purpose of this chapter is to give a general introduction to polygonal meshes and especially triangle meshes. We introduce the notation and basic concepts required for understanding how mesh data structures are implemented. In addition, this chapter contains a discussion of some data sources for polygonal meshes and some primitive manipulation operations which form the basic operations for many algorithms discussed in later chapters.

Polygonal and especially triangle meshes are a very important shape representation, and their importance is arguably increasing. There are several reasons for this: first of all, the ability of graphics hardware to render an ever increasing number of triangles in real-time, secondly, the fact that ever larger triangle meshes are acquired from devices such as laser scanners and structured light scanners. Thirdly, many algorithms for manipulation of triangle meshes have been developed recently. Thus, we are now able to perform direct manipulations of shapes represented as triangle meshes where we might previously have had to convert the shape to some other representation.

In the following section, we give a brief overview of primitives for shape representation. This is followed by a more in-depth treatment of polygonal meshes in Sect. 5.2. In Sect. 5.3 we discuss various common sources of polygonal mesh data in order to motivate our interest in this representation and also issues which often arise when dealing with polygonal meshes—to motivate much of the material in this book. Section 5.4 deals with primitive operations for mesh manipulation. Finally, Sect. 5.5 contains descriptions of a number of common data structures for polygonal mesh representation.

5.1   Primitives for Shape Representation

However, other types of polygon are frequently more useful than triangles. In particular quadrilaterals (quads for short) often align better with the symmetries of an object than triangles would. In other types of application, polygons with even more edges are useful, and for yet other types of application, we might prefer entirely other representations: polygonal meshes are by nature non-smooth, and it does

J.A. Bærentzen et al., *Guide to Computational Geometry Processing*,
DOI 10.1007/978-1-4471-4075-7_5, © Springer-Verlag London 2012

take many polygons to produce a fair approximation of a smooth surface. Consequently, the *de facto* industry standard for CAD surfaces remains NURBS patches (Sect. 6.6). However, for models which are animated, subdivision surfaces are very popular (Chap. 7). Finally, implicit surfaces (cf. Sect. 3.10, Chaps. 18 and 16) easily lend themselves to modeling of complex surfaces with very few primitives.

Some years ago, there was a big interest in points as both a primitive for modeling and rendering. It was observed in the seminal paper by Grossman and Dally [1] that points are simpler than even triangles, and in cases where we deal with fairly high resolution meshes, triangles often project to no more than a single pixel when the model is displayed. This led to a number of papers about point rendering, e.g. [2–5] to name a few and, subsequently, also a novel interest in how to manipulate point clouds. Many algorithms from image analysis and polygonal mesh manipulation were applied to point clouds [6, 7]. However, the lack of connectivity information remains a problem. In the presence of very thin structures, it is a problem that the distance between point samples may be greater than the thickness of the structure. Moreover, points are an even more memory demanding representation of surfaces than triangles.

To sum up, there are reasons to use other representations for geometry for many particular applications, but it seems fair to say that the polygonal mesh representation is the representation most widely useful in geometry processing, and in reality other representations are often converted to polygons (triangles) before rendering.

5.2   Basics of Polygonal and Triangle Meshes

Definition 5.1 Polygonal Mesh Entities:
- A polygonal mesh is simply a set of *faces* $\mathcal{F}$, a set of *edges* $\mathcal{E}$, and a set of *vertices* $\mathcal{V}$ where it is understood that $\mathcal{E}$ and $\mathcal{V}$ are precisely the edges and vertices of the faces in $\mathcal{F}$, respectively. We will assume a numbering of the faces, edges, and vertices.
- The point in space corresponding to vertex i will be denoted $\mathbf{p}_i$.
- The edge from vertex i to vertex j is $\mathbf{e}_{ij} = \mathbf{p}_j - \mathbf{p}_i$. Just as for vertices, we may refer to edge i as $\mathbf{e}_i$.
- The total number of vertices (the cardinality of $\mathcal{V}$) will be denoted $|\mathcal{V}|$ and likewise for edges and faces.
- It will be useful to refer to the set of *neighbors*, $\mathcal{N}_i$, of a vertex i, i.e. the set j such that ij is an edge in the triangle mesh.
- The number of neighbors of a vertex i, $|\mathcal{N}_i|$, is called the *valence* of vertex i.
- The *1-ring* of a vertex is the set of polygons sharing that vertex, see Fig. 5.1. The neighbors of a vertex are also sometimes referred to as the 1-ring neighbors since they belong to the 1-ring.

Many algorithms require the mesh to correspond to our intuitive notion of a surface which can be formalized by requiring the mesh to represent a *manifold.*

Fig. 5.1 A vertex and its 1-ring

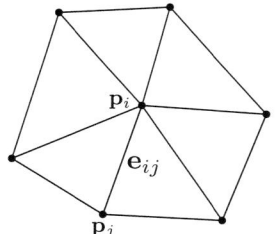

Fig. 5.2 Two examples of triangle meshes which are not 2-manifold

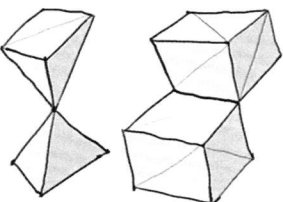

Definition 5.2 A shape is a manifold if any sufficiently small patch is *homeomorphic* (cf. Definition 2.39) to a disk.

In other words, this definition means that any sufficiently small patch has an invertible (i.e., one-to-one and onto) mapping of that patch onto a unit disk. This would be true, for instance, for a point on an edge shared by two triangles, but it would *not* be true if three triangles met at the edge. Two triangles meeting at a vertex is another example of a non-manifold configuration, see Fig. 5.2. We state without proof (but see [8]) the following theorem:

Theorem 5.1 *Triangle mesh manifoldness conditions: faces must meet only along edges or vertices, and an edge must be shared by either one or two faces, depending on whether it is an interior edge or along the boundary. Secondly, the faces around a vertex must form a single cycle.*

These conditions entail a geometric property, namely that the mesh does not self-intersect since that would violate the condition that faces meet only along edges of the mesh. Also invalid are shapes like two boxes glued together along a common edge or two pyramids sharing an apex.

For many purposes it is convenient to deal only with triangles. Triangles have the nice property that they are planar. If we know that a mesh is composed only of triangles, it simplifies both storage and a number of algorithms. On the other hand, it is generally very easy to convert polygons with more than three edges to triangles—especially if they are all planar and convex or approximately so [9].

5.3   Sources and Scourges of Polygonal Models

Most popular modeling packages allow the users to directly model polygonal meshes. In this case, the polygon of choice is usually the quadrilateral. In many other cases, the polygons are the result of some form of acquisition. Typically, the acquisition process leads to data in one of the three forms discussed below.

- Scattered points in 2D. These could be terrain points measured by a surveyor. Such data have the nice feature that they can be projected into 2D and triangulated in the plane, typically using Delaunay triangulation (see Chap. 14). In this and other cases where the data can be considered to be samples of a height function $f : \mathbb{R}^2 \to \mathbb{R}$ the situation is much simpler than in the following case.

- Scattered points in 3D. We may have point samples of a complex 3D surface that cannot easily be flattened to a planar surface. This is a harder case, in particular when the topology of the object cannot be assumed known. Frequently, in fact, such data are converted to the implicit representation discussed below. Chapters 16 and 17 discuss methods for precisely that.

- Implicit surfaces (cf. Sect. 3.10). In many cases, we have a function $\Phi : \mathbb{R}^3 \to \mathbb{R}$ with some associated isovalue τ such that the surface $S = \Phi^{-1}(\tau)$. In 2D the contour of Φ corresponding to the isovalue τ would generally be a closed curve, and in 3D it is a watertight surface. In later chapters we will discuss both various representations for implicit surfaces and techniques for extracting the contour surface as a polygonal mesh. Another well known source of implicit surfaces is volumetric data which may be acquired using *computed tomography* scanners or *magnetic resonance imagery*. These scanning modalities are often used in medical imaging.

Polygonal meshes which were produced as described above have a tendency to be fraught with a number of problems which are completely analogous to those faced when dealing with other types of signal: photographs, sounds or any other type of signal, which has been acquired from a real world data source. We will return to this viewpoint that a polygonal mesh produced by some sort of acquisition process is a *discrete geometric signal*. However, meshes have some additional problems, since the mesh connectivity and the overall object topology may both suffer from noise. To sum up, the most important issues we have to deal with are as follows:

- *Oversampling*. Most acquisition techniques do not distinguish between regions of high or low degree of geometric detail. Consequently a mesh produced in this fashion will have a tendency to be oversampled in places leading to unnecessarily large geometric models.

- *Undersampling*. For the same reason tiny features and sharp edges or corners are rarely sampled well. Consequently, e.g. a sharp edge is typically jagged in a mesh produced by some acquisition technique.

- *Irregularity*. In a completely regular triangle mesh, all vertices have valence 6, but only a torus or a planar patch admit a completely regular mesh. However, in a highly regular mesh, we can have relatively few vertices with valencies other than six. Since many algorithms tend to work better on meshes which are relatively regular, it is often desirably to improve the regularity of a mesh. Particularly nasty

Fig. 5.3 Many mesh
manipulation algorithms, for
instance smoothing
algorithms, only modify the
vertex positions

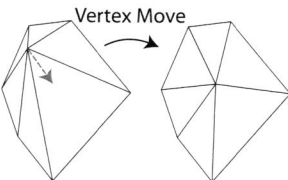

triangles are often classified as *needles* and *caps*. A needle is a triangle with one
edge that is extremely short and a cap is an obtuse triangle with an angle of close
to 180°. A needle may also be a cap and vice versa.

- Topological issues. The mesh may consist of disconnected patches of polygons.
  Since many laser scanners produce data in patches, this issue frequently arises.
  In more extreme cases we may be dealing with a soup of disconnected triangles
  where we have no knowledge of shared vertices and edges—typically because
  that information has been misplaced. Finally, we may have a manifold mesh but
  of a different genus than expected—either because a handle was introduced or
  removed.

5.4   Polygonal Mesh Manipulation

Many of the above problems have spawned a lot of work in the geometry processing research community. For instance, there is a large body of literature on mesh
smoothing. Most of this concerns how to remove noise by attenuating high frequency variations in the mesh with the underlying assumption that the surface is
smooth. However, in recent years many authors have focused on anisotropic smoothing where sharp edges (in particular) are preserved while noise is removed (cf.
Chap. 9).

Mesh smoothing algorithms preserve mesh connectivity. In fact, vertex positions
(see Fig. 5.3) are typically the only mesh attribute which is modified.

Except for *vertex move*, all primitive operations for mesh manipulation somehow
change the vertex connectivity. In the following, we will present four often used
mesh manipulation operations, and discuss their uses briefly but leave their precise
applications to later chapters.

The first of these primitive operations is *edge collapse* which is shown in Fig. 5.4.
This operation removes a single edge and (in the case of triangle meshes) the two
adjacent faces. Since it reduces mesh complexity it is often used for simplification—
especially since the introduction of the very popular QSlim algorithm [10] by Garland and Heckbert, which is discussed in Sect. 11.1.

Collapsing an edge can also be seen as welding its two endpoint vertices, and,
depending on the mesh representation, these two vertices need not be connected. If
we allow the merging of unconnected vertices non-manifold situations may arise. If
we weld two vertices from different components, the result is a single vertex with
two disconnected 1-rings. In the original paper by Garland and Heckbert, this was
seen as an advantage since it allows different components to be joined. On the other

Fig. 5.4 Collapsing an edge
and the inverse operation
known as vertex split

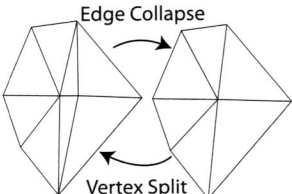

hand, some mesh representations do not allow non-manifold meshes. If we wish to prevent non-manifold situations from occurring, only vertices connected by an edge can be collapsed, and we need to perform the following tests.

1. We find all vertices connected to both endpoint vertices. If a face adjacent to the edge is a triangle, the corner not on the edge will obviously be connected to both endpoint vertices, and that is not a problem since the triangle disappears when the edge is collapsed which means that the two connections are joined. In all other cases where a vertex is connected to both endpoints of an edge, we need to disallow the edge collapse since the result would be a vertex with two edges to the welded vertex.

2. If all faces around the edge being collapsed are triangles and the endpoints of the edge have valence three, the object is a tetrahedron, and collapsing the edge will collapse the tetrahedron.

3. If a face adjacent to the edge being collapsed is a triangle, it will disappear, and the valence of the final vertex belonging to this edge will be reduced by one. If the valence of this vertex is three it will be reduced to two. If the remaining two faces in the 1-ring of this vertex are triangles, they will become coplanar. If the faces are not triangles we may wish to allow the collapse since valence two vertices could be considered legal.

4. We may not want to collapse an edge which connects two boundary loops, because in that case we join the two loops in a figure eight construct. However, in reality it may be fairly easy for most mesh representations to handle this situation.

All of these tests are required to enforce a manifold constraint on the mesh, but, of course, we may not always wish to do so. However, if the representation can only handle manifolds, failure to enforce the constraints would lead to a degenerate mesh.

The inverse of an edge collapse is a *vertex split*. While edge collapse reduces geometric detail, vertex split introduces detail, and in fact multi-resolution hierarchies based on these two operators are often used in mesh processing.

Irregular connectivity in triangle meshes can be improved with a simple operation known as *edge flip* or edge swap illustrated in Fig. 5.5. An edge flip removes an edge, producing a quad and forms two new triangles by introducing the other diagonal of the quad as a new edge. This simple operation can greatly improve the regularity of the mesh. For 3D meshes, it also changes the geometry since the triangles are usually non planar. Again, we must be a little careful when applying this operation. If a vertex has valence three, it is impossible to flip any incident edge since this would reduce the valence to two causing a non-manifold situation.

Fig. 5.5 Flipping an edge

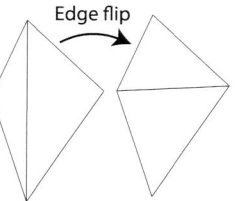

Fig. 5.6 Splitting an edge or
a face

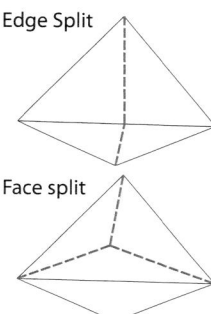

If we wish to introduce more detail, for instance produce a smoother model from
a coarse mesh, techniques for introducing detail are needed. While vertex split is
one option, other techniques work by inserting vertices in edges or in the interior of
polygons as illustrated in Fig. 5.6. However, regular subdivision schemes which will
be discussed later usually work by splitting a triangle into four new triangles and a
quad into four new quads. Repeated application of these operations combined with
a smoothing operation typically leads to very fair surfaces. Subdivision is discussed
in detail in Chap. 7. In contrast, the two operations shown in the margin are usually
combined with edge flips to ensure a proper distribution of valences. Their main
advantage is that they introduce detail in a highly localized manner.

5.5 Polygonal Mesh Representations

Till now, we have simply assumed that there is some mesh representation without
considering in detail how to actually store the polygons. In fact, there are many
representations, and which we choose has great bearing on what we can easily do
with the mesh.

The simplest solution is to store each polygon with the geometric position of each
of its vertices. However, doing so means that we have no connectivity information—
effectively, we have just a polygon soup which can be rendered but little manipulation is possible since all the primitive operations assume that we have some knowledge about how polygons relate to each other.

A more useful and very simple representation is an *indexed face set*. An indexed
face set stores the mesh in two arrays. The first one contains all vertices indexed
by number (typically just in a linear array). For each vertex we store its attributes,

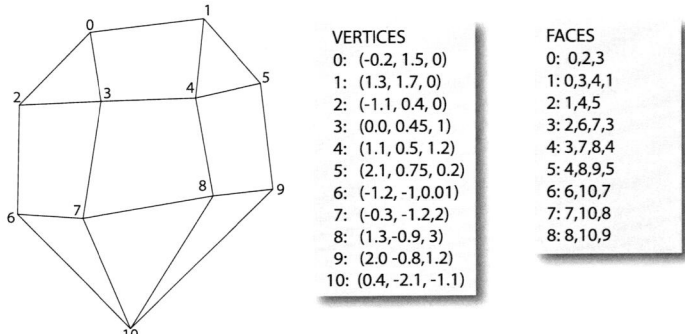

Fig. 5.7 An indexed face set with the actual geometry on the *left*, the list of vertices in the *middle*, and the face list on the *right*

which means at least its position in space but frequently also other information such as a vertex normal, and possibly some attributes needed for specific purposes such as rendering. The second array is a list of faces which contains at least a list of indices where each index refers to the vertex array. An example is shown below in Fig. 5.7.

While indexed face sets allow us to deduce for instance which polygons share edges, this information is not explicitly encoded. In general, it takes a loop over all faces to find a face sharing an edge with a given other face. Similarly we have to visit all faces to find the set of vertices sharing an edge with a given vertex. We can easily bring down this time to a constant time query if we create an additional array with adjacency information. For each vertex we could store the faces containing that vertex, and the vertices which share an edge with that vertex. Another way to accomplish roughly the same goal is to store with each polygon the adjacent polygon across each edge. In this case, we also need to store the index of the corresponding edge in the adjacent polygon. Zorin advocates this representation for subdivision surfaces [11].

Given such an augmented representation, we can perform many algorithms efficiently, but the auxiliary data structures need to be updated if the mesh is changed. While this is certainly feasible, many researchers and professional choose instead to use *edge-based* data structures. Edge-based data structures are in general also very flexible with regard to what types of polygon can be stored.

The fundamental idea is to represent connectivity by explicitly storing how edges relate to each other. For each face of the mesh, we simply store a pointer to a single one of its edges. Likewise, we store for each vertex just a pointer to a single edge. If we want the next edge in the edge loop which defines a face, we simply follow a pointer from the current edge. Consequently, we can circulate a face, visiting all its edges even though the face only has a pointer to one of these. Since edge-based representations are based on edges knowing their incident faces, only manifolds are usually representable using such data structures.

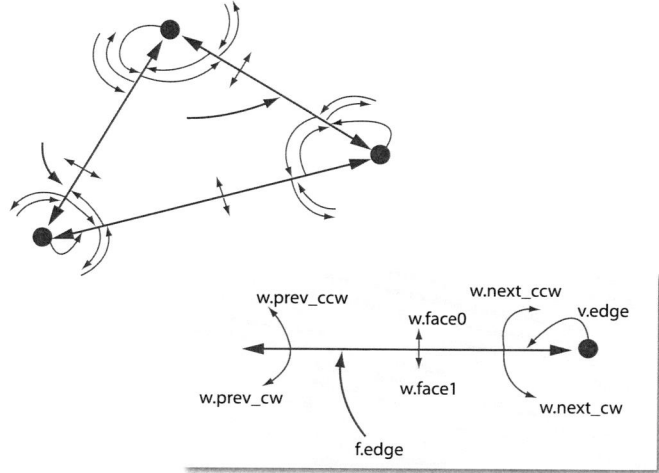

Fig. 5.8   The classic winged edge data structure. While widely used, it is probably fair to say that it is now superseded by the half edge data structure

The classic winged edge data structure [12] is shown below in Fig. 5.8. The winged edge is an oriented edge connecting two vertices. For both of its two face loops, it has pointers to the next edges in both clockwise and counter clockwise direction. This allows us to move around a face in the fashion described above. Unfortunately, the winged edge representation is a bit complicated, and it has another flaw: the edge itself is bidirectional, but the winged edge is oriented. If we want to move from an edge to the next edge, which pointer we have to use depends on the orientation of the edge. This means that to circulate a face we need to use conditionals (if statements) which is s potentially a problem with today's CPUs which have very deep pipelines. If a branch is mispredicted many instructions already partially processed in the pipeline must be discarded which is detrimental to performance.

5.6   The Half Edge Data Structure

The *half edge* representation [13] solves this problem by having two representations of a given edge. Half edges come in pairs, and each member of the pair represents the edge from the point of view of one of the two polygons sharing the edge. Thus, there is no ambiguity. Each half edge has a next pointer which points to the next edge in the loop that corresponds to *its* face. By simply following the next pointers, we can visit all the edges in the edge loop of a face. See Fig. 5.9

It is also possible to visit all edges around a vertex since the vertex knows one of its (outgoing) half edges. To visit all vertices sharing an edge with a given vertex, v, we can proceed as follows:

1. Follow the pointer which v holds to an outgoing half edge h which we also label h0.

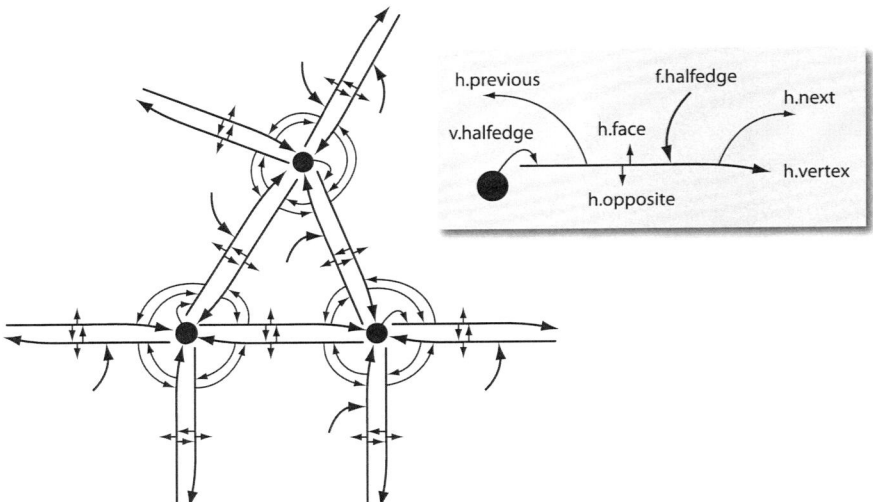

Fig. 5.9 This is an illustration of the half edge data structure. To avoid further overloading of the crowded diagram, the arrows are only named in the example in the *box*. The half edge (h) itself points to one incident vertex, to its opposite, and to the next and previous half edges in the counter clockwise loop around the face. A vertex (v) points to one outgoing half edge, and a face (f) points to one half edge in its loop. *Arrows* corresponding to all these pointers are shown

2. h holds a pointer to a vertex w which we visit.
3. Follow the pointer to the edge opposite h and its next pointer. This yields the next outgoing half edge in a clockwise loop around v. Name this half edge h.
4. If h is identical to h0 we are done, else go to 2.

The reader should be able to follow the steps above by referring to Fig. 5.9. In practice it is rarely necessary to directly deal with the pointers in the half edge data structure unless implementing it. Usually a library such as OpenMesh [14] or GEL [15] will provide so called *circulators*. They usually come in two flavors: face circulators and vertex circulators. Face circulators allow the user to visit all edges in a loop around the face, and vertex circulators do the same for vertices. In other words, a circulator simply maintains a pointer to a current half edge in a face or vertex loop, and it can be incremented in order to move to the next in the loop.

There is a number of variations of the theme discussed above. Three concrete examples are the quad-edge data structure [16], the G-map [17] and the lath data structure [18]. All three emphasize ease of obtaining the dual mesh (where each face is a vertex and vice versa) but fundamentally, the same topological (i.e. spatial connectivity) information is encoded. Another variation is the directed edge data structure [19] which is a triangle mesh specialized edge-based data structure.

5.6.1   The Quad-edge Data Structure

The quad-edge data structure operates on pairs of subdivisions, for which there exist two one-to-one mappings that send vertices of one subdivision to faces of the other subdivision and vice versa, and edges of one subdivision to edges of the other one and vice versa, where the term subdivision is defined as follows.

Definition 5.3 A subdivision of a manifold[1] M is [16] a partition S of M into three finite collections of disjoint parts, the vertices (denoted by $\mathcal{V}$), the edges (denoted by $\mathcal{E}$ or E) and the faces (denoted by $\mathcal{F}$) with the following properties:
- every vertex is a point of M,
- every edge is a line of M,
- every face is a disk of M,
- the boundary of every face is a closed path of edges and vertices.

A directed edge of a subdivision P is an edge of P together with a direction along it and a given orientation (see page 80 in [16]). Since directions and orientations can be chosen independently, for every edge of a subdivision there are four different directed and oriented edges [16]. For any oriented directed edge e we can define unambiguously its vertex of *origin* Org(e), its *destination*, Dest(e), its *left face*, Left(e), and its *right face*, Right(e). The flipped version Flip(e) of an edge e is the same unoriented edge taken with *opposite orientation* and same direction. The *symmetric* of e, Sym(e) corresponds to the same undirected edge with the *opposite direction* but the same orientation as e.

The *next edge with the same origin*, Onext(e) is defined as the one immediately following e (counterclockwise) in the ring of edges out of the origin of e (see Fig. 5.10). The *next counterclockwise edge with the same left face*, denoted by Lnext(e), is defined as the first edge we encounter after e when moving along the boundary of the face $F = $ Left(e) in the counterclockwise sense as determined by the orientation of F.

Definition 5.4 Two subdivisions S and S^* are said to be the *dual* [16] of each other if for every directed and oriented edge e of either subdivision, there is another edge Dual(e) (which is defined as the dual of e) of the other subdivision such that:
- the dual of Dual(e) is e: Dual(Dual(e)) = e,
- the dual of the symmetric of e is the symmetric of Dual(Sym(e)): Dual(Sym(e)) = Sym(Dual(e)),
- the dual of the flipped version of e is the symmetric of the flipped version of Dual(e): Flip(Dual(e)) = Sym(Flip(Dual(e)))
- moving counterclockwise around the left face of e in one subdivision is the same as moving clockwise around the origin of Dual(e) in the other subdivision: Lnext(Dual(e)) = Onext^{-1}(Dual(e)).

---

[1]A two-dimensional manifold is a topological space with the property that every point has an open neighborhood which is a disk.

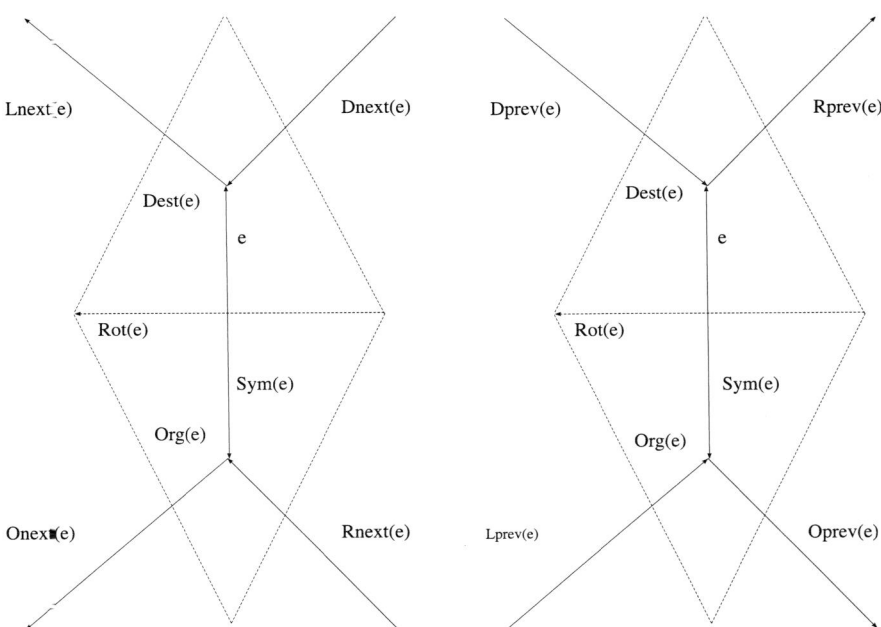

Fig. 5.10 The edge functions

The corresponding subdivisions of the quad-edge data structure are called dual. The perfect example of such a pair of dual subdivisions in computational geometry is the Voronoi diagram and the Delaunay triangulation (or more generally the Delaunay graph) of a point set, and the quad-edge data structure allows one to traverse concurrently the Voronoi diagram and the Delaunay triangulation.

The dual of an edge e is the edge of the dual subdivision that goes from the (vertex corresponding to the) left face of e to the (vertex corresponding to the) right face of e but taken with orientation opposite to that of e.

The quad-edge data structure [16] is in fact a convenient mathematical structure for representing the topological relationships among edges of any pair of dual subdivisions on a two-dimensional manifold. Edge functions (see Fig. 5.10) allow the traversal of the pair of dual subdivisions at the same time, and it can be used to construct a pair of dual subdivisions at the same time, as in the case with the incremental construction of the Delaunay triangulation and the Voronoi diagram, which we will see in Chap. 14.

As shown in the top part of Fig. 5.11, each branch of the quad-edge is part of a loop around a Delaunay vertex/Voronoi face, or around a Delaunay triangle/Voronoi vertex. The lower part of Fig. 5.11 shows the corresponding Delaunay/Voronoi structure, where (a, b, c) are quad-edges, and (1, 2, 3) are Delaunay vertices.

The topology of the subdivision is completely determined by its edge algebra, and vice versa. This allows all the edge functions to be expressed using three basic primitives, Flip, Rot, and Onext described above [16]. The quad-edge traversal operations are based on the edge algebra $(E, E*, \text{Onext}, \text{Flip}, \text{Rot})$, and their ex-

Fig. 5.11 A simple Voronoi
diagram and its
corresponding quad-edge

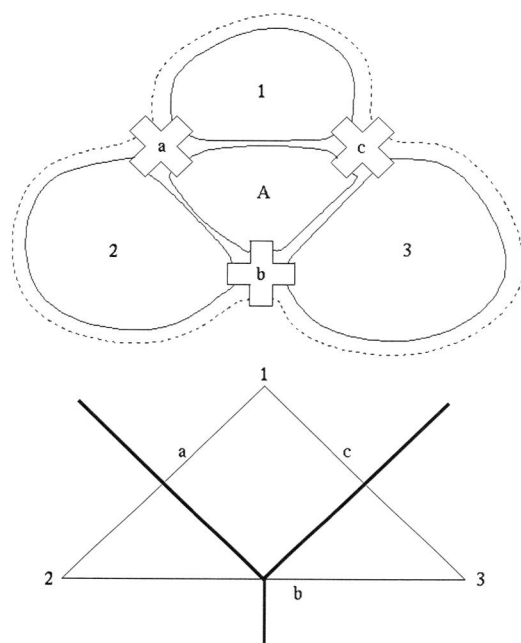

Table 5.1 Basic quad-edge
topological operators

Operation	Description
e := MakeEdge[]	Creates an edge e to a newly created data structure representing an empty manifold
Splice[a, b]	Joins or separates the two edge rings a Org and b Org, and independently, the two dual edge rings a Left and b Left (see Fig. 5.12)

pression as composition of the basic primitives [16].The main advantage of the
quad-edge data structure is that all the construction and modification of planar
graphs can be done using two basic topological operators (see Table 5.1), and
the complex topological operations built from these two basic topological operators.

In summary, a reasonably complete mesh library should (independent of implementation) contain functionality allowing the user to iterate through all entities, vertices, edges (or half edges) and faces and to locally move from one entity to the
next—say from a vertex to any of its outgoing edges. These local operators could
be implemented via circulators. It should also provide a (rich) set of primitive manipulation operations such as those discussed in Sect. 5.4.

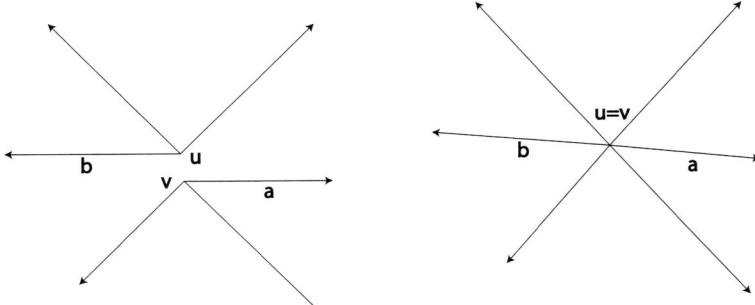

Fig. 5.12 The Splice topological operator

5.7   Exercises

Exercise 5.1 Write a computer program that loads and displays a polygonal mesh from a file in one of the common mesh formats (e.g. OBJ, OFF, PLY, or X3D).

[**GEL Users**] Such a program has been written for GEL users. Download it and try it out on some of the 3D models distributed with the example program.

Exercise 5.2 Based on the program from the exercise above, your goal is to write a function which computes the dual of a polygonal mesh. A dual mesh has a vertex for every face and a face for every vertex. Simply create a vertex for each face and a face for each vertex. The vertex created for a face should be at the center of the face.

Hint: probably the easiest way of solving this problem is to create a new mesh as an indexed face set and then to convert that indexed face set to your representation of choice. You will need to be able to iteratively visit all vertices and faces, a method to associate an attribute with a face, and a method to circulate around a vertex.

[**GEL Users**] If you base your work on the example program, you simply need to fill in the `compute_dual` function in the program from the previous exercise.

Once you have filled in this function, if you run the program with an example model, pressing 'd' should produce the dual. Note that pressing 'd' repeatedly will smooth and shrink the mesh.

Exercise 5.3 Implement the quad-edge data structure using GEL. Use two different manifolds, and add for each edge in one manifold, add the pointer to its dual edge in the other manifold. Keep your implementation for further use.

References

1. Grossman, J.P., Dally, W.J.: Point sample rendering. In: Proceedings of the 9th Eurographics Workshop on Rendering, June pp. 181–192 (1998)
2. Pfister, H., Zwicker, M., van Baar, J., Gross, M.: Surfels: surface elements as rendering primitives. In: Proceedings of SIGGRAPH 2000, pp. 335–342 (2000)

3. Zwicker, M., Pfister, H., van Baar, J., Gross, M.: EWA splatting. IEEE Trans. Vis. Comput. Graph. **8**(3), 223–238 (2002)
4. Dachsbacher, C., Vogelgsang, C., Stamminger, M.: Sequential point trees. ACM Trans. Graph. **22**(3), 657–662 (2003). doi:10.1145/882262.882321
5. Botsch, M., Hornung, A., Zwicker, M., Kobbelt, L.: High-quality surface splatting on today's GPUs. In: Proceedings of Symposium on Point-Based Graphics, pp. 17–24 (2005)
6. Pauly, M., Keiser, R., Kobbelt, L.P., Gross, M.: Shape modeling with point-sampled geometry. ACM Trans. Graph. **22**(3), 641–650 (2003)
7. Pauly, M., Gross, M.: Spectral processing of point-sampled geometry. In: Proceedings of the 28th Annual Conference on Computer Graphics and Interactive Techniques, pp. 379–386 (2001)
8. Hoffmann, C.M.: Geometric and Solid Modeling. Morgan Kaufmann, San Mateo (1989)
9. Chazelle, B.: Triangulating a simple polygon in linear time. Discrete Comput. Geom. **6**, 485–524 (1991)
10. Garland, M., Heckbert, P.S.: Surface simplification using quadric error metrics. In: Proceedings of the 24th Annual Conference on Computer Graphics and Interactive Techniques, pp. 209–216 (1997)
11. Zorin, D., Schröder, P., DeRose, T., Kobbelt, L., Levin, A., Sweldens, W.: Subdivision for modeling and animation. Technical Report, SIGGRAPH 2000 Course Notes (2000)
12. Baumgart, B.G.: Winged edge polyhedron representation. Technical Report, Stanford University (1972)
13. Mantyla, M.: Introduction to Solid Modeling. Freeman, New York (1988)
14. Botsch, M., Steinberg, S., Bischoff, S., Kobbelt, L.: OpenMesh-a generic and efficient polygon mesh data structure. In: OpenSG Symposium (2002)
15. Bærentzen, A.: GEL. http://www2.imm.dtu.dk/projects/GEL/
16. Guibas, L., Stolfi, J.: Primitives for the manipulation of general subdivisions and the computation of Voronoi diagrams. ACM Trans. Graph. **4**, 74–123 (1985)
17. Lienhardt, P.: Extension of the notion of map and subdivisions of a three-dimensional space. In: STACS 88 (Bordeaux, 1988). Lecture Notes in Comput. Sci., vol. 294, pp. 301–311. Springer, Berlin (1988)
18. Joy, K.I., Legakis, J., MacCracken, R.: Data structures for multiresolution representation of unstructured meshes. In: Farin, G., Hagen, H., Hamann, B. (eds.) Hierarchical Approximation and Geometric Methods for Scientific Visualization. Springer, Heidelberg (2002)
19. Campagna, S., Kobbelt, L., Seidel, H.P.: Directed edges—a scalable representation for triangle meshes. J. Graph. Tools **3**(4), 1–12 (1998)

Splines

6

In this chapter, we describe the type of curves and surfaces often used in modern CAD systems. The de facto industry standard here is NURBS which stands for *Non Uniform Rational B-Splines*. Even though the animation industry has largely switched to subdivision surfaces (cf. Chap. 7) B-Splines are still relevant: They are a very flexible tool for the representation of smooth surfaces, they allow for exact representation of conic surfaces, and the CAD business has a lot of software and know-how pertaining to B-Splines. For further reading and proofs we refer to the vast literature on the subject, e.g. [1–4].

The main drawback of B-splines (and other parametric surfaces) compared to subdivision is the fact that continuity across surface patches is hard to maintain when the surface deforms. Thus, subdivision surfaces are preferred for animation. On the other hand, subdivision surfaces have issues with irregular vertices which also occur near the seams between patches.

6.1 Parametrization

Often a surface cannot be parametrized by a single regular parametrization but need to be specified by several patches. This is certainly true for CAD models of complex objects like cars, ships, and aeroplanes. They consist of thousands of patches. In differential geometry the individual parametrizations are normally required to be defined on open sets. Then a smoothness conditions like C^1, C^2 etc., see Example 2.6 in Chap. 2, on the surface can be formulated as the same smoothness condition on the individual parametrizations. In a CAD system the parametrizations are typically defined on a closed rectangle and the individual patches on the surface only overlap at the boundary,

$$\mathbf{x}_1(U_1) \cap \mathbf{x}_2(U_2) \subseteq \mathbf{x}_1(\partial U_1) \cap \mathbf{x}_2(\partial U_2). \tag{6.1}$$

Now a smoothness condition on the individual parametrization does not secure the same smoothness on the surface. Certain boundary conditions have to be satisfied. The terminology regarding smoothness is also slightly different. In the

J.A. Bærentzen et al., *Guide to Computational Geometry Processing*,
DOI 10.1007/978-1-4471-4075-7_6, © Springer-Verlag London 2012

Fig. 6.1 Parametrization

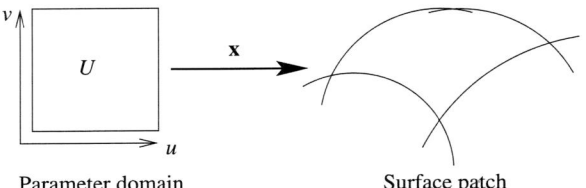

Parameter domain   Surface patch

CAGD (Computer Aided Geometric Design) community two patches join C^k if the parametrization meet in a C^k fashion along the common boundary, while G^k or GC^k means that there at each point on the common boundary locally exists a regular C^k parametrization. In differential geometry this will be called a C^k surface. This is the same as the tangent planes agree and the projection from the common tangent plane is C^k.

6.2 Basis Functions and Control Points

The parametrization (illustrated in Fig. 6.1) $\mathbf{x} : U \to \mathbb{R}^3$ is of the form

$$\mathbf{x}(u, v) = \sum_{\ell=1}^{n} \mathbf{c}_\ell F_\ell(u, v).$$

The functions $F_\ell : U \to \mathbb{R}$ are called the *basis functions* and the points $\mathbf{c}_\ell \in \mathbb{R}^3$ are called the *control points*. The definition of the parametrization $\mathbf{x}$ is independent of the coordinate system if and only if the basis functions form a partition of unity, i.e.,

$$\sum_{\ell=1}^{n} F_\ell(u, v) = 1, \quad \text{for all } (u, v) \in U. \tag{6.2}$$

The basis functions, F_ℓ, are normally products, $F_\ell(u, v) = G_\ell(u)H_\ell(v)$, of univariate polynomials, rational functions, piecewise polynomials, or piecewise rational functions. So even though we are interested in surfaces we start with functions of one variable.

6.3 Knots and Spline Spaces on the Line

We consider functions $f : [a, b] \to \mathbb{R}$ that are piecewise polynomials of degree d. That is, we have a sequence of break points, called *knots*, where the different pieces of polynomials meets with a certain degree of differentiability.

Fig. 6.2 A spline of degree 2 consisting of four polynomial pieces, parabolas

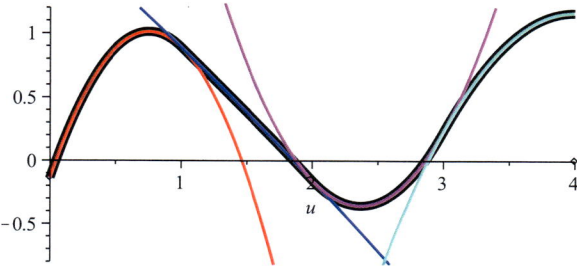

Definition 6.1 A *knot sequence* or *knot vector* is a non-decreasing sequence, denoted **u**,

$$\underbrace{u_1 \leq \cdots \leq u_{d+1}}_{\text{boundary knots}} < \underbrace{u_{d+2} \leq \cdots \leq u_{d+n}}_{\text{inner knots}} < \underbrace{u_{d+n+1} \leq \cdots \leq u_{2d+n+1}}_{\text{boundary knots}}, \qquad (6.3)$$

where $a = u_{d+1}$ and $b = u_{n+d+1}$. If $u_\ell < u_{\ell+1} = \cdots = u_{\ell+\nu} < u_{\ell+\nu}$ then we say that the knot $u_{\ell+1} = \cdots = u_{\ell+\nu-1}$ has *multiplicity* ν.

If the multiplicity of a knot is one, i.e., $\nu = 1$, then we call the knot *simple*. If the multiplicity is equal to the degree, i.e., $\nu = d$, we say that the knot has *full multiplicity*. Often the boundary knots have multiplicity $d + 1$, i.e., $u_1 = \cdots = u_{d+1} = a$ and $u_{n+d+1} = \cdots = u_{n+2d+1} = b$.

Definition 6.2 The *spline space* of degree d on the knot vector **u** is the space

$$\mathcal{S}_{\mathbf{u}}^d = \mathcal{S}_{\mathbf{u}}^d\big([a, b]\big)$$

$$= \left\{ f : [a, b] \to \mathbb{R} \;\middle|\; \begin{array}{l} f|_{]u_\ell, u_{\ell+1}[} \text{ is a polynomial of degree } d \text{ and} \\ f \text{ is } C^{d-\nu} \text{ at a knot with multiplicity } \nu \end{array} \right\}. \qquad (6.4)$$

Notice that with the knot vector (6.3) we have at most n knot intervals, $[u_{d+1}, u_{d+2}], \ldots, [u_{d+n}, u_{d+n+1}]$, and consequently each spline in the spline space consists of n polynomial pieces. To be precise, if all knots are simple then we have n pieces and each time the multiplicity of a knot is increased by one the number of pieces drops by one. In Fig. 6.2 we have plotted a spline of degree two consisting of four polynomial pieces.

If a knot is simple then two polynomial pieces meet with at least C^{d-1} differentiability; if a knot has full multiplicity then two polynomial pieces meet with continuity only, and finally if the multiplicity is greater than the degree, i.e., $\nu > d$ then two polynomial pieces meet discontinuously.

We can also consider functions $f : \mathbb{R} \to \mathbb{R}$ defined on all of the reals. In that case we will have a bi-infinite knot sequence, e.g. $\mathbb{Z}$.

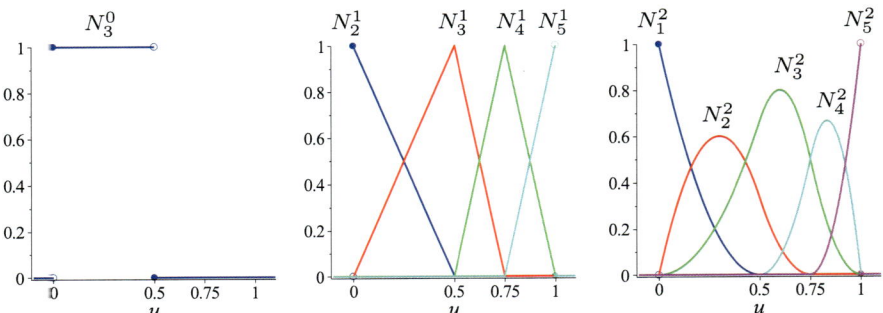

Fig. 6.3 B-splines of degree 0, 1, and 2 on the knot vector $\mathbf{u} = 0, 0, 0, \frac{1}{2}, \frac{3}{4}, 1, 1, 1$. The boundary knots have multiplicity 3 so $N_1^0 = N_2^0 = N_6^0 = N_7^0 = N_1^1 = N_6^1 = 0$

6.4   B-Splines

B-splines (B for basis, i.e., basis splines) form a basis with minimal support for a spline space.

Definition 6.3 The *B-splines* $N_{\mathbf{u},\ell}^d(u)$ of degree d on the knot vector $\mathbf{u}$ can be defined recursively,

$$N_{\mathbf{u},\ell}^0(u) = \begin{cases} 1 & \text{if } u_\ell \leq u < u_{\ell+1}, \\ 0 & \text{otherwise}, \end{cases}$$

$$N_{\mathbf{u},\ell}^d(u) = \frac{u - u_{\ell-1}}{u_{\ell+d-1} - u_{\ell-1}} N_{\mathbf{u},\ell}^{d-1}(u) + \frac{u_{\ell+d} - u}{u_{\ell+d} - u_\ell} N_{\mathbf{u},\ell+1}^{d-1}(u).$$

With the knot vector (6.3) we have $n + d$ functions, but some of them could be the zero function. Indeed, if a knot has multiplicity $v > d$, then $d - v$ of the functions N_ℓ^d are the zero function. In Fig. 6.3 the case $d = 0, 1, 2$ is illustrated for the knot sequence $\mathbf{u} = 0, 0, 0, \frac{1}{2}, \frac{3}{4}, 1, 1, 1$, and in Fig. 6.4 some cubic B-splines on different knot vectors are plotted.

We give the following theorem without proof, for that we refer to [1–3] or Exercise 6.1.

Theorem 6.1 *The B-splines on a knot vector $\mathbf{u}$ have the following properties.*
1. *The restriction of the B-spline $N_{\mathbf{u},\ell}^d(u)$ to the open interval $]u_\ell, u_{\ell+1}[$ is a polynomial of degree d.*
2. *The B-spline $N_{\mathbf{u},\ell}^d(u)$ is C^{d-v} at a knot with multiplicity v.*
3. *The support of the B-spline $N_{\mathbf{u},\ell}^d(u)$ is $[u_\ell, u_{\ell+d+1}]$.*
4. *The collection of B-splines $N_{\mathbf{u},\ell}^d(u)$ forms a basis with minimal support for the spline space $\mathcal{S}_{\mathbf{u}}^d$.*
5. *The B-splines form a partition of unity, i.e., $\sum_\ell N_{\mathbf{u},\ell}^d(u) = 1$.*

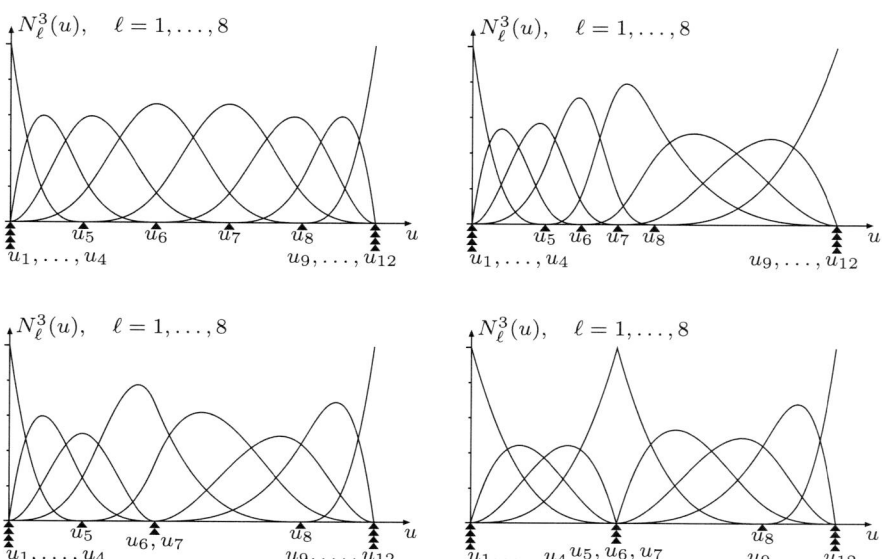

Fig. 6.4 Cubic B-splines on different knot vectors

6. *If the knot sequence is* $\mathbb{Z}$, *then the B-splines of a fixed degree are the translates of each other*

$$N_{\mathbb{Z},\ell}^d(u) = N_{\mathbb{Z},1}^d(u + 1 - \ell),\qquad(6.5)$$

and they satisfy the following refinement equation:

$$N_{\mathbb{Z},0}^d(u) = \sum_{\ell} N_{\mathbb{Z},0}^d(2u - \ell).\qquad(6.6)$$

A *spline curve* is a curve of the form

$$\mathbf{r}(u) = \sum_{\ell=1}^{n+d} \mathbf{c}_\ell N_{\mathbf{u},\ell}^d(u),\qquad(6.7)$$

and the points $\mathbf{c}_\ell$ are called the *control points*. The control points are normally not points on the curve but if a knot $u_{\ell+1} = \cdots = u_{\ell+d}$ has full multiplicity then $\mathbf{c}_\ell = \mathbf{r}(u_{\ell+d})$, cf. Exercise 6.2. As a corollary to Property 3 above we can describe the influence of a single control point on a spline curve.

Corollary 6.1 *Let* $\mathbf{r}(u) = \sum_\ell \mathbf{c}_\ell N_{\mathbf{u},\ell}^d(u)$ *be a spline curve on a knot vector* $\mathbf{u}$. *Then the control point* $\mathbf{c}_\ell$ *only influences the curve on the interval* $[u_\ell, u_{\ell+d+1}]$. *Conversely, the piece of the curve given by the knot interval* $[u_\ell, u_{\ell+1}]$ *is only influenced by the* $d + 1$ *control points* $\mathbf{c}_{\ell-d-1}, \ldots, \mathbf{c}_\ell$.

6.4.1   Knot Insertion and de Boor's Algorithm

If we insert a knot $u^* \in [u_k, u_{k+1})$ in a knot sequence, $\mathbf{u} = \cdots \leq u_k < u_{k+1} \leq \cdots$, then we obtain a refined knot sequence $\mathbf{u}^* = \cdots \leq u_k \leq u^* < u_{k+1} \leq \cdots$ and we clearly have $\mathcal{S}_\mathbf{u}^d \subseteq \mathcal{S}_{\mathbf{u}^*}^d$. That is, any spline, $\mathbf{r}(u) = \sum c_k N_{\mathbf{u},k}^d(u)$, with knot sequence $\mathbf{u}$ can be written as a spline, $\mathbf{r}(u) = \sum c_\ell^* N_{\mathbf{u}^*,\ell}^d(u)$, with knot sequence $\mathbf{u}^*$. The new control points, c_ℓ^*, are given by

$$
\mathbf{c}_\ell^* = \begin{cases} \mathbf{c}_\ell, & \ell \leq k - d, \\ (1 - \alpha_\ell)\mathbf{c}_{\ell-1} + \alpha_\ell \mathbf{c}_\ell, & \ell = k - d + 1, \ldots, k - 1, \\ \mathbf{c}_{\ell-1}, & \ell \geq k, \end{cases} \tag{6.8}
$$

where

$$
\alpha_\ell = \frac{u^* - u_{\ell-1}}{u_{\ell+d-1} - u_{\ell-1}}. \tag{6.9}
$$

The *de Boor's algorithm* is the process of repeated insertion of a knot until we have full multiplicity and obtain a point on the curve. We formulate it as a theorem.

Theorem 6.2 *Let* $\mathbf{r}(u) = \sum_k \mathbf{c}_k N_{\mathbf{u},k}^d(u)$ *be a spline curve of degree d with knots* $\mathbf{u} = u_1, \ldots, u_{2d+n+1}$ *and control points* $\mathbf{c}_1, \ldots, \mathbf{c}_{n+d}$. *If* $u \in [u_\ell, u_{\ell+1}]$ *then the point* $\mathbf{r}(u)$ *on the curve can be determined by de Boor's algorithm. Put* $\mathbf{c}_i^0 = \mathbf{c}_i$ *for* $i = \ell - d, \ldots, \ell$ *and calculate for* $k = 1, \ldots, d$

$$
\begin{aligned}
\alpha_i^k &= \frac{u - u_\ell}{u_{\ell+d} - u_\ell}, \\
\mathbf{c}_i^k &= \left(1 - \alpha_i^k\right)\mathbf{c}_i^{k-1} + \alpha_i^k \mathbf{c}_i^{k-1}, \quad i = \ell - d + k, \ldots, \ell.
\end{aligned} \tag{6.10}
$$

Then $\mathbf{r}(u) = \mathbf{c}_\ell^d$.

6.4.2   Differentiation

The derivative of a spline of degree d is clearly a spline of degree $d - 1$.

Theorem 6.3 *The derivative of the B-spline* $N_{\mathbf{u},\ell}^d$ *is*

$$
\begin{aligned}
\frac{d}{du} N_{\mathbf{u},1}^d(u) &= -\frac{d}{u_{2+d} - u_2} N_{\mathbf{u},2}^{d-1}(u), \\
\frac{d}{du} N_{\mathbf{u},\ell}^d(u) &= \frac{d}{u_{\ell+d} - u_\ell} N_{\mathbf{u},\ell}^{d-1}(u) - \frac{d}{u_{\ell+d+1} - u_{\ell+1}} N_{\mathbf{u},\ell+1}^{d-1}(u), \\
& \qquad\qquad \ell = 2, \ldots, n + d - 1, \\
\frac{d}{du} N_{\mathbf{u},n+d}^d(u) &= \frac{d}{u_{n+2d} - u_{n+d}} N_{\mathbf{u},n+d}^{d-1}(u).
\end{aligned} \tag{6.11}
$$

Now consider a spline curve $\mathbf{r}(u) = \sum_\ell \mathbf{c}_\ell N^d_{\mathbf{u},\ell}(u)$. Using Theorem 6.3 we see that

$$\mathbf{r}'(u) = \sum_{\ell=1}^{n+d} \mathbf{c}_\ell \frac{\mathrm{d}}{\mathrm{d}u} N^d_{\mathbf{u},\ell}(u) = -\mathbf{c}_1 \frac{d}{u_{2+d} - u_2} N^{d-1}_{\mathbf{u},2}(u)$$

$$+ \sum_{\ell=2}^{n+d-1} \mathbf{c}_\ell \left(\frac{d}{u_{\ell+d} - u_\ell} N^{d-1}_{\mathbf{u},\ell}(u) - \frac{d}{u_{\ell+d+1} - u_{\ell+1}} N^{d-1}_{\mathbf{u},\ell+1}(u) \right)$$

$$+ \mathbf{c}_{n+d} \frac{d}{u_{n+2d} - u_{n+d}} N^{d-1}_{\mathbf{u},n+d}(u)$$

$$= \left(\sum_{\ell=2}^{n+d} \mathbf{c}_\ell \frac{d}{u_{\ell+d} - u_\ell} N^{d-1}_{\mathbf{u},\ell}(u) - \sum_{\ell=2}^{n+d} \mathbf{c}_{\ell-1} \frac{d}{u_{\ell+d} - u_\ell} N^{d-1}_{\mathbf{u},\ell}(u) \right)$$

$$= \sum_{\ell=2}^{n+d} \frac{d(\mathbf{c}_\ell - \mathbf{c}_{\ell-1})}{u_{\ell+d} - u_\ell} N^{d-1}_{\mathbf{u},\ell}(u), \tag{6.12}$$

where we have used $N^{d-1}_{\mathbf{u},1} = N^{d-1}_{\mathbf{u},n} = 0$ on the interval $[u_{d+1}, u_{d+n+1}]$. This gives us the following.

Theorem 6.4 *The derivative of a spline curve $\mathbf{r}(u) = \sum_\ell \mathbf{c}_\ell N^d_{\mathbf{u},\ell}(u)$ with knot vector $\mathbf{u} = u_1, \ldots, u_{2d+1+n}$ is a spline curve with knot vector $u_2, \ldots, u_{2d+n}$ and with control points*

$$\mathbf{d}_\ell = \frac{d(\mathbf{c}_\ell - \mathbf{c}_{\ell-1})}{u_{\ell+d} - u_\ell}. \tag{6.13}$$

6.5 NURBS

The acronym NURBS stands for Non Uniform Rational Basis Spline. We again have a knot vector, $\mathbf{u} = u_1, \ldots, u_{2d+1+n}$, and the B-splines, $N^d_{\mathbf{u},k}$. But now we introduce *weights*, $\mathbf{w} = w_1, \ldots, w_{d+n}$, and the weighted B-splines, $w_k N^d_{\mathbf{u},k}$. The weighted B-splines are no longer a partition of unity, so to restore affine invariance they are normalized and we arrive at NURBS,

$$R^d_{\mathbf{u},\mathbf{w},k}(u) = \frac{w_k N^d_{\mathbf{u},k}(u)}{\sum_{\ell=1}^{n+d} w_\ell N^d_{\mathbf{u},\ell}(u)}. \tag{6.14}$$

A *rational spline curve* is a curve of the form

$$\mathbf{r}(u) = \sum_{k=1}^{n+d} \mathbf{c}_k R^d_{\mathbf{u},\mathbf{w},k}(u) = \sum_{k=1}^{n+d} \mathbf{c}_k \frac{w_k N^d_{\mathbf{u},k}(u)}{\sum_{\ell=1}^{n+d} w_\ell N^d_{\mathbf{u},\ell}(u)}$$

$$= \frac{\sum_{k=1}^{n+d} w_k \mathbf{c}_k N^d_{\mathbf{u},k}(u)}{\sum_{k=1}^{n+d} w_k N^d_{\mathbf{u},k}(u)}. \tag{6.15}$$

We immediately see that the weights are not unique. First of all, if all the weights are multiplied by a common factor then the NURBS, $R^d_{\mathbf{u},\mathbf{w},k}$, are obviously left unchanged. Secondly, a rational linear reparametrization will change the weights without changing the shape of a rational spline curve. More precisely, we have the following theorem, see [5].

Theorem 6.5 *Let $R^d_{\mathbf{u},\mathbf{w},k}$ be NURBS with knots $\mathbf{u}$ and weights $\mathbf{w}$ and consider the rational linear transformation*

$$\psi(u) = \frac{\alpha u + \beta}{\gamma u + \delta}, \quad \text{with inverse } \phi(t) = \frac{-\delta t + \beta}{\gamma t - \alpha}. \tag{6.16}$$

We then have $R^d_{\mathbf{u},\mathbf{w},k}(\phi(t)) = R^d_{\widehat{\mathbf{u}},\widehat{\mathbf{w}},k}(t)$, where $\widehat{\mathbf{u}} = \psi(\mathbf{u})$ and

$$\widehat{w}_k = \frac{w_k}{\prod_{j=1}^{d}(\gamma u_{k+j} + \delta)}. \tag{6.17}$$

Observe that the parameters α, β, γ, and δ, in the rational transformations, are only determined up to a common factor, but that does not matter because multiplying the weights with a common factor does not change the NURBS.

We still have two degrees of freedom left in the rational linear transformation. They can be determined by the position of two parameter values, e.g., $\psi(u_{d+1}) = u_{d+1}$ and $\psi(u_{d+n+1}) = u_{d+n+1}$. It is always possible to choose γ and δ such that $w_1 = w_{n+d} = 1$; in that case we say that the NURBS are in *standard form*.

If $\mathbf{r}(u)$ is a rational spline curve in $\mathbb{R}^m$ then we can define an ordinary polynomial spline curve $\widehat{\mathbf{r}}$ in $\mathbb{R}^{m+1}$ by letting

$$\widehat{\mathbf{r}}(u) = \left[\sum_{k=1}^{n+d} w_k \mathbf{c}_k N^d_{\mathbf{u},k}(u), \sum_{k=1}^{n+d} w_k N^d_{\mathbf{u},k}(u) \right]$$

$$= \sum_{k=1}^{n+d} [w_k \mathbf{c}_k, w_k] N^d_{\mathbf{u},k}(u). \tag{6.18}$$

We see that we can get $\mathbf{r}$ from $\widehat{\mathbf{r}}$ by dividing the first m coordinates with the last. That is, $\mathbf{r}$ is the central projection of $\widehat{\mathbf{r}}$. That gives us the following.

Theorem 6.6 *A rational spline curve on some knot vector with weights w_k and control points $\mathbf{c}_k$ is the central projection of a polynomial spline curve on the same knot vector and with control points $[w_k \mathbf{c}_k, w_k]$.*

The theorem implies that we can obtain any conic section as a rational spline, see Fig. 6.5 and Sect. 6.7.1. In the figure a part of an ellipse is obtained as the central projection of a piece of a parabola. By taking several parabolas it is possible to obtain a full ellipse, but if the parabolas are written as a quadratic spline it is only C^0, not C^1. If a rational spline with smoothness C^n is requested then it has to have degree $2n + 2$, see [6]. Hence the inner knots have multiplicity $n + 2$.

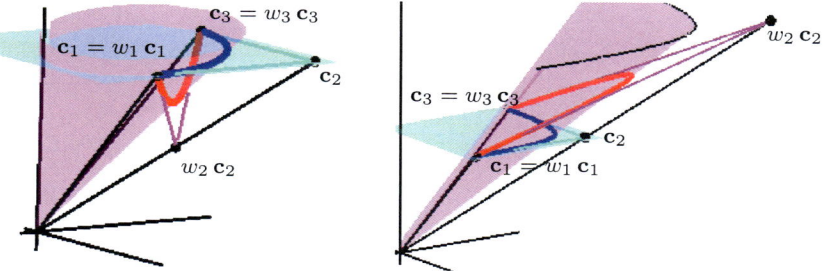

Fig. 6.5 On the *left* an ellipse and to the *right* a hyperbola both colored blue and obtained as the central projection of a parabola, colored *red*

6.6 Tensor Product Spline Surfaces

We finally are ready to define tensor product spline surfaces. The basis functions are simply products of B-splines, i.e., we have basis functions

$$R_{k,\ell}^{d,e}(u, v) = N_{\mathbf{u},k}^d(u) N_{\mathbf{v},\ell}^e(v), \tag{6.19}$$

where $\mathbf{u} = u_1, \ldots, u_{2d+m+1}$ and $\mathbf{v} = v_1, \ldots, v_{2e+n+1}$ are knot vectors. They give a basis with minimal support for the spline space

$$\left\{ f : \mathbb{R}^2 \to \mathbb{R} \middle| \begin{array}{l} f|_{]u_k, u_{k+1}[\times]v_\ell, v_{\ell+1}[} \text{ is a polynomial of degree } d \times e \\ f \text{ is } C^{d-\mu} \text{ at a knot line } u = u_k \text{ with multiplicity } \mu \\ f \text{ is } C^{e-\nu} \text{ at a knot line } v = v_\ell \text{ with multiplicity } \nu \end{array} \right\}.$$

A *tensor product spline surface* is a surface of the form

$$\mathbf{x}(u, v) = \sum_{k=1}^{m+d} \sum_{\ell=1}^{n+d} \mathbf{c}_{k,\ell} R_{k,\ell}^{d,e}(u, v)$$

$$= \sum_{k=1}^{m+d} \sum_{\ell=1}^{n+d} \mathbf{c}_{k,\ell} N_{\mathbf{u},k}^d(u) N_{\mathbf{v},\ell}^e(v). \tag{6.20}$$

Just as in the case of curves we introduce weights $w_{k,\ell}$ and obtain rational basis functions

$$R_{k,\ell}^{d,e}(u, v) = \frac{w_{k,\ell} N_{\mathbf{u},k}^d(u) N_{\mathbf{v},\ell}^e(v)}{\sum_{k=1}^{m+d} \sum_{\ell=1}^{n+d} w_{k,\ell} N_{\mathbf{u},k}^d(u) N_{\mathbf{v},\ell}^e(v)}. \tag{6.21}$$

A *rational tensor product spline surface* is a surface of the form

$$\mathbf{x}(u, v) = \sum_{k=1}^{m+d} \sum_{\ell=1}^{n+d} \mathbf{c}_{k,\ell} R_{k,\ell}^{d,e}(u, v)$$

$$= \frac{\sum_{k=1}^{m+d} \sum_{\ell=1}^{n+d} w_{k,\ell} \mathbf{c}_{k,\ell} N_{\mathbf{u},k}^d(u) N_{\mathbf{v},\ell}^e(v)}{\sum_{k=1}^{m+d} \sum_{\ell=1}^{n+d} w_{k,\ell} N_{\mathbf{u},k}^d(u) N_{\mathbf{v},\ell}^e(v)}. \tag{6.22}$$

We can again define an ordinary tensor product spline surface $\widehat{\mathbf{x}}$ in a space of one higher dimension by letting

$$\widehat{\mathbf{x}}(u, v) = \left[\sum_{k=1}^{m+d} \sum_{\ell=1}^{n+d} w_{k,\ell} \mathbf{c}_{k,\ell} N_{\mathbf{u},k}^d(u) N_{\mathbf{v},\ell}^e(v), \sum_{k=1}^{m+d} \sum_{\ell=1}^{n+d} w_{k,\ell} N_{\mathbf{u},k}^d(u) N_{\mathbf{v},\ell}^e(v) \right]$$

$$= \sum_{k=1}^{m+d} \sum_{\ell=1}^{n+d} [w_{k,\ell} \mathbf{c}_{k,\ell}, w_{k,\ell}] N_{\mathbf{u},k}^d(u) N_{\mathbf{v},\ell}^e(v). \tag{6.23}$$

Once more we see that we can obtain the rational surface $\mathbf{x}$ by dividing all coordinates but the last in $\widehat{\mathbf{x}}$ by the last coordinate in $\widehat{\mathbf{x}}$. So we also have the corresponding theorem.

Theorem 6.7 *A rational tensor product spline surface with weights $w_{k,\ell}$ and control points $\mathbf{c}_{k,\ell}$ is the central projection of a polynomial tensor product spline surface on the same knot vectors and with control points $[w_{k,\ell}\mathbf{c}_{k,\ell}, w_{k,\ell}]$.*

Just as we can represent pieces of conic section exactly as rational spline curves, we can represent pieces of quadric surfaces exactly by rational tensor product surfaces, see Sect. 6.7.1.

6.7   Spline Curves and Surfaces in Practice

Probably, NURBS curves and surfaces are largely used through CAD systems where the implementation is hidden. One example of a completely NURBS-based modeler is Rhinoceros™ from Robert McNeel & Associates.

Even people who need NURBS curves or surfaces in software rarely need to implement them from scratch. For people using OpenGL, the OpenGL utility library (GLU) has a set of functions for dealing with both NURBS curves and surfaces. An example of NURBS surfaces produced by OpenGL and glu is shown in Fig. 6.6. The GLU API supports direct rendering of NURBS curves and surfaces and through callback functions it is also possible to retrieve the polygons produced by the tessellation.

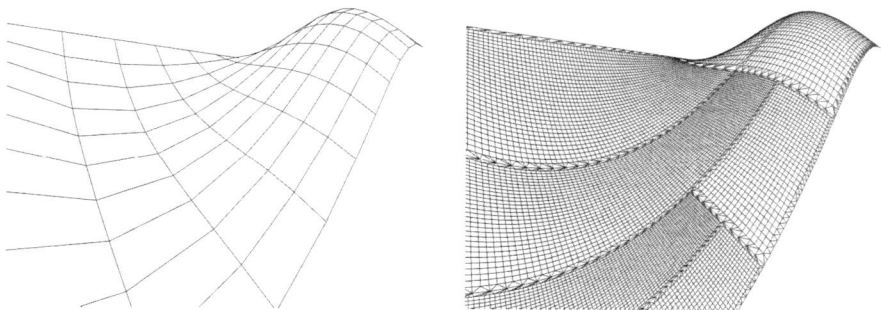

Fig. 6.6 An example of a NURBS surface rendered with the OpenGL utility library. On the *left* the surface was tessellated by dividing the parameter domain into equally large areas. On the *right* a tolerance in pixel units was imposed on the edge lengths

6.7.1 Representation of Conics and Quadrics

One of the advantages of rational spline curves and surfaces is that it is possible to represents conic curves and quadric surfaces exactly. Another advantage is that the curves are invariant under projective transformations. A thorough treatment requires some projective geometry and the use of homogeneous coordinates. But we stick with Euclidean coordinates.

Definition 6.4 Let X and Y be affine spaces. A projective map is a map $P : U \to V$ of the form

$$P : \mathbf{v} \mapsto \frac{f(\mathbf{v})}{\omega(\mathbf{v})}, \tag{6.24}$$

where $f : X \to Y$ and $\omega : U \to \mathbb{R}$ are affine maps.

If $X = Y$ then P is called a projective transformation.

If we are given a rational spline curve $\mathbf{x}$ with control points $\mathbf{c}_k$ and weights w_k and subject it to a projective transformation then we obtain

$$P(\mathbf{x}) = P\left(\frac{\sum_k w_k \mathbf{c}_k N_k}{\sum_k w_k N_k}\right) = \frac{\sum_k w_k f(\mathbf{c}_k) N_k}{\sum_k w_k N_k} \bigg/ \frac{\sum_k w_k \omega(\mathbf{c}_k) N_k}{\sum_k w_k N_k}$$

$$= \frac{\sum_k w_k f(\mathbf{c}_k) N_k}{\sum_k w_k \omega(\mathbf{c}_k) N_k} = \frac{\sum_k w_k \omega(\mathbf{c}_k) \frac{f(\mathbf{c}_k)}{\omega(\mathbf{c}_k)} N_k}{\sum_k w_k \omega(\mathbf{c}_k) N_k}. \tag{6.25}$$

We see that $P(\mathbf{x})$ is rational spline curve with control points $\widehat{\mathbf{c}}_k = P(\mathbf{c}_k)$ and weights $\widehat{w}_k = w_k \omega(\mathbf{c}_k)$. We similarly have the following.

Theorem 6.8 *Suppose* $P : \mathbf{v} \mapsto f(\mathbf{v})/\omega(\mathbf{v})$ *is a projective transformation and*

$$\mathbf{x}(u, v) = \left(\sum_{k,\ell} w_{k,\ell} \mathbf{c}_{k,\ell} N_{\mathbf{u},k}^d(u) N_{\mathbf{v},\ell}^e(v)\right) \bigg/ \left(\sum_{k,\ell} w_{k,\ell} N_{\mathbf{u},k}^d(u) N_{\mathbf{v},\ell}^e(v)\right)$$

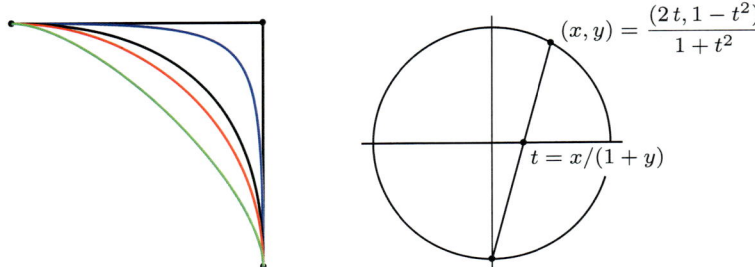

Fig. 6.7 On the *left* different conics. In *blue* a hyperbola with $w = 3$, in *black* a parabola with $w = 1$, in *red* a *circle* with $w = \sqrt{(2)}/2$, and in *green* an *ellipse* with $w = 1/3$. On the *right* a stereographic projection which gives a rational parametrization of the *circle*

is a rational spline surface. Then $P(\mathbf{x}(u, v))$ *is rational spline surface too, with control points*

$$\widehat{\mathbf{c}}_{k,\ell} = P(\mathbf{c}_{k,\ell}) = \frac{f(\mathbf{c}_{k,\ell})}{\omega(\mathbf{c}_{k,\ell})} \tag{6.26}$$

and weights

$$\widehat{w}_{k,\ell} = w_{k,\ell}\omega(\mathbf{c}_{k,\ell}). \tag{6.27}$$

Ellipses and hyperbolas can be written as the image of a parabola under a projective transformation. So using such a transformation and the formulas above we can write ellipses and hyperbolas as a quadratic rational spline curve. We will not give a general treatment but only a simple example.

Consider a quadratic curve with knot vector $0, 0, 0, 1, 1, 1$. Then there is only one polynomial segment and the B-splines are simply the Bernstein polynomials of degree two, $B_0^2(t) = (1-t)^2$, $B_1^2(t) = 2(1-t)t$, and $B_2^2(t) = t^2$, see Exercise 6.4. We now consider the control points $(1, 0)$, $(1, 1)$, and $(0, 1)$ and the weights 1, w, and 1. If $w > 1$ then we obtain a piece of a hyperbola, if $w = 1$ then we obtain a piece of a parabola, and if $0 < w < 1$ then we obtain a piece of an ellipse. A particular case of the latter is $w = \sqrt{2}/2$ in which case we obtain a quarter of a circle, see Fig. 6.7. Another way to get the rational parametrization of the circle is to consider stereographic projection, see Fig. 6.7.

We can also use a stereographic projection to give a rational parametrization of the sphere. On the sphere stereographic projection from the south pole and its inverse is given as

$$(x, y, z) \mapsto \frac{(x, y)}{1 + z} \quad \text{and} \quad (u, v) \mapsto \frac{(2u, 2v, 1 - u^2 - v^2)}{1 + u^2 + v^2}, \tag{6.28}$$

respectively. If we compose a rational parametrization of a domain in the uv-plane with stereographic projection then we obtain a rational parametrization of the corresponding domain on the sphere, but the degree in u and v is doubled, see Fig. 6.8.

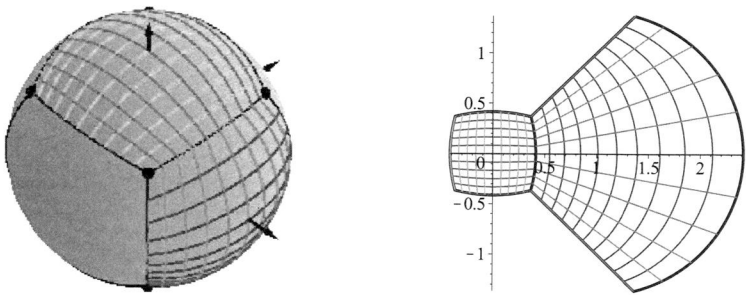

Fig. 6.8 On the *left* two different patches on the sphere. One, with *red* and *green* parameter *lines*, has degree (4, 4). The other, with *blue* and *magenta parameter lines*, has degree (2, 4), both covers one sixth of sphere. On the *right* the image of the two patches under stereographic projection. Here the patches has degree (2, 2) and (1, 2), respectively

Any quadrilateral in the plane can be parametrized by a bilinear patch, i.e., a polynomial patch of degree (1, 1). If we compose with stereographic projection then we obtain rational patches on the sphere with degree (2, 2). But it is not possible to partition the sphere into patches of degree (2, 2). If we use higher degree patches it becomes possible.

One way is first to project a cube, co-centered with the sphere, from the center onto the sphere. This partitions the sphere into six equally sized patches, corresponding to the six faces of the cube, see Fig. 6.8. The boundaries of these six patches are arcs of great circles and stereographic projection maps these patches onto six domains in the plane who's boundaries are circles or lines. There is a central "square-like" domain with four arcs of circles as boundaries, four equally sized "fan-like" domains with two arcs of circles and two line segments as boundaries, and finally one unbounded domain, see Fig. 6.8.

The central planar domain has a boundary consisting of four circles and it can be parametrized by a rational patch of degree (2, 2). The knot vectors are without inner knots so the B-splines reduces to the Bernstein polynomials of degree 2, cf., Exercise 6.4. That is, we have a parametrization $\mathbf{x} : [0, 1]^2 \to \mathbb{R}^2$ of the form

$$\mathbf{x}(u, v) = \frac{\sum_{k,\ell=0}^{2} w_{k,\ell} \mathbf{c}_{k,\ell} B_k^2(u) B_\ell^2(v)}{\sum_{k,\ell=0}^{2} w_{k,\ell} B_k^2(u) B_\ell^2(v)}. \tag{6.29}$$

If the weights are in standard form then the boundary control points and weights are uniquely given. We are then left with one inner control point and one inner weight that both are far from unique. One symmetric choice is the following set of control points and weights:

$$\mathbf{c}_{k,\ell} = \begin{bmatrix} (a_1, a_1) & (0, b_1) & (-a_1, a_1) \\ (b_1, 0) & (0, 0) & (-b_1, 0) \\ (a_1, -a_1) & (0, -b_1) & (-a_1, -a_1) \end{bmatrix},$$

$$w_{k,\ell} = \begin{bmatrix} 1 & w_1 & 1 \\ w_1 & 1 & w_1 \\ 1 & w_1 & 1 \end{bmatrix}$$

(6.30)

where

$$a_1 = \frac{\sqrt{3}}{3 + \sqrt{3}}, \qquad b_1 = \frac{6\sqrt{3}}{(3 + \sqrt{3})^2}, \qquad w_1 = \frac{\sqrt{6} + \sqrt{2}}{4}. \qquad (6.31)$$

Another natural choice for the inner weight could be w_1^2 instead of 1. In any case, when we compose with stereographic projection we obtain a patch of degree $(4, 4)$ on the sphere. It has the form

$$\widehat{\mathbf{x}} = \frac{\sum_{k,\ell=0}^{4} \widehat{w}_{k,\ell} \widehat{\mathbf{c}}_{k,\ell} B_k^4(u) B_\ell^4(v)}{\sum_{k,\ell=0}^{4} \widehat{w}_{k,\ell} B_k^4(u) B_\ell^4(v)}, \qquad (6.32)$$

where the control points and weights can be found from the equation

$$\widehat{\mathbf{x}}(u, v) = \frac{(2\mathbf{x}(u, v), 1 - \|\mathbf{x}(u, v)\|^2)}{1 + \|\mathbf{x}(u, v)\|^2}, \qquad \text{all } u, v \in [0, 1], \qquad (6.33)$$

or equivalently

$$\sum_{k,\ell=0}^{4} \left[\widehat{w}_{k,\ell} \widehat{\mathbf{c}}_{k,\ell}, \widehat{w}_{k,\ell} \right] B_k^4(u) B_\ell^4(v)$$

$$= \left[2q(u, v)\mathbf{p}(u, v), q(u, v)^2 - \|\mathbf{p}(u, v)\|^2, q(u, v)^2 + \|\mathbf{p}(u, v)\|^2 \right], \quad (6.34)$$

where $\mathbf{p}(u, v)$ and $q(u, v)$ are the numerator and the denominator in (6.29), respectively.

The "fan-like" planar domain has two circles and two straight lines as the boundary and can be parametrized by a patch of degree $(1, 2)$ of the form

$$\mathbf{x}(u, v) = \frac{\sum_{k=0}^{1} \sum_{\ell=0}^{2} w_{k,\ell} \mathbf{c}_{k,\ell} B_k^1(u) B_\ell^2(v)}{\sum_{k=0}^{1} \sum_{\ell=0}^{2} w_{k,\ell} B_k^1(u) B_\ell^1(v)}, \qquad (6.35)$$

where the control points and weights are

$$\mathbf{c}_{k,\ell} = \begin{bmatrix} (a_1, a_1) & (b_1, 0) & (a_1, -a_1) \\ (a_2, a_2) & (b_2, 0) & (a_2, -a_2) \end{bmatrix}, \qquad w_{k,\ell} = \begin{bmatrix} 1 & w_1 & 1 \\ 1 & w_2 & 1 \end{bmatrix}, \qquad (6.36)$$

and

$$a_2 = \frac{\sqrt{3}}{3 - \sqrt{3}}, \qquad b_2 = \frac{6\sqrt{3}}{(3 - \sqrt{3})^2}, \qquad w_2 = \frac{\sqrt{6} - \sqrt{2}}{4}. \qquad (6.37)$$

When we compose with stereographic projection we obtain a patch of degree $(2, 4)$ on the sphere. The control points and weights can be found like in the case of the patch of degree $(4, 4)$.

On the sphere the two domains look exactly the same, so it is of course possible to parametrize both with a patch of degree $(2, 4)$. It is in fact possible to parametrize any domain on the sphere bounded by four circles by a patch of degree $(2, 4)$, see [7].

6.7.2   Interpolation and Approximation

We first consider the univariate case. Here we are given parameter values $\mu_\ell \in \mathbb{R}$ and corresponding data points $\mathbf{x}_\ell$ that belong to some affine space, (it could be just real numbers). We are now looking for a spline $\mathbf{r}(u) = \sum_k \mathbf{c}_k N_k^d(\mathbf{u}, u)$ such that $\mathbf{r}(\mu_\ell) = \mathbf{x}_\ell$ for all ℓ. If we have L points then we get L linear equations

$$\sum_{k=1}^{n+d} \mathbf{c}_k N_{\mathbf{u},k}^d(\mu_\ell) = \mathbf{x}_\ell, \quad \ell = 1, \dots, L, \qquad (6.38)$$

in the control points $\mathbf{c}_k$. In order to have a unique solution we certainly need to have the same number of variables and equations, i.e., we need to have $L = n + d$. If this is the case then the following theorem gives a necessary and sufficient condition for when (6.38) has a unique solution, see [1, p. 200].

Theorem 6.9 (Schoenberg–Whitney) *Let* $\mathbf{u} = u_1 \leq \cdots \leq u_{2d+n+1}$ *be a knot sequence where all inner knots have at most full multiplicity, i.e., all B-splines of degree* d *are continuous. Given* $n + d$ *strictly increasing parameter values,* $\mu_1 < \cdots < \mu_{n+d}$, *the matrix* $N_{\mathbf{u},k}^d(\mu_\ell)$ *of the system* (6.38) *is regular if and only if* $N_{\mathbf{u},k}^d(\mu_k) \neq 0$ *for all* $k = 1, \dots, n + d$, *i.e., if and only if* $u_{d+k} < \mu_k < u_{d+2k}$, *all* k.

So given some univariate data it is not hard to find an interpolating spline. The most difficult case is when the parameter values are not given. Then they have to be chosen and the resulting interpolating spline is sensitive to this choice.

When it comes to bivariate interpolation the situation is much more difficult. Unless the data have a nice rectangular structure it is not possible to find an interpolating tensor product spline. So we give up on exact interpolation and use approximation instead. We are still given parameter values $(\mu_j, v_j) \in \mathbb{R}^2$ and data points $\mathbf{x}_j, \ell = 1, \dots, N$, but now we solve the equations

$$\mathbf{x}(\mu_j, v_j) = \sum_{k,\ell} \mathbf{c}_{k,\ell} N_{\mathbf{u},k}^d(\mu_j) N_{\mathbf{v},\ell}^e(v_j) = \mathbf{x}_j, \quad j = 1, \dots, N, \qquad (6.39)$$

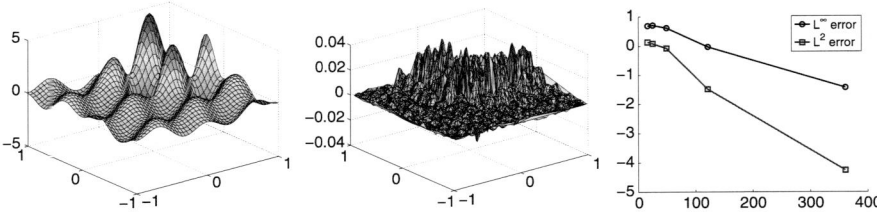

Fig. 6.9 A tensor product spline surface approximating 10,000 data points randomly sampled from the function depicted to the *left*. In the *middle* the error when using cubic splines and $19 \times 19 = 361$ control points. On the *right* $\log_{10}$ of the L^2 and L^∞ error as a function of the number of control points

in the least square sense. That is, we minimize the L^2 error

$$\text{error} = \frac{1}{N} \sum_{j=1}^{N} \left\| \mathbf{x}(\mu_j, \nu_j) - \mathbf{x}_j \right\|^2. \tag{6.40}$$

If we choose an ordering (k_i, ℓ_i), $i = 1, \ldots, (m + d)(n + e) = M$ of the indices (k, ℓ) and define the $N \times M$ matrix

$$\mathbf{A} = \begin{bmatrix} N_{k_1}^d(\mathbf{u}|\widetilde{u}_1) N_{\ell_1}^e(\mathbf{v}|\widetilde{v}_1) & \cdots & N_{k_M}^d(\mathbf{u}|\widetilde{u}_1) N_{\ell_M}^e(\mathbf{v}|\widetilde{v}_1) \\ \vdots & \ddots & \vdots \\ N_{k_1}^d(\mathbf{u}|\widetilde{u}_N) N_{\ell_1}^e(\mathbf{v}|\widetilde{v}_N) & \cdots & N_{k_M}^d(\mathbf{u}|\widetilde{u}_N) N_{\ell_M}^e(\mathbf{v}|\widetilde{v}_N) \end{bmatrix},$$

the $M \times K$ matrix $\quad \mathbf{C} = \begin{bmatrix} \mathbf{c}_{k_1,\ell_1} & \cdots & \mathbf{c}_{k_M,\ell_M} \end{bmatrix}^T$,

and the $N \times K$ matrix $\quad \mathbf{X} = \begin{bmatrix} \mathbf{x}_1 & \cdots & \mathbf{x}_N \end{bmatrix}^T$,

where K is the dimension of the space where the data points, $\mathbf{x}_j$, and the control points, $\mathbf{c}_{k,\ell}$, are lying in, then minimizing the L^2 error (6.40) can be formulated as

$$\text{minimize} \|\mathbf{AC} - \mathbf{X}\|^2,$$

cf. Example 2.56. To illustrate this we have taken 10,000 random samples from the function (6.43) depicted to the left in Fig. 6.9. We have used cubic splines in both the u and v direction, i.e., $d = e = 3$. As the number of knot intervals we have tried $m = n = 1, 2, 4, 16$. That is, we have $K = 1$, $N = 10,000$, and $M = 16, 25, 49, 361$. The result is shown in Fig. 6.9, where we have plotted the graph of the function from which the data were sampled, the error at all data points in the case of the highest resolution, and finally the logarithm of L^2 error and the L^∞ error, and the $\max_j \|\mathbf{x}(\mu_j, \nu_j) - \mathbf{x}_j\|^2$ are plotted as a function of the number of control points. It should be noted that already with 49 control points we obtain a visually good approximation.

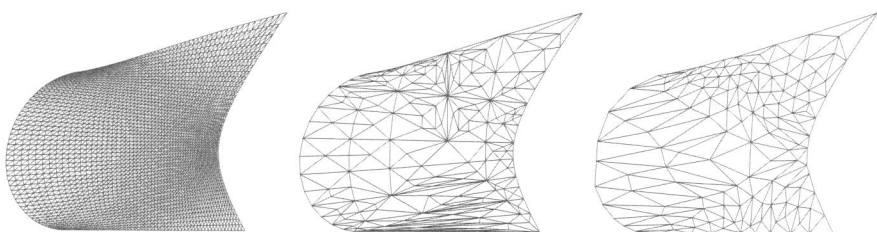

Fig. 6.10 An example of a NURBS surface tessellated uniformly (*left*) with Velho's adaptive scheme (*center*) and uniformly with subsequent triangle reduction (*right*)

The corresponding problem for rational tensor product splines is linear in the control points but non-linear in the weights. If the weights $w_{k,\ell}$ are fixed then we have the same results; we simply replace $N^d_{\mathbf{u},k}(\mu_j) N^e_{\mathbf{v},\ell}(\nu_j)$ with the NURBS, $R^{d,e}_{k,\ell}(\mu_j, \nu_j)$, defined in (6.21).

6.7.3   Tessellation and Trimming of Parametric Surfaces

Parametric surfaces can be either uniformly or adaptively *tessellated*, i.e., converted from a smooth parametric surface to a polygonal mesh. Moreover, adaptive tessellation can be done in a multitude of ways. Velho et al. present a unified and hierarchical approach to tessellation [8], which is shown applied to a NURBS patch in Fig. 6.10 (middle). This adaptive tessellation scheme starts from an initial triangulation and then refines it adaptively where needed. Compared to a uniform tessellation, much fewer triangles are spent to achieve a given fidelity. The GLU interface also affords some adaptivity in the tessellation, but the method used is somewhat cruder, see Fig. 6.6.

In many cases, the quadrilateral domain over which a NURBS patch is defined does not exactly match the boundary which is needed for some design. For this reason, most APIs (including GLU) for NURBS rendering support trimming.

The principle is simply that we use curves (typically also NURBS curves) to define a closed path. Everything (arbitrarily) to the right of the path is then left out. Thus, if our trimming curve is clockwise, we leave out the interior, making a hole, and if it is counterclockwise, we keep the interior. Typically, trimming is implemented in a very general fashion allowing us to cut out multiple pieces, to compose closed trim loops of several curves, and to have trimming curves within trimming curves, for instance to create an annular region [9]. In Fig. 6.11 a very simple example is shown where a circular region of the domain has been cut out of a NURBS patch using a closed NURBS curve.

6.8   Exercises

Exercise 6.1 Prove the B-spline properties in Theorem 6.1.

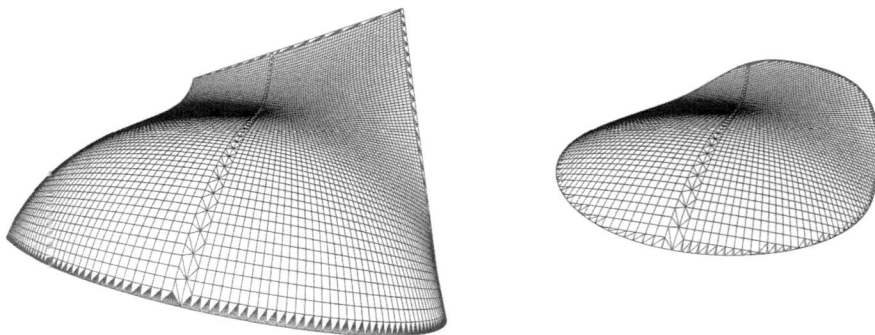

Fig. 6.11 An example of a NURBS surface with and without trimming. Note that the trimming curve is actually a perfect circle in the parameter domain created as a quadratic 2D NURBS curve

Exercise 6.2 Show that if a knot $u_{l+1} = \cdots = u_{l+d}$ has full multiplicity and $\mathbf{x}(u) = \sum_{\ell=1}^{n-d} \mathbf{c}_\ell N_{\mathbf{u},\ell}^d(u)$, then $\mathbf{c}_\ell = \mathbf{x}(u_{\ell+d})$.

Exercise 6.3 Prove Corollary 6.1.

Exercise 6.4 Show that if there is only one knot interval and the knot vector is $\mathbf{u} = \underbrace{0, \ldots, 0}_{d+1}, \underbrace{1, \ldots, 1}_{d+1}$, then the B-splines are the *Bernstein polynomials*

$$N_{\mathbf{u},\ell+1}^d(u) = B_\ell^d(u) = \binom{d}{\ell}(1-u)^{d-\ell}u^\ell, \quad \ell = 0, \ldots, d. \tag{6.41}$$

Exercise 6.5 Prove the knot insertion algorithm (6.8).

Exercise 6.6 Let

$$h(t) = \frac{1}{2}\left(1 + \frac{t}{\sqrt{1+t^2}}\right), \tag{6.42}$$

and sample the function

$$f(u, v) = h(5x)h(5y)\sin\big(\pi(x+y)\big)\cos\big(2\pi(x-y)\big), \tag{6.43}$$

densely on the square $[-1, 1]^2$. Choose degrees d and e, knot vectors

$$\mathbf{u} : \underbrace{-1, \ldots, -1}_{d+1} < u_{d+1} \leq \cdots \leq u_{d+m} < \underbrace{1, \ldots, 1}_{d+1},$$

$$\mathbf{v} : \underbrace{-1, \ldots, -1}_{e+1} < v_{e+1} \leq \cdots \leq v_{e+n} < \underbrace{1, \ldots, 1}_{e+1},$$

and minimize the L^2 error

$$\sum_{j=1}^{N}\left(\sum_{k=1}^{d+m}\sum_{\ell=1}^{e+n}c_{k,\ell}N_{\mathbf{u},k}^{d}(\mu_j)N_{\mathbf{v},\ell}^{e}(\nu_j)-f_j\right)^2,$$

where $f_j = f(\mu_j, \nu_j)$, $j = 1, \ldots, N$, are the samples.

References

1. de Boor, C.: A Practical Guide to Splines. Springer, New York (1978)
2. Farin, G.: Curves and surfaces for computer-aided geometric design. In: Computer Science and Scientific Computing, 4th edn., Academic Press, San Diego (1997). A practical guide, Chap. 1 by P. Bézier; Chaps. 11 and 22 by W. Boehm, with 1 IBM-PC floppy disk (3.5 inch; HD)
3. Piegl, L., Tiller, W.: The NURBS Book. Springer, New York (1987)
4. Farin, G., Hoschek, J., Kim, M.-S. (eds.): Handbook of Computer Aided Geometric Design. Elsevier Science, North-Holland (2002)
5. Lee, E.T.Y., Lucian, M.L.: Möbius reparametrizations of rational B-splines. Comput. Aided Geom. Des. **8**(3), 213–215 (1991). doi:10.1016/0167-8396(91)90004-U
6. Bangert, C., Prautzsch, H.: Circle and sphere as rational splines. Neural Parallel Sci. Comput. **5**(1–2), 153–161 (1997). Computer aided geometric design
7. Dietz, R., Hoschek, J., Jüttler, B.: Rational patches on quadric surfaces. Comput. Aided Des. **27**(1), 27–40 (1995). doi:10.1016/0010-4485(95)90750-A
8. Velho, L., de Figueiredo, L.H., Gomes, J.: A unified approach for hierarchical adaptive tessellation of surfaces. ACM Trans. Graph. **18**, 329–360 (1999). doi:10.1145/337680.337717
9. Woo, M., Neider, J., Davis, T.: OpenGL Programming Guide, 2nd edn. Addison Wesley, Reading (1998)

Subdivision

Subdivision curves and surfaces are defined as the limit of a sequence of successive refinements of a control polygon or mesh. There is in many cases a close connection to spline curves with a uniform knot vector and uniform tensor product surfaces.

In many ways, subdivision curves and surfaces provide nearly the same possibilities as spline curves and surfaces. However, in some ways, subdivision is simpler. There are no issues with stitching together the patches (even if the surface is animated) since the starting point is a polygonal mesh. This is of great practical importance: if a character is modeled with spline patches, ensuring that these join smoothly as the character is animated is a tricky problem.

On the other hand, for subdivision surfaces, we need to deal with extraordinary vertices where the subdivided mesh may be less smooth than elsewhere.

There are also some other important differences. Subdivision surfaces can be parametrized which is convenient because it means that they can be used where a parametric form is required, but we normally tessellate subdivision surfaces precisely by subdividing the mesh. Hence, one would not normally use an adaptive tessellation algorithm for parametric surfaces on subdivision surfaces. Nor would one often trim a subdivision surface since we may instead trim the initial mesh.

In Sect. 7.1, we study subdivision curves. Curve subdivision is simple to express using matrix multiplication, and we discuss the relation to spline curves and how an eigenanalysis can be used to find points on the limit curve (after infinitely many subdivision steps). In Sect. 7.2, we present a similar discussion but now for subdivision surfaces. Here the matrix representation is somewhat more difficult but still highly useful for analysis of the schemes. In Sect. 7.2.1 the characteristic map is presented. The characteristic map is a tool for analysis of whether subdivision schemes are tangent plane continuous in the limit.

In Sect. 7.3 we turn to concrete subdivision schemes and discuss the Loop, Catmull–Clark, modified butterfly, sqrt(3), and Doo–Sabin schemes. Finally, some advanced techniques and recent methods are discussed briefly in Sect. 7.3.4.

J.A. Bærentzen et al., *Guide to Computational Geometry Processing*,
DOI 10.1007/978-1-4471-4075-7_7, © Springer-Verlag London 2012

7.1   Subdivision Curves

A B-spline curve of degree n on the knot vector consisting of the integers, i.e., a curve which is piecewise polynomial of degree n and is C^{n-1} at the integers, can be written as $\sum_{k\in\mathbb{Z}} \mathbf{c}_k^0 B^n(t-k)$, where $B^n(t) = N_{\mathbb{Z},1}^n(t)$ and the *control points* $\mathbf{c}_k^0$ form the *control polygon*. In the plane we have $\mathbf{c}_k^0 = (x_k^0, y_k^0)$ and in space we have $\mathbf{c}_k^0 = (x_k^0, y_k^0, z_k^0)$. The curve is of course also C^{n-1} (in fact C^∞) at the half integers so it is also a spline curve on the knot vector consisting of the half integers. Thus, it can be written $\sum_{k\in\mathbb{Z}} \mathbf{c}_k^1 B^n(2t-k)$ where $\mathbf{c}_k^1$ are new control points. This process can be repeated and we can write the spline curve as $\sum_{k\in\mathbb{Z}} \mathbf{c}_k^\ell B^n(2^\ell t - k)$. The above process gives us linear mappings (*subdivision*)

$$\left[\mathbf{c}_k^0\right] \mapsto \left[\mathbf{c}_k^1\right] \mapsto \cdots \mapsto \left[\mathbf{c}_k^\ell\right] \mapsto \left[\mathbf{c}_k^{\ell+1}\right] \mapsto \cdots, \tag{7.1}$$

and it turns out that this sequence of control polygons converges to the spline curve $\sum_{k\in\mathbb{Z}} \mathbf{c}_k^0 B^n(t-k)$.

Let us as a simple example find the uniform quadratic B-spline $B^2(t)$. The B-splines can be found recursively, cf. Sect. 6.4,

$$B^0(t) = \begin{cases} 1 & t \in [0,1) \\ 0 & t \notin [0,1) \end{cases} \tag{7.2}$$

Using $B^1(t) = t B^0(t) + (2-t) B^0(t-1)$, we obtain

$$B^1(t) = \begin{cases} t & t \in [0,1) \\ 2-t & t \in [1,2) \\ 0 & t \notin [0,2) \end{cases} \tag{7.3}$$

and performing this once more, $B^2(t) = \frac{t}{2} B^1(t) + \frac{3-t}{2} B^1(t-1)$ yields

$$B^2(t) = \begin{cases} \frac{1}{2}t^2 & t \in [0,1) \\ \frac{1}{4}(3-(3-2t)^2) & t \in [1,2) \\ \frac{1}{2}(t-3)^2 & t \in [2,3) \\ 0 & t \notin [0,3) \end{cases} \tag{7.4}$$

Theorem 7.1 *The uniform B-splines (cf., Fig. 7.1) satisfy the* refinement equation

$$B^k(t) = \frac{1}{2^k} \sum_{\ell=0}^{k+1} \binom{k+1}{\ell} B^k(2t-\ell). \tag{7.5}$$

Proof The case $k=0$ is easy, $B^0(t) = B^0(2t) + B^0(2t-1)$. For the general case we first note that

Fig. 7.1 The B-splines of degree 0, 1, and 2

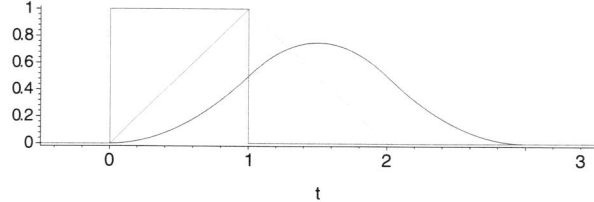

Fig. 7.2 The uniform B-spline of degree 2 is a linear combination of dilated and translated copies of itself with weights $\frac{1}{4}$, $\frac{3}{4}$, $\frac{3}{4}$, and $\frac{1}{4}$

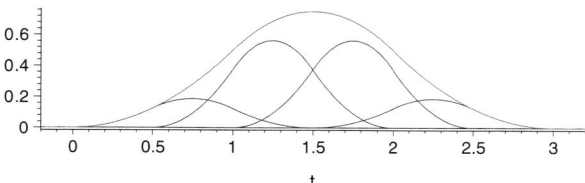

$$B^\ell(2t) * B^0(2t) = \int B^\ell(2s) B^0(2t - 2s)\, \mathrm{d}s$$
$$= \frac{1}{2} \int B^\ell(s) B^0(2t - s)\, \mathrm{d}s = \frac{1}{2} B^{\ell+1}(2t)$$

and

$$B^\ell(2t) * B^0(2t - 1) = \int B^\ell(2s) B^0(2t - 1 - 2s)\, \mathrm{d}s$$
$$= \frac{1}{2} \int B^\ell(s) B^0(2t - 1 - s)\, \mathrm{d}s = \frac{1}{2} B^{\ell+1}(2t - 1).$$

Now we write B^k as a $k+1$ fold convolution of B^0 with itself,

$$B^k(t) = \underbrace{B^0(t) * \cdots * B^0(t)}_{k+1}$$
$$= \underbrace{\left(B^0(2t) + B^0(2t - 1)\right) * \cdots * \left(B^0(2t) + B^0(2t - 1)\right)}_{k+1}$$
$$= \sum_{\ell=0}^{k+1} \binom{k+1}{\ell} \underbrace{B^0(2t) * \cdots * B^0(2t)}_{k+1-\ell} * \underbrace{B^0(2t - 1) * \cdots * B^0(2t - 1)}_{\ell}$$
$$= \frac{1}{2^k} \sum_{\ell=0}^{k+1} \binom{k+1}{\ell} B^k(2t - \ell). \qquad \square$$

In particular, see Fig. 7.2,

$$B^2(t) = \frac{1}{4} B^2(2t) + \frac{3}{4} B^2(2t - 1) + \frac{3}{4} B^2(2t - 2) + \frac{1}{4} B^2(2t - 3). \qquad (7.6)$$

If we substitute this into a uniform quadratic spline curve, then we obtain

$$\sum_{k \in \mathbb{Z}} \mathbf{c}_k^0 B^2(t - k)$$

$$= \sum_{k \in \mathbb{Z}} \mathbf{c}_k^0 \left(\frac{1}{4} B^2(2t - 2k) + \frac{3}{4} B^2(2t - 2k - 1) + \frac{3}{4} B^2(2t - 2k - 2) \right.$$

$$\left. + \frac{1}{4} B^2(2t - 2k - 3) \right)$$

$$= \frac{1}{4} \sum_{k \in \mathbb{Z}} \mathbf{c}_k^0 B^2(2t - 2k) + \frac{3}{4} \sum_{k \in \mathbb{Z}} \mathbf{c}_k^0 B^2(2t - 2k - 1)$$

$$+ \frac{3}{4} \sum_{k \in \mathbb{Z}} \mathbf{c}_{k+2}^0 B^2(2t - 2k) + \frac{1}{4} \sum_{k \in \mathbb{Z}} \mathbf{c}_{k+2}^0 B^2(2t - 2k - 1)$$

$$= \sum_{k \in \mathbb{Z}} \frac{\mathbf{c}_k^0 + 3\mathbf{c}_{k+2}^0}{4} B^2(2t - 2k) + \sum_{k \in \mathbb{Z}} \frac{3\mathbf{c}_k^0 + \mathbf{c}_{k+2}^0}{4} B^2(2t - 2k - 1)$$

$$= \sum_{k \in \mathbb{Z}} \mathbf{c}_k^1 B^2(2t - k),$$

where

$$\mathbf{c}_{2k-1}^1 = \frac{3}{4} \mathbf{c}_k^0 + \frac{1}{4} \mathbf{c}_{k+2}^0 \quad \text{and} \quad \mathbf{c}_{2k}^1 = \frac{1}{4} \mathbf{c}_k^0 + \frac{3}{4} \mathbf{c}_{k+2}^0. \tag{7.7}$$

This is the *subdivision rules* for the quadratic spline curve.

All this can also be seen another way. If we write the basis functions as a row vector

$$\mathbf{B}(t) = \begin{bmatrix} \dots & B^2(t + 2) & B^2(t + 1) & B^2(t) & B^2(t - 1) & B^2(t - 2) & \dots \end{bmatrix} \tag{7.8}$$

and the control polygon as a column matrix

$$\mathbf{C}^0 = \begin{bmatrix} \dots & \mathbf{c}_{-2}^0 & \mathbf{c}_{-1}^0 & \mathbf{c}_0^0 & \mathbf{c}_1^0 & \mathbf{c}_2^0 & \dots \end{bmatrix}^T, \tag{7.9}$$

then the spline curve can be written as a matrix product

$$\sum_{k \in \mathbb{Z}} \mathbf{c}_k^0 B^2(t - k) = \mathbf{B}(t) \mathbf{C}^0. \tag{7.10}$$

The refinement equation (7.5) can be written

$$\mathbf{B}(t) = \mathbf{B}(2t) \mathbf{S} \tag{7.11}$$

where the *subdivision matrix* $\mathbf{S}$ is a bi-infinite matrix with entries $\mathbf{S}_{2\ell+k,\ell} = \frac{1}{4}\binom{3}{k}$,

$$
\mathbf{S} = \frac{1}{4}
\begin{bmatrix}
\ddots & & & \overset{\ell}{\underset{\downarrow}{}} & & & \\
& 3 & 0 & 0 & & & \\
2\ell \rightarrow & 3 & 1 & 0 & & & \\
& 1 & 3 & 0 & 0 & & \\
& 0 & 3 & 1 & 0 & & \\
& 0 & 1 & 3 & 0 & & \\
& 0 & 0 & 3 & 1 & & \\
& & 0 & 1 & 3 & & \\
& & 0 & 0 & 3 & & \\
& & & & & \ddots &
\end{bmatrix}.
\tag{7.12}
$$

Observe that the rows of $\mathbf{S}$ sum to one, and this will be important later. Notice that the columns are shifted down by two when we go from one column to the next.

If we consider cubic splines then the sequence $\frac{1}{4}, \frac{3}{4}, \frac{3}{4}, \frac{1}{4}$ in the columns are replaced with the sequence $\frac{1}{8}, \frac{4}{8}, \frac{6}{8}, \frac{4}{8}, \frac{1}{8}$ and if the degree is n then the sequence becomes $\frac{1}{2^n}, \ldots, \frac{1}{2^n}\binom{k}{n}, \ldots, \frac{1}{2^n}$. In all cases the down shift by two is the same and the rows sum to one. By taking other numbers than the binomial coefficients we get other subdivision schemes. In order for the rows to sum to one we need that both the even and odd elements in the sequence sum to one.

We can now write

$$
\sum_{k \in \mathbb{Z}} \mathbf{c}_k^0 B^2(t-k) = \mathbf{B}(t)\mathbf{C}^0 = \mathbf{B}(2t)\mathbf{S}\mathbf{C}^0 = \mathbf{B}(2t)\mathbf{C}^1
$$

$$
= \sum_{k \in \mathbb{Z}} \mathbf{c}_k^1 B^2(2t-k),
\tag{7.13}
$$

and we see that subdivision can be written $\mathbf{C}^1 = \mathbf{S}\mathbf{C}^0$. Repeated subdivision is given by

$$
\mathbf{C}^k = \mathbf{S}\mathbf{C}^{k-1} = \cdots = \mathbf{S}^k \mathbf{C}^0.
\tag{7.14}
$$

Using the same matrix $\mathbf{S}$ at each level is called *stationary subdivision*. In Fig. 7.3 the first two levels of subdivision are illustrated.

Suppose we want to investigate the local behavior at $t = 0$. In a small neighborhood of $t = 0$ only the four basis functions $B^2(t)$, $B^2(t+1)$, $B^2(t+2)$, and $B^2(t+3)$ are non-zero. So we need only four control points to determine the local behavior, and we say that the *invariant neighborhood* has size 4. We can also see that if we start with four points and perform repeated subdivision then that piece of the control polygon converges to two segments of the final curve, cf. Fig. 7.4.

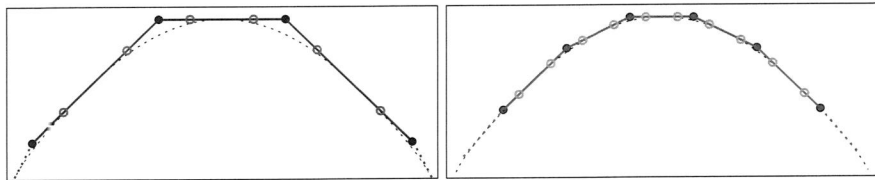

Fig. 7.3 The first two levels of subdivision. If the subdivision continues the refined polygon converges towards the *black dotted* spline curve

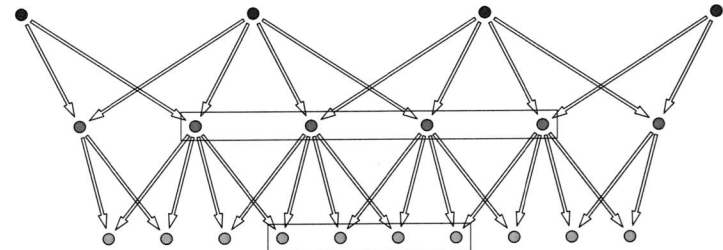

Fig. 7.4 The invariant neighborhood of a quadratic spline has size 4. In each row the middle four points determine the local behavior at $t = 0$. The outermost points in each row converges from the outside to the midpoints of the outermost polygon legs

7.1.1 Eigenanalysis

Now (the right) four consecutive points of the control polygon determine the local behavior at $t = 0$, but so do the middle four consecutive points, see Fig. 7.4, when the control polygon is subdivided. That means that we do not need the full infinite subdivision matrix (7.12) but only the small bold 4×4 sub-matrix. In the following $\mathbf{S}$ denotes this *local subdivision matrix*,

$$\mathbf{S} = \frac{1}{4} \begin{bmatrix} 1 & 3 & 0 & 0 \\ 0 & 3 & 1 & 0 \\ 0 & 1 & 3 & 0 \\ 0 & 0 & 3 & 1 \end{bmatrix}. \tag{7.15}$$

The local subdivision matrix $\mathbf{S}$ has eigenvalues

$$\lambda_1 = 1, \qquad \lambda_2 = \frac{1}{2}, \qquad \lambda_3 = \frac{1}{4}, \qquad \lambda_4 = \frac{1}{4}, \tag{7.16}$$

with corresponding eigenvectors

$$\mathbf{x}_1 = \begin{bmatrix} 1 \\ 1 \\ 1 \\ 1 \end{bmatrix}, \qquad \mathbf{x}_2 = \begin{bmatrix} 1 \\ \frac{1}{3} \\ -\frac{1}{3} \\ -1 \end{bmatrix}, \qquad \mathbf{x}_3 = \begin{bmatrix} 1 \\ 0 \\ 0 \\ 0 \end{bmatrix}, \qquad \mathbf{x}_4 = \begin{bmatrix} 0 \\ 0 \\ 0 \\ 1 \end{bmatrix}. \tag{7.17}$$

We can now write the column of four control points as

$$\mathbf{C}^0 = \sum_{i=1}^{4} \mathbf{x}_i \mathbf{a}_i = \mathbf{XA},$$

where the coefficients $\mathbf{a}_i$ are points or vectors,

$$\mathbf{X} = \begin{bmatrix} 1 & 1 & 1 & 0 \\ 1 & \frac{1}{3} & 0 & 0 \\ 1 & -\frac{1}{3} & 0 & 0 \\ 1 & -1 & 0 & 1 \end{bmatrix} \quad \text{and} \quad \mathbf{a} = \begin{bmatrix} \mathbf{a}_1 \\ \mathbf{a}_2 \\ \mathbf{a}_3 \\ \mathbf{a}_4 \end{bmatrix}.$$

We clearly have

$$\mathbf{A} = \mathbf{X}^{-1}\mathbf{C}^0 \quad \text{or} \quad \begin{bmatrix} \mathbf{a}_1 \\ \mathbf{a}_2 \\ \mathbf{a}_3 \\ \mathbf{a}_4 \end{bmatrix} = \begin{bmatrix} 0 & \frac{1}{2} & \frac{1}{2} & 0 \\ 0 & \frac{3}{2} & -\frac{3}{2} & 0 \\ 1 & -2 & 1 & 0 \\ 0 & 1 & -2 & 1 \end{bmatrix} \begin{bmatrix} \mathbf{c}_{-3}^0 \\ \mathbf{c}_{-2}^0 \\ \mathbf{c}_{-1}^0 \\ \mathbf{c}_0^0 \end{bmatrix}.$$

The result of repeated subdivision is now easy to calculate

$$\mathbf{C}^k = \mathbf{S}^k \mathbf{C}^0 = \mathbf{S}^k \sum_{i=1}^{4} \mathbf{x}_i \mathbf{a}_i = \sum_{i=1}^{4} \mathbf{S}^k \mathbf{x}_i \mathbf{a}_i = \sum_{i=1}^{4} \lambda_i^k \mathbf{x}_i \mathbf{a}_i \to \mathbf{x}_1 \mathbf{a}_1. \tag{7.18}$$

Furthermore,

$$\mathbf{x}_1 \mathbf{a}_1 = \begin{bmatrix} 1 \\ 1 \\ 1 \\ 1 \end{bmatrix} \begin{bmatrix} 0 & \frac{1}{2} & \frac{1}{2} & 0 \end{bmatrix} \begin{bmatrix} \mathbf{c}_{-3}^0 \\ \mathbf{c}_{-2}^0 \\ \mathbf{c}_{-1}^0 \\ \mathbf{c}_0^0 \end{bmatrix} = \begin{bmatrix} 1 \\ 1 \\ 1 \\ 1 \end{bmatrix} \frac{\mathbf{c}_{-2}^0 + \mathbf{c}_{-1}^0}{2}.$$

A similar calculation can be made for any stationary subdivision, and when the rows sum to one, we always find that 1 is an eigenvalue with an eigenvector consisting of all ones. This is important; if the largest eigenvalue of the local subdivision matrix is different from one then repeated subdivisions diverge. If we do not have a basis of eigenvectors then the analysis is more complicated.

We now continue our analysis of the local behavior by subtracting the limit,

$$\frac{\mathbf{C}^k - \mathbf{x}_1 \mathbf{a}_1}{\lambda_2^k} = \sum_{i=2}^{4} \left(\frac{\lambda_i}{\lambda_2} \right)^k \mathbf{x}_i \mathbf{a}_i \to \mathbf{x}_2 \mathbf{a}_2 \tag{7.19}$$

and

$$\mathbf{x}_2 \mathbf{a}_2 = \begin{bmatrix} 1 \\ \frac{1}{3} \\ -\frac{1}{3} \\ -1 \end{bmatrix} \begin{bmatrix} 0 & \frac{3}{2} & -\frac{3}{2} & 0 \end{bmatrix} \begin{bmatrix} \mathbf{c}_{-3}^0 \\ \mathbf{c}_{-2}^0 \\ \mathbf{c}_{-1}^0 \\ \mathbf{c}_0^0 \end{bmatrix} = \begin{bmatrix} 3 \\ 1 \\ -1 \\ -3 \end{bmatrix} \frac{\mathbf{c}_{-2}^0 - \mathbf{c}_{-1}^0}{2}.$$

Fig. 7.5 Subdivision of
quadratic tensor product
B-spline surface

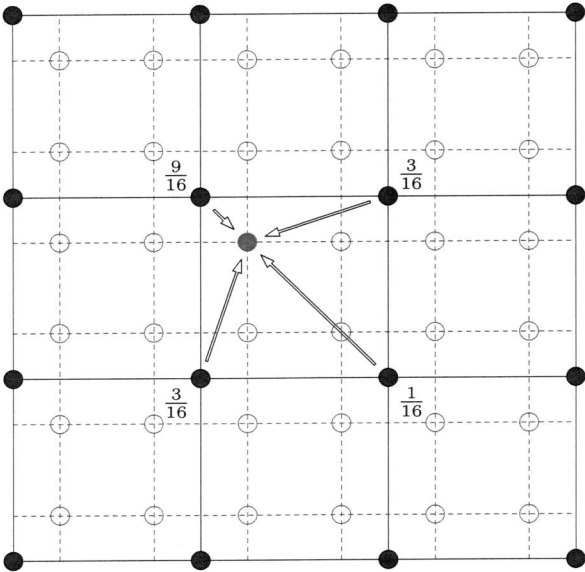

Recalling that the second eigenvalue is $1/2$ and the third eigenvalue is $1/4$, we have
$\mathbf{C}^k = a_1\mathbf{x}_1 + 2^{-k}a_2\mathbf{x}_2 + O(4^{-k})$ or

$$
\begin{bmatrix} \mathbf{c}^k_{-3} \\ \mathbf{c}^k_{-2} \\ \mathbf{c}^k_{-1} \\ \mathbf{c}^k_{0} \end{bmatrix} = \begin{bmatrix} 1 \\ 1 \\ 1 \\ 1 \end{bmatrix} \frac{\mathbf{c}^0_{-2} + \mathbf{c}^0_{-1}}{2} + \begin{bmatrix} 3 \\ 1 \\ -1 \\ -3 \end{bmatrix} \frac{\mathbf{c}^0_{-2} - \mathbf{c}^0_{-1}}{2^{k+1}} + O\left(4^{-k}\right).
$$

In the limit when $k \to \infty$ we obtain a well defined tangent in the direction of $\mathbf{a}_2$
or equivalently in the direction $\mathbf{c}^0_{-1} - \mathbf{c}^0_{-2}$. Again the same kind of analysis can be
done for any stationary subdivision scheme and the conclusion is that we want the
following properties of the local subdivision matrix.

- The first (dominant) eigenvalue λ_1 should be 1.
- The next (sub dominant) eigenvalue λ_2 should be strictly less than 1.
- All other eigenvalues should be strictly less than λ_2.

7.2   Subdivision Surfaces

Just as for curves subdivision of surfaces is performed by calculating new points by
taking convex combination of the old points, i.e., each new point is a weighted average of (some) old points. In Fig. 7.5 the case of a quadratic tensor product B-spline
surface is illustrated. There is also a local subdivision matrix for subdivisions surfaces. It is not hard to find the 16×16 subdivision matrix for the quadratic tensor
product B-spline surface in Fig. 7.5, but the precise result depends of course on how

the control points are numbered. One particular numbering gives

$$\mathbf{S} = \frac{1}{16} \begin{bmatrix} 1 & 3 & 0 & 0 & 3 & 9 & 0 & 0 & 0 & 0 & 0 & 0 & 0 & 0 & 0 & 0 \\ 0 & 3 & 1 & 0 & 0 & 9 & 3 & 0 & 0 & 0 & 0 & 0 & 0 & 0 & 0 & 0 \\ 0 & 1 & 3 & 0 & 0 & 3 & 9 & 0 & 0 & 0 & 0 & 0 & 0 & 0 & 0 & 0 \\ 0 & 0 & 3 & 1 & 0 & 0 & 9 & 3 & 0 & 0 & 0 & 0 & 0 & 0 & 0 & 0 \\ 0 & 0 & 0 & 0 & 3 & 9 & 0 & 0 & 1 & 3 & 0 & 0 & 0 & 0 & 0 & 0 \\ 0 & 0 & 0 & 0 & 0 & 9 & 3 & 0 & 0 & 3 & 1 & 0 & 0 & 0 & 0 & 0 \\ 0 & 0 & 0 & 0 & 0 & 3 & 9 & 0 & 0 & 1 & 3 & 0 & 0 & 0 & 0 & 0 \\ 0 & 0 & 0 & 0 & 0 & 0 & 9 & 3 & 0 & 0 & 3 & 1 & 0 & 0 & 0 & 0 \\ 0 & 0 & 0 & 0 & 1 & 3 & 0 & 0 & 3 & 9 & 0 & 0 & 0 & 0 & 0 & 0 \\ 0 & 0 & 0 & 0 & 0 & 3 & 1 & 0 & 0 & 9 & 3 & 0 & 0 & 0 & 0 & 0 \\ 0 & 0 & 0 & 0 & 0 & 1 & 3 & 0 & 0 & 3 & 9 & 0 & 0 & 0 & 0 & 0 \\ 0 & 0 & 0 & 0 & 0 & 0 & 3 & 1 & 0 & 0 & 9 & 3 & 0 & 0 & 0 & 0 \\ 0 & 0 & 0 & 0 & 0 & 0 & 0 & 0 & 3 & 9 & 0 & 0 & 1 & 3 & 0 & 0 \\ 0 & 0 & 0 & 0 & 0 & 0 & 0 & 0 & 0 & 9 & 3 & 0 & 0 & 3 & 1 & 0 \\ 0 & 0 & 0 & 0 & 0 & 0 & 0 & 0 & 0 & 3 & 9 & 0 & 0 & 1 & 3 & 0 \\ 0 & 0 & 0 & 0 & 0 & 0 & 0 & 0 & 0 & 9 & 3 & 0 & 0 & 3 & 1 \end{bmatrix},$$

the eigenvalues are $1, \frac{1}{2}, \frac{1}{2}, \frac{1}{4}, \frac{1}{4}, \frac{1}{4}, \frac{1}{4}, \frac{1}{4}, \frac{1}{8}, \frac{1}{8}, \frac{1}{8}, \frac{1}{16}$ and this result is independent of the numbering. Just as in the curve case the dominant eigenvalue is 1 and has multiplicity 1. Now the sub-dominant eigenvalue (still $\frac{1}{2}$) has multiplicity 2. We can write the original configuration of control points as $\mathbf{C}^0 = \sum \mathbf{x}_i \mathbf{a}_i$ where $\mathbf{x}_i$ are the eigenvectors and the coefficients $\mathbf{a}_i$ are points or vectors. The configuration after k steps of subdivision is now

$$\mathbf{C}^k = \mathbf{S}^k \mathbf{C}^0 = \mathbf{S}^k \sum \mathbf{x}_i \mathbf{a}_i = \sum \mathbf{S}^k \mathbf{x}_i \mathbf{a}_i = \sum \lambda_i^k \mathbf{x}_i \mathbf{a}_i \to \mathbf{x}_1 \mathbf{a}_1, \qquad (7.20)$$

and we see that $\mathbf{a}_1$ is the limit point. Performing the same calculation as in the curve case gives

$$\frac{\mathbf{C}^k - \mathbf{x}_1 \mathbf{a}_1}{\lambda_2^k} = \sum \left(\frac{\lambda_i}{\lambda_2} \right)^k \mathbf{x}_i \mathbf{a}_i \to \mathbf{x}_2 \mathbf{a}_2 + \mathbf{x}_3 \mathbf{a}_3, \qquad (7.21)$$

and we see that $\mathbf{a}_2$ and $\mathbf{a}_3$ spans the tangent plane. So now we want the following properties of the local subdivision matrix.

- The first (dominant) eigenvalue λ_1 should be 1.
- The next two (sub dominant) eigenvalues λ_2, λ_3 should be equal and strictly less than 1.
- All other eigenvalues should be strictly less than $\lambda_2 = \lambda_3$.

Unfortunately, the picture is slightly complicated by *extraordinary vertices*. An extraordinary vertex is a vertex whose valency is not regular. In a mesh of identical regular quads (i.e., squares), which tile the plane, all vertices are incident on four such quads. In a mesh of regular triangles, six triangles meet at a vertex. For a quad-based scheme the extraordinary vertices are those of valency different from four, and, in a triangle-based scheme, the extraordinary vertices are those of valency different from six.

If the original control mesh has extraordinary vertices and/or faces we cannot use the standard rules. The generalization of the quadratic tensor product B-spline to the case of non-quadrilateral is the Doo–Sabin scheme. After one step of subdivision all vertices have valence four and the number of extraordinary faces are constant. So we need special rules for extraordinary faces, and the corresponding subdivision matrix is different in the vicinity of an extraordinary face.

One generalization of the cubic tensor product B-spline to vertices with valence different from four is known as the Catmull–Clark scheme. After one step of subdivision all faces are quadrilaterals and the number of extraordinary vertices are constant. Now we need special rules for extraordinary vertices and the subdivision matrix once more changes in the vicinity of an extraordinary vertex.

Most subdivision schemes are as above extensions of a tensor product (or box-spline surface), so we know the degree of differentiability away from extraordinary vertices and faces. As subdivision does not create new extraordinary points or faces by repeated subdivision will separate two extraordinary point or faces by an arbitrary number of regular points and faces. So when we analyze the surface at an extraordinary vertex or face, we may assume it is the only one. The eigenanalysis is not enough though, we also need to consider the so called characteristic map, which we now proceed to define.

7.2.1 The Characteristic Map

Given an extraordinary point or face we can then find a ring around it that we can divide into a number of sectors where each sector is a piece of an ordinary tensor product B-spline surface (or box spline surface), parametrized over a fixed domain Ω, see Fig. 7.6. That is, the jth sector can be written

$$\mathbf{x}(u, v, j) = \mathbf{B}_j(u, v)\mathbf{C}^0 = \mathbf{B}_j(u, v)\left(\sum \mathbf{x}_i \mathbf{a}_i\right), \quad (u, v) \in \Omega, \tag{7.22}$$

where $\mathbf{B}_j(u, v)$ is a row vector containing the basis functions, $\mathbf{C}^0$ is a column matrix containing the control points, the columns $\mathbf{x}_1, \mathbf{x}_2, \dots$ are the eigenvectors for the local subdivision matrix and the coefficients $\mathbf{a}_i$ are points or vectors. When we now subdivide the mesh the regular part becomes larger and larger. Thus, we obtain a sequence of rings, parametrized as above, that shrink to the point we want to investigate. The jth sector in the kth ring is parametrized as

$$\mathbf{x}^k(u, v, j) = \mathbf{B}_j(u, v)\mathbf{C}^k = \mathbf{B}_j(u, v)\left(\mathbf{S}^k \sum \mathbf{x}_i \mathbf{a}_i\right)$$
$$= \mathbf{B}_j(u, v)\left(\sum \lambda_i^k \mathbf{x}_i \mathbf{a}_i\right), \tag{7.23}$$

where $\mathbf{S}$ is the local subdivision matrix. When $k \to \infty$ we see that

$$\mathbf{x}^k(u, v, j) \to \mathbf{B}_j(u, v)\mathbf{x}_1 \mathbf{a}_1 = \mathbf{a}_1,$$

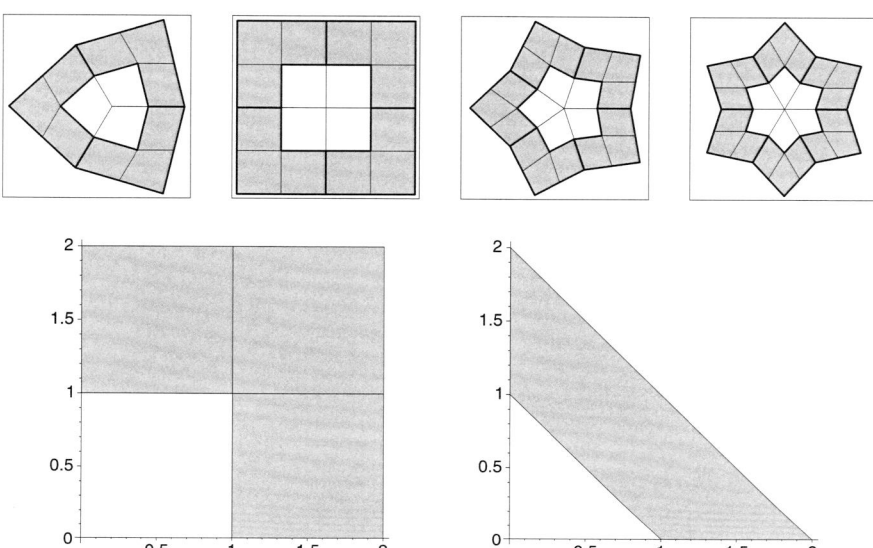

Fig. 7.6 Above the ring of a quadrilateral mesh for a point of valence 3, 4, 5, and 6, respectively. Below the domain Ω for the standard sector. On the left for a quadrilateral control mesh, on the right for a triangular mesh

since, by affine invariance, all rows of $\mathbf{S}$ must sum to one, which entails that $\mathbf{x}_1 = [1\ 1\ \ldots\ 1]^T$, and the sum of the elements in $\mathbf{B}_j(u, v)$ is also one. Consequently, $\mathbf{B}_j(u, v)\mathbf{x}_1 = 1$. If $\lambda_2 = \lambda_3 > \lambda_4 \geq \cdots$, and we furthermore have

$$\frac{\mathbf{x}^k(u, v, j) - \mathbf{a}_1}{\lambda_2^k} = \mathbf{B}_j(u, v)\left(\mathbf{x}_2\mathbf{a}_2 + \mathbf{x}_3\mathbf{a}_3 + \sum_{i \geq 4}\left(\frac{\lambda_i}{\lambda_2}\right)^k \mathbf{x}_i\mathbf{a}_i\right), \qquad (7.24)$$

so the local behavior is approximately given by $\mathbf{B}_j(u, v)(\mathbf{x}_2\mathbf{a}_2 + \mathbf{x}_3\mathbf{a}_3)$.

Definition 7.1 For a local subdivision matrix with eigenvalues $\lambda_1 = 1 > \lambda_2 = \lambda_3 > |\lambda_4| \geq \cdots$, and corresponding eigenvectors $\mathbf{x}_i$ the *characteristic map* is on the jth sector defined by

$$\Psi(u, v, j) = \mathbf{B}_j(u, v)[\mathbf{x}_2, \mathbf{x}_3], \qquad (u, v) \in \Omega. \qquad (7.25)$$

Recall that $\mathbf{x}_2, \mathbf{x}_3$ are column vectors of length n, where n is the size of the invariant neighborhood, and $\mathbf{B}_j(u, v)$ is a row vector of length n. Thus, $\mathbf{B}_j(u, v)[\mathbf{x}_2, \mathbf{x}_3]$ is a row vector of length two and $\Psi(\cdot, \cdot, j)$ is a map $\Omega \to \mathbb{R}^2$.

Definition 7.2 The characteristic map is *regular* if

$$\det \frac{\partial \Psi(u, v, j)}{\partial(u, v)} \neq 0, \quad \text{for all } (u, v) \in \Omega \text{ and for all sectors } j.$$

Fig. 7.7 Many subdivision
schemes split triangles or
quadrilaterals as shown in this
figure

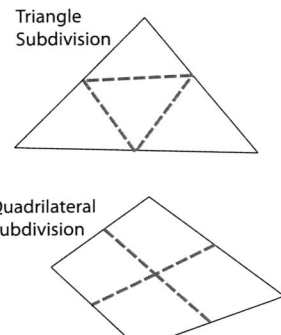

Theorem 7.2 *If the eigenvalues* λ_i *for the local subdivision matrix satisfy i* $\lambda_1 = 1 > \lambda_2 = \lambda_3 > |\lambda_4| \geq \cdots$, *and the characteristic map is regular, then the subdivision surface is a tangent plane continuous for all initial control meshes* $\mathbf{c}^0 = \sum \mathbf{x}_i a_i$ *where the coefficients* a_2, a_3 *are linearly independent.*

7.3   Subdivision Surfaces in Practice

Subdivision curves are very easy to implement. From a control polygon subdivision recursively produces a new polygon with twice as many vertices. Vertex positions are conveniently stored in an array, and if we are given a particular vertex occupying a given entry in this array, the next vertex is known to be stored in the subsequent entry in this vector.

For surfaces the principle is almost the same. The control polyhedron is refined by insertion of new vertices, which are computed as linear combinations (summing to one) of vertices in the coarser polyhedron. However, the connectivity is more complicated. We can store our vertices in a single array, but there is no longer a simple connection between the position of the vertex in the mesh and its array index.

Instead, to define a subdivision scheme, we must specify how more refined polygons are produced from the coarse polygons, and how the positions of the resulting new vertices are computed. The two most common strategies for the former task are shown in Fig. 7.7. As shown in the marginal figure, triangles are split by inserting a new vertex on each edge. The result is four new triangles: a center triangle and three triangles which each share one corner with the old triangle. In a quad mesh, we use a slightly different strategy, and split each edge by inserting a vertex, but we also add a vertex to the center of each face and form new faces by connecting the *edge vertices* with the *face vertex*. Note that a general N-gon is turned into N quads by this procedure.

These operations are sometimes referred to as 1–4 splits. Note that in a triangle mesh, only valence six vertices are introduced by the triangle splitting scheme and only valence four vertices by the quadrilateral splitting scheme. Thus, *extraordinary vertices* whose valence is different from the regular valence are not introduced by these splits (cf. Sect. 7.3.2).

Fig. 7.8 The Loop masks for old vertices and vertices inserted on edges. Note that there is no rule for face vertices since vertices are not inserted in faces in the Loop scheme. v denotes valence

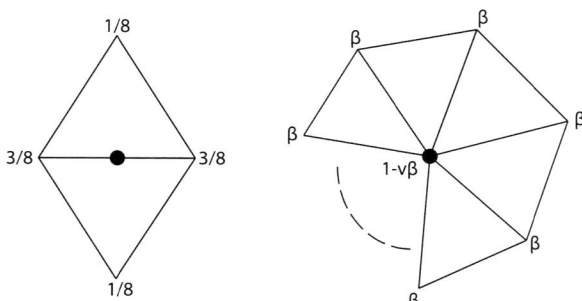

From the above discussion, it is clear that we are dealing with three classes of vertices when we refine a mesh by splitting polygons. The first class consists of the vertices which are shared by the coarse and the refined mesh. The second class consists of edge vertices, i.e. vertices which we insert on edges, and the final class consists of face vertices.

These classes are convenient when it comes to specifying how vertex positions are computed. It is not quite practical to use a matrix notation. Instead subdivision schemes are often specified via so called masks which specify how a new vertex is computed from the old vertices. For instance, the Loop masks are shown in Fig. 7.8. The fat dots indicate what vertex position we are computing. Say, we are computing a vertex on an edge. The left hand image shows the mask for an edge vertex. It tells us that we should compute the position of the new vertex by taking three eighths of the vertices at either end of the edge and one eighth of the vertices opposite the edge. Note that these weights sum to one. The right image tells us how to update the position of an existing vertex. The gap in the 1-ring illustrates that the rule is used for vertices of arbitrary valence. To apply the rule, β needs to be computed which is done according to

$$\beta = \frac{1}{v}\left(\frac{5}{8} - \frac{(3 + 2\cos(2\pi/v))^2}{64}\right) \tag{7.26}$$

where v is the valency of the vertex [1].

Thus, we now know the Loop scheme almost well enough to implement it. The only thing left to cover is a practical concern. The normal procedure for splitting faces is to visit every face and split it into four new faces, but a new vertex created on an edge will be shared by a set of faces on either side of that edge. If our implementation is based on an indexed face set representation, there is no natural way to store the new edge vertex between the point in the time that we visit the first face and the second face sharing that edge. A common solution is to create an auxiliary edge data structure (for instance a hash table) to store the new vertices. However, if we are using an edge-based representation it is somewhat simpler, since we already have explicit edge representations, and we can probably store the positions of new vertices as a datum associated with the edges.

Fig. 7.9 The tetrahedron on the *left* was subdivided three times producing the sequence of images from *left* to *right*. Observe that Loop shrinks the models a great deal

Another implementation concern is whether to subdivide "in place" or create a new mesh. The latter is simpler. If our mesh is represented using an edge-based representation, it is clearly possible to split as we go along, but the practical difficulties are greater than if we simply build a new mesh. There is less house-keeping because we do not have to keep track of which vertices are new, which vertices are old, edges which have been split, etc.

7.3.1 Subdivision Schemes

In this chapter, we deal exclusively with stationary subdivision schemes based on linear combinations of vertices. We can classify such subdivision schemes according to several criteria. The most common are as follows.
- The type of polygon the scheme is designed for: triangles, quadrilaterals or some other type.
- Whether it is primal or dual. Primal schemes are those which work as described above. In a dual (also known as vertex splitting) scheme, we split each vertex into a number of new vertices. The number is equal to the number of faces incident on that vertex.
- Whether it is interpolating or approximating. In other words whether the original vertices are interpolated or not.

In the following, we will discuss a number of concrete subdivision schemes. Similar discussions can be found in a number of books, e.g. Zorin's SIGGRAPH course notes [1], Warren and Weimer's book [2], and Real-Time Rendering [3].

7.3.1.1 Loop
The Loop scheme [4], which we have just used as an example, is a triangle-based, primal, and approximating scheme. In regular regions, Loop subdivision reproduces C^2 *quartic triangular box splines* in the limit. At extraordinary vertices the limit surface is C^1. Loop subdivision is useful because triangle meshes are so common. Loop subdivision is shown in Fig. 7.9.

7.3.1.2 Catmull–Clark
However, subdivision-based modeling systems often employ a scheme designed for quadrilaterals, because quads align better with the symmetries of most objects. A commonly used scheme is known as Catmull–Clark (see Fig. 7.11). Catmull–Clark is a quad-based, primal, and approximating scheme. In regular regions,

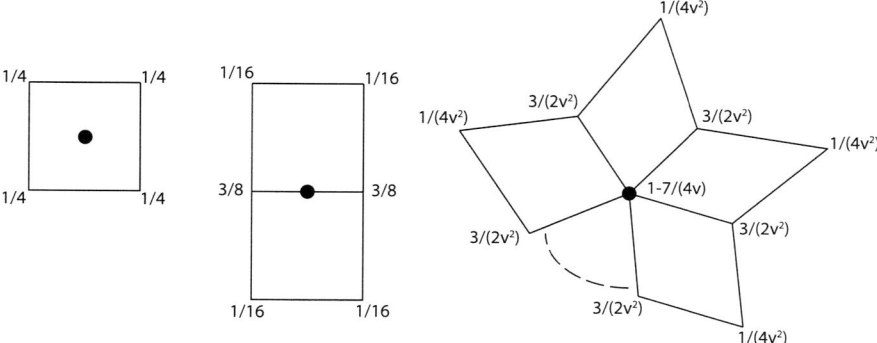

Fig. 7.10 The Catmull–Clark masks for old vertices and vertices inserted on edges, and vertices inserted in faces. In the vertex mask, v is the valency

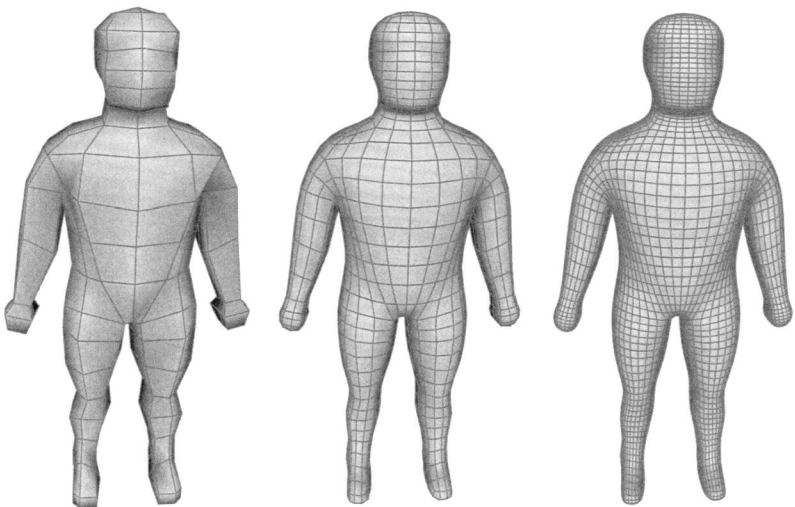

Fig. 7.11 The simple mesh on the *left* was subdivided two times to produce the *center* and *right* meshes using Catmull–Clark subdivision

the Catmull–Clark limit surface is C^2 and it reproduces a bicubic tensor product B-spline patch, as mentioned earlier. At extraordinary vertices the limit surface is C^1 [1].

All faces are split by inserting a vertex at the center and on each edge and, subsequently, connecting the face vertex with all the edge vertices as shown in Fig. 7.7. The Catmull–Clark masks are shown in Fig. 7.10, and an example of Catmull–Clark subdivision is shown in Fig. 7.11.

Fig. 7.12 The Doo–Sabin scheme is an example of a dual scheme where a vertex is split into multiple vertices—one for each incident face. On *top* a cube after 1, 2, and 3 steps of Doo–Sabin. *Below* the cube subdivided using 1, 2, and 3 steps of Catmull–Clark

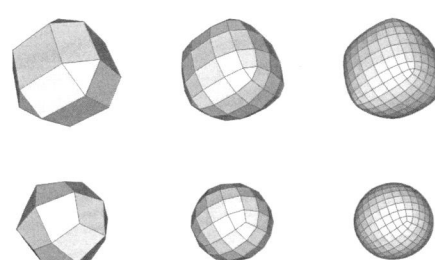

7.3.1.3 Doo–Sabin

Duality is easiest to explain with an example: Doo–Sabin is a quad-based, dual, and approximating scheme which reproduces quadratic tensor product B-spline surfaces in the limit.

The scheme is a dual scheme, which means that, conceptually, we split vertices rather than faces. Specifically, we produce a new vertex for each corner of each face. The position of the new vertex is the average of [1]

- the centroid of the face,
- the midpoints of the two edges adjacent to the old corner, and
- the position of the old corner.

The face is retained in a slightly shrunk version, and we create a new face for each edge and each vertex. Edges become quads and vertices become faces with the same number of edges as the valency of the vertex. Three steps of Doo–Sabin subdivision are shown in Fig. 7.12.

Intuitively, the Doo–Sabin scheme can be seen as a *chamfering* of the corners and edges of the old model. From a different point of view, it can also be seen as a generalization of the Chaikin curve subdivision scheme where we simply cut off the corners of a polygon until a smooth surface is obtained. In the Doo–Sabin scheme, we repeatedly cut the edges and corners until we obtain a smooth surface. The method was originally described in [5].

7.3.1.4 Modified Butterfly

So far, we have only discussed approximating schemes. When subdivision surfaces are used for design, we typically want approximating schemes because if we force a surface to interpolate a set of points, the surfaces have a tendency to develop wrinkles. However, approximating schemes tend to cause considerable shrinkage. Consequently, it is best to use an interpolating scheme if we just want to refine a coarse model slightly with a few steps of subdivision.

One well known method for creating interpolating surfaces is the Modified Butterfly, which is a primal, triangle-based subdivision method. Triangles are split in the same 1–4 fashion as in the Loop scheme. However, the limit surface is only C^1 even in regular regions. Since the scheme is interpolating, there is no vertex rule. However, there is a rule for edge vertices which applies to edges adjacent to two regular vertices and one which applies to edges where one of the vertices is irregular.

Fig. 7.13 The Modified Butterfly scheme

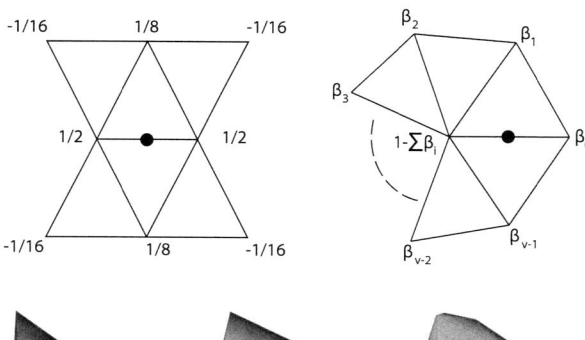

Fig. 7.14 The tetrahedron on the *left* was subdivided three times producing the sequence of images from *left* to *right*. Unlike the Loop, Modified Butterfly does not shrink the mesh. On the other hand, the resulting surface is a little less fair

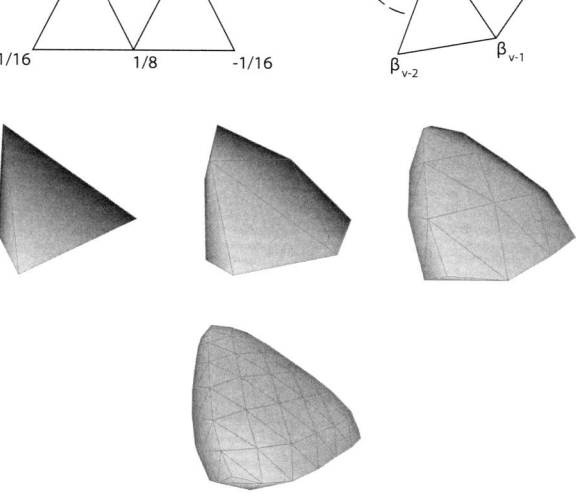

The masks for the Modified Butterfly are shown in Fig. 7.13, and the β_i are computed as follows:

$$v = 3: \quad \beta_0 = \frac{5}{12}, \qquad \beta_1 = -\frac{1}{12}, \qquad \beta_2 = -\frac{1}{12}$$

$$v = 4: \quad \beta_0 = \frac{3}{8}, \qquad \beta_1 = 0, \qquad \beta_2 = -\frac{1}{8}, \qquad \beta_3 = 0$$

$$v \geq 5: \quad \beta_i = \frac{\frac{1}{4} + \cos(2\pi i/v) + \frac{1}{2}\cos(4\pi i/v)}{v}$$

where v denotes valency as usual [1].

In the first step (and only in the first step), we may have an edge where both endpoints are irregular. In this case, we can simply compute the edge vertex using the irregular rule for both endpoints and average the result. An example of a Modified Butterfly subdivision is shown in Fig. 7.14.

7.3.1.5 $\sqrt{3}$ Subdivision

The number of triangles grows exponentially when subdividing. For real-time rendering applications this can be a problem, since we wish to keep the amount of geometry down in order to keep the frame rate up. The $\sqrt{3}$ scheme [6] addresses

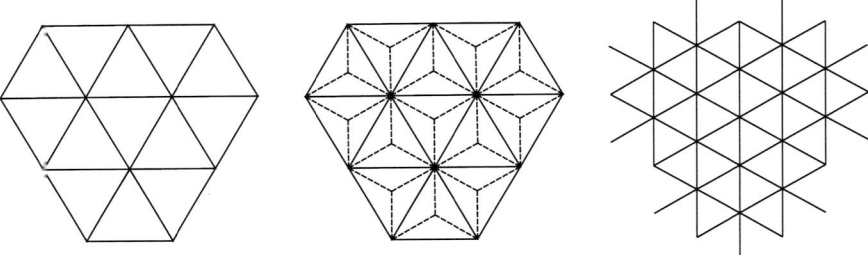

Fig. 7.15 One iteration of the $\sqrt{3}$ subdivision scheme. First we insert vertices at the centers of all faces (*center image*), and then all the old edges are flipped producing the final result on the *right*

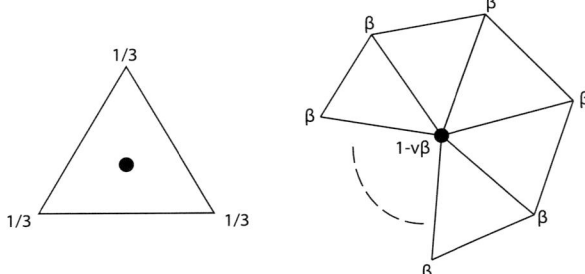

Fig. 7.16 The masks for the $\sqrt{3}$ scheme

this problem in two ways. First of all, it is a triangle-based approximating scheme, and arguably primal, but it does not employ the usual 1–4 split. Instead, a vertex is inserted at the center of every triangle, and then the old edges are flipped as illustrated in Fig. 7.15. Thus, the number of triangles is only tripled in each iteration, and *two iterations* of the algorithm lead to a regular subdivision of each triangle where the original edges are tri-sected. This is the reason for the unusual name. In $\sqrt{3}$ subdivision, the face vertex is simply inserted in the center of the old face (weighting each old vertex by $1/3$). The masks are shown in Fig. 7.16. β is computed:

$$\beta = \frac{4 - 2\cos(2\pi/v)}{9v}. \tag{7.27}$$

The $\sqrt{3}$ scheme has a very nice feature when it comes to adaptive subdivision. In adaptive subdivision, we selectively subdivide where needed. Typically, when a region is deemed sufficiently flat, we stop the subdivision process. Unfortunately, problems arise where faces at different levels of subdivision are adjacent. Consider a face which is refined using a 1–4 based scheme which is adjacent to an unrefined face. In this case, we have a T-junction on the edge which is split from one side only. Usually this is fixed by inserting an edge connecting the junction to the opposite

Fig. 7.17 Adaptive subdivision using the $\sqrt{3}$ scheme above and the normal 1–4 scheme below. On the *left*, we have shown how the triangles are subdivided in the two schemes. On the *right*, the edge flips in the second step of $\sqrt{3}$ are shown, but notice that the old edge separating the subdivided triangle from the unsubdivided is left unflipped. Note also the addition of an edge to avoid a T-junction in the 1–4 scheme shown below

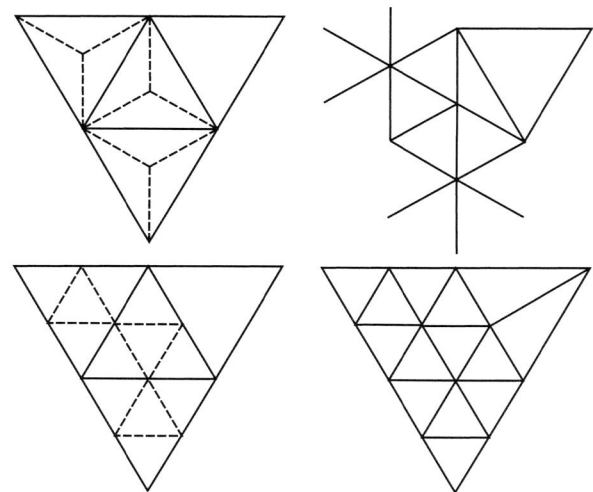

face. This problem does not exist in the $\sqrt{3}$ scheme since the edges are not split. All we have to do is omit an edge flip. The differences are illustrated in Fig. 7.17.

Finally, the slower rate of increase in the number of triangles makes transitions between subdivided and not subdivided areas more gradual.

7.3.2 The Role of Extraordinary Vertices

Extraordinary vertices (cf. Sect. 7.2) are simply vertices of irregular valence, but what does that mean in practice? Regular triangles, quadrilaterals, and hexagons, and only those, tile the plane. However, only an infinite plane or a torus admits a tiling with *only* regular vertices. Thus, any non-trivial mesh contains extraordinary vertices, but the number is constant during subdivision because subdivision algorithms never create or remove extraordinary vertices. In general, however, subdivision surfaces tend to become a little less smooth in the vicinity of an extraordinary vertex, and if the valence is very high, it usually leads to artifacts.

Thus, extraordinary vertices seem to be something the designer should avoid, but that is neither possible nor desirable. Extraordinary vertices should be placed at the corners of the model: Observe that if we assume that the quads are nearly regular, a valence three vertex in a quad mesh has an angle sum of roughly 270 degrees and a valence five vertex a sum of 450 degrees. In other words, if we consider (8.5) we should expect that a valency three vertex is at a point of positive and a valency five vertex at a point of negative Gaussian curvature. Conversely, a vertex which is intended to be nearly flat in the limit surface should be of valence four.

Finally, note that if a triangle mesh is subdivided with, say, Catmull–Clark, the result is a very unusual pattern because in the first step nearly every vertex is extraordinary. Thus, subdivision schemes should only be used on meshes of the intended type.

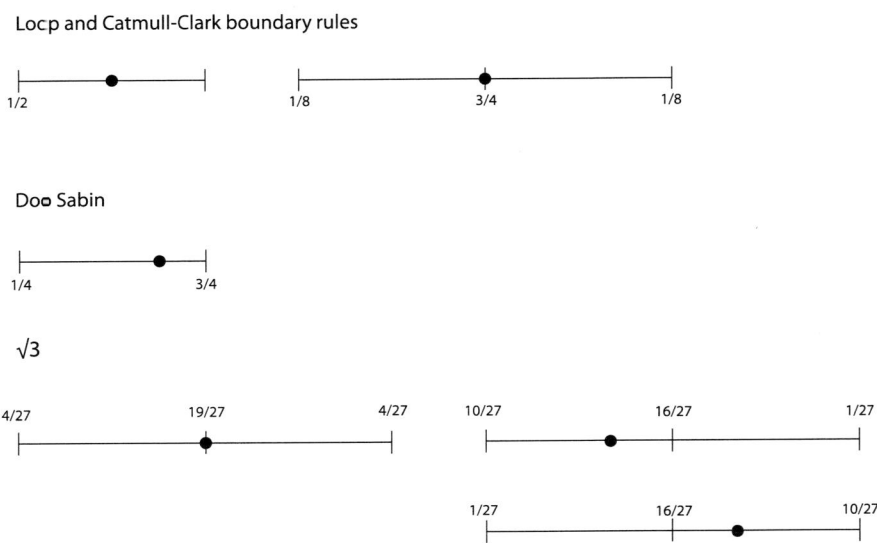

Fig. 7.18 Boundary rules for Loop, Catmull–Clark, Doo–Sabin, and $\sqrt{3}$. The vertex we are computing is indicated by a *fat dot*, and the other vertices of the edges by a *vertical line*

7.3.3 Boundaries and Sharp Edges

So far, we have assumed the mesh to be a closed 2-manifold. In many cases, this assumption is not acceptable, and we need to deal with boundaries. This is usually done by making special masks for boundary edges and vertices which only include other edge vertices. By letting new boundary vertices depend only on other boundary vertices, we ensure that the same smooth curve is obtained when two subdivision surfaces meet along a common boundary. Of course, boundary rules are simply subdivision rules for subdivision curves.

Loop and Catmull–Clark both use boundary rules for curves whose limit curves are cubic splines. The boundary rules for Doo–Sabin lead to quadratic splines. The boundary rules in the case of the Modified Butterfly scheme are fairly complex, and we refer the reader to [7] for a discussion.

In the case of $\sqrt{3}$ subdivision, it is a little harder to deal with boundary edges, since these cannot be flipped. However, after two subdivision steps, the boundary edge has been trisected. Consequently, the boundary rules are only applied in every other step, and we have three types of rule: one for each of the two types of edge vertex and one for the original vertices.

The boundary rules for the various schemes are shown diagrammatically in Fig. 7.18.

The boundary edges are also useful in other situations where we want a sharp edge. If a sequence of edges interior to the surface needs to be sharp, we simply subdivide them using the boundary rules instead of the face rules.

7.3.4 Advanced Topics

It is possible to *push* a vertex to its position on the limit surface without performing an infinite number of subdivisions. Recall that the limit position is given by the first eigenvalue of the subdivision matrix. Thus, what we need to do is to decompose the vertex positions inside the mask into a linear combination of the eigenvectors of the matrix. It then follows from (7.20) that the limit position is simply the first eigenvector times its weight in this linear combination. Similarly, we can find the tangent frame from the second and third eigenvectors. In practice, this work has been done in a generic fashion for the most popular schemes, and it is possible to find closed form formulas for the computation of these limit positions and tangent vectors, e.g. in [3].

It is fairly convenient to implement subdivision in terms of the rules encoded in the masks shown in Sect. 7.3.1 if we are dealing with data structures that directly encode the mesh connectivity. However, both Loop and Catmull–Clark admit a different implementation in two loops where edges and faces are split by inserting a vertex in the middle during the first iteration. This produces a topologically split mesh with the same geometry. In the second loop, a simple smoothing updates the vertex positions. Although we still need to treat extraordinary vertices in a special fashion, this is arguably simpler than using the traditional approach [8].

In a seminal paper [9], Jos Stam pointed out that it is possible to parametrize Catmull–Clark surfaces using a method (that also generalizes to other schemes), which is based on analyzing the eigenstructure of the subdivision matrix. In fact, it is, of course, easy away from the extraordinary vertices where the limit surface is a parametric surface, but if just one vertex in a patch is extraordinary, more work has to be done. If we have performed two iterations of subdivision, we can divide the mesh into patches where only a single vertex of any quad is extraordinary. It is possible to write down the subdivision matrices for these patches, i.e. the matrices that will produce a new patch at finer scale around the same extraordinary vertex. Somewhat simplified, to evaluate the parametrization of the subdivision surface near an extraordinary point, we can simply subdivide until the vicinity of the point is a regular patch and then evaluate the surface as a bicubic B-spline. Subdivision is costly, but since we know the subdivision matrix, we do not have to do the subdivision. We can find the control points for the spline surface from the eigenvector decomposition of the vertices by multiplying with a sufficiently high power of the eigenvalue. We cannot avoid irregular vertices, and as pointed out by Myles and Peters [10], it is a bit ironical that popular subdivision schemes are motivated by the ability to handle extraordinary vertices while having lower degree of continuity precisely at those vertices. The C^2PS scheme is C^2 at the so called *polar vertices* which may be irregular vertices. Polar vertices are placed at the tips of protrusions and surrounded by a ring of triangles. A polar configuration (see Fig. 7.19) is characterized by edge sequences flowing away from the pole in a radial manner and edge loops circling the pole. Subdivision with C^2PS preserves the polar structure by doubling the number of radial edge sequences and loops as shown in Fig. 7.19 (middle). Notice that the valency of the pole doubles for each step of subdivision: nonetheless, the limit

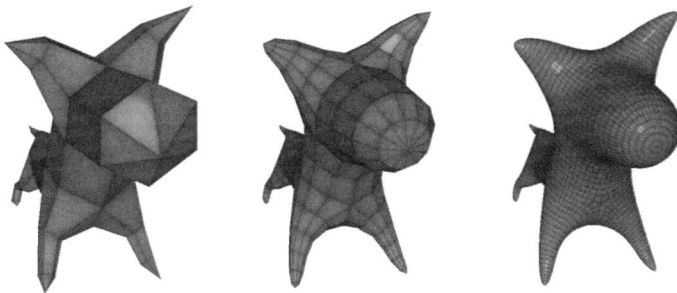

Fig. 7.19 A polar configuration and how it looks after one and two iterations of subdivision

surface is C^2 smooth. Only the pole and the two innermost loops of edges are dealt with in a special way. The remaining vertices are subdivided using Catmull–Clark. Unfortunately, the C^2PS is scheme is not widely adopted in commercial modeling packages, and, in fact, the images in Fig. 7.19 show our own slightly simplified implementation. However, it seems natural to model a wide range of shapes using polar configurations, so we believe the scheme holds significant promise.

Specifically, we look at the center of a polar configuration. With the advent of programmable graphics hardware, many algorithms have been transferred to the graphics processing unit. Subdivision has been a problem in this respect because it is not possible to subdivide a triangle (or quad) in isolation. We always need neighboring vertices. Fortunately, this is becoming easier since recent graphics cards are equipped with so called geometry shaders which allow computations to be carried out on a per-triangle level with access to neighboring vertices. The current trend also seems to go in the direction of adding *tessellation units* to graphics hardware. This means that smooth surface primitives whether formulated as subdivision surfaces or parametric surfaces are making their way into the graphics hardware, and we may, finally, be moving away from the polygon as the standard primitive for rendering.

7.4   Exercises

Exercise 7.1 Implement Doo–Sabin subdivision. Hint: This scheme can be implemented as a very simple subdivision where each quad is split into four quads by inserting a vertex in the center of each face and midpoint of each edge. This is followed by taking the dual of the mesh (where the dual vertices are averages of the corners of the corresponding face).

Exercise 7.2 Implement the $\sqrt{3}$ scheme.

  [**GEL Users**] GEL and many other subdivision schemes allow you to easily flip edges making this scheme straightforward to implement.

Exercise 7.3 Implement Catmull–Clark. Hint: This is hard to do if you change the connectivity of the original It is easiest to first compute the new vertices created for edges and faces and the new positions of the old vertices. Subsequently, a new mesh is created with this information.

Exercise 7.4 Implement Loop subdivision. Hint: see above.

Exercise 7.5 Implement the Modified Butterfly subdivision. Hint: same connectivity as Loop.

Exercise 7.6 Implement boundary rules for the implemented schemes.

References

1. Zorin, D., Schroder, P., et al.: Subdivision for modeling and animation. In: SIGGRAPH 99 Course Notes, vol. 2 (1999)
2. Warren, J., Weimer, H.: Subdivision Methods for Geometric Design. Morgan Kaufmann, San Francisco (2002)
3. Moller, T., Haines, E.: Real-Time Rendering, 2nd edn. AK Peters, Wellesley (2002)
4. Loop, C.T.: Smooth subdivision surfaces based on triangles. Master's thesis, Dept. of Mathematics, University of Utah (1987)
5. Doo, D., Sabin, M.: Behaviour of recursive division surfaces near extraordinary points. Comput. Aided Des. **10**(6), 356–360 (1978). doi:10.1016/0010-4485(78)90111-2
6. Kobbelt, L.: $\sqrt{3}$ subdivision. In: Proceedings of SIGGRAPH 2000, pp. 103–112 (2000)
7. Zorin, D., Schröder, P., DeRose, T., Kobbelt, L., Levin, A., Sweldens, W.: Subdivision for modeling and animation. Technical report, SIGGRAPH 2000 Course Notes (2000)
8. Warren, J., Schaefer, S.: A factored approach to subdivision surfaces. IEEE Comput. Graph. Appl. **24**(3), 74–81 (2004)
9. Stam, J.: Exact evaluation of Catmull-Clark subdivision surfaces at arbitrary parameter values. In: Proceedings of the 25th Annual Conference on Computer Graphics and Interactive Techniques, pp. 395–404 (1998)
10. Myles, A., Peters, J.: Bi-3 C^2 polar subdivision. ACM Trans. Graph. **28**(3), 1–12 (2009)

Curvature in Triangle Meshes

8

Algorithms for effective manipulation and visualization of triangle meshes often require some knowledge about the differential properties of the mesh. A trivial example is the surface normal, which is dealt with in Sect. 8.1. The surface normal is only defined for triangle faces, but when a mesh is drawn as a smooth surface, one generally defines the normal at each vertex and interpolates it across faces. To do so, some technique for normal estimation is required. For many other applications such as shape smoothing, shape analysis, artistic rendering, etc. information about curvature is needed.

This chapter explains how the notion of curvature can be extended to triangle meshes. We will assume that the mesh is a triangle mesh and not a general polygonal mesh. This is not a limitation since any polygonal mesh can be triangulated in order to produce a triangle mesh.

As regards notation: perhaps, it would be more stringent to use different symbols for the discrete versions of differential geometry, say, mean curvature, for which we use H both in the discrete and continuous case. Going further, we could have used various notational devices to distinguish between mean curvature computed using different methods. We have done neither and use H for mean curvature both in the continuous and discrete setting. We believe that it can be seen from the context to what precisely the symbol refers.

Since curvature is not, strictly speaking, defined on meshes, computing curvature from meshes must be approached by one of the following approaches:

- Compute the integral curvature of a small area instead of the pointwise curvature. This is the basic strategy used in the formulas for the computation of the mean curvature normal in Sect. 8.2 and the Gaußian curvature in Sect. 8.3.
- Find a smooth surface that is very close to the triangle mesh: if every edge in a triangle mesh is replaced by a cylindrical patch and every vertex is replaced by a spherical cap, we can compute the curvature of this smooth approximation. This basic strategy is employed in the methods discussed in Sect. 8.4.
- Compute a smooth approximation. This is the classic way of estimating curvature from triangle meshes is based on almost the opposite outlook. The triangle vertices are assumed to be sampled from some smooth surface, and we fit a smooth

J.A. Bærentzen et al., *Guide to Computational Geometry Processing*,
DOI 10.1007/978-1-4471-4075-7_8, © Springer-Verlag London 2012

surface to the vertices in the neighborhood of the point where we wish to find the curvature. This method is discussed in Sect. 8.5.

In Sect. 8.6 we discuss how to compute the directions of principal curvature from the shape operator, which has been discussed in the preceding sections.

Finally, it should be noted that there are many other strategies for estimating curvature than those mentioned here. Section 8.7 provides some pointers to additional literature. The Appendix contains a derivation of the Cotan formula [1] for the mean curvature normal [2].

For references to books on differential geometry as well as an introduction to the fundamental notions, the reader is referred to Chap. 3.

8.1   Estimating the Surface Normal

All curvature measures are basically functions of the derivatives of the normal, and information about the surface normal in itself is used for many purposes, e.g. for shading when rendering meshes. However, we frequently want to compute the surface normal at vertices where the normal is, strictly speaking, not defined. This section explains a sound method for computing what is termed a *pseudo normal* at mesh vertices. By pseudo normal we understand a vector, associated with a point on a triangle mesh, that is imbued with some properties analogous to those of the normal of a smooth surface.

In Chap. 3, we introduced the notion of a parametrization of a smooth surface. A parametrization $S : \mathbb{R}^2 \supseteq U \to \mathbb{R}^3 : (u, v) \mapsto S(u, v)$ is a map from a 2D domain to a smooth surface in 3D. As discussed in Sect. 3.3, the normal to a smooth surface is the cross product of the derivative of S with respect to u and v,

$$\mathbf{n} = \frac{S_u \times S_v}{\| S_u \times S_v \|}.$$

For a triangular mesh, the normal is clearly defined on the faces, where it is just a vector perpendicular to the face, but what do we do at vertices and edges? A starting point is to ask which properties we should require of a vertex normal. An obvious requirement is that the normal should be similar to the normal of a smooth surface, if the mesh is very close to a smooth surface. In other words, we should require that the normal computed for a mesh converges to the normal of a smooth surface, if we refine the mesh (divide the faces into smaller and smaller faces) until it converges to the smooth surface.

If the faces are refined in a sound fashion such that the radii of their circumcircles tend to zero, all the normals around a vertex will tend to the same normal as we refine the mesh, and the average of the face normals of faces incident on a given vertex would satisfy this property. Indeed, a common solution for computing the normal at a vertex is to simply average the normals of the incident faces. However, this is not the best solution. To see why, we consider two additional properties that we need to require of a *good vertex normal*.

Fig. 8.1 The angle weighted
normal is the average of the
face normals weighted by the
angle α_i which the face
makes at the vertex

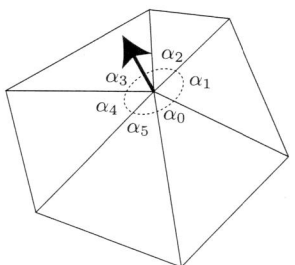

(1) The normal should be independent of the tessellation and depend only on the
     geometry of the mesh. Put differently, if one were to add an outgoing edge to a
     vertex by splitting an incident triangle into two triangles, this should not affect
     the normal estimate at that vertex.
(2) The normal should allow us to determine whether a point is inside or outside the
     mesh. This is more subtle, but for an arbitrary point near but *not* on a smooth
     surface, we know that the normal at the closest point on the nearby smooth
     surface will point towards that point. A similar property would be desirable for
     vertex normals in meshes.
Having this in mind, we arrive at the following.

Definition 8.1 The angle weighted pseudo normal AWPN at a vertex is

$$\mathbf{n}_\alpha = \frac{\sum \alpha_i \mathbf{n}_i}{\| \sum \alpha_i \mathbf{n}_i \|}, \tag{8.1}$$

where i runs over all faces incident on the vertex in question.

The AWPN has both these desirable properties. It is almost the same as just
averaging the normals of incident faces, but now the normals are weighted by the
angle which the face makes at the vertex as illustrated in Fig. 8.1. It is obvious that
simply splitting a face does not affect the AWPN since if we split a face by a line
going through a vertex, we merely obtain two faces both with the normal of the old
face, and the angles sum to the angle of the old face [3].
The AWPN also has the second property. Let $\mathbf{p}_i \in \mathbb{R}^3$ be the point in space associated with vertex i. If we are given a second point $\mathbf{q}$ such that $\mathbf{p}_i$ is the point on the
mesh closest to $\mathbf{q}$, we have

$$\mathbf{n}_\alpha \cdot (\mathbf{q} - \mathbf{p}_i) > 0, \tag{8.2}$$

where $\mathbf{n}_\alpha$ is the normal at vertex i, precisely if $\mathbf{q}$ is outside the mesh. Of course,
we can insert a vertex on a face or an edge without changing the geometry so the
AWPN is defined everywhere: on a face it is simply the face normal. On an edge the
AWPN is the average of the normals of the two incident faces. For a proof of (8.2)
the reader is referred to [4].

8.2   Estimating the Mean Curvature Normal

In the case of plane curves, curvature is unambiguously defined: it is the rate of change of the unit tangent vector. For surfaces, the situation is more involved, but, as discussed in Chap. 3, we can define the *normal curvature* at a point as the curvature of a *normal section*. A normal section is the intersection of the surface and a plane containing the normal. Rotating the plane around the normal, we obtain a different normal curvature for each angle. For two perpendicular directions we obtain, respectively, the smallest and the greatest normal curvature. These are denoted the minimum and maximum *principal curvatures*. The mean curvature is the average of the principal curvatures.

The mean curvature is also related to the notion of a *minimal surface* [5], i.e., a surface that has the smallest area among all surfaces with the same boundary. If the mean curvature is zero everywhere on a surface, any small local change to the surface will increase the area. Thus, $H = 0$ everywhere on a minimal surface.

To be a bit more precise, imagine a small patch around a point $\mathbf{p}$. We can think of $\mathbf{p}$ as a control point which can be moved freely, and the patch (a small neighborhood of $\mathbf{p}$) will deform smoothly as this happens. Note that the patch is also assumed to join the rest of the surface in a smooth fashion and that the rest of the surface is not affected by the deformation. Let the area of the patch be A. The area A can be seen as a function of the position of $\mathbf{p}$, and, hence, we can define the gradient of A.

The mean curvature is defined as the following limit:

$$2H\mathbf{n} = 2\mathbf{H} = \lim_{A \to 0} \frac{\nabla A}{A}, \tag{8.3}$$

where the vector $\mathbf{H}$ is the *mean curvature normal* and with the condition that the greatest distance between two points on the curve enclosing the area A also tends to zero. This is a slightly different formulation from the one in Sect. 3.7, because we consider the change of the area, A, to be a function of a deformation of the local patch in any direction and not just the normal direction. The salient difference is that (8.3) yields the mean curvature normal, and (3.40) is a formula for the scalar mean curvature.

In the case of discrete surfaces, we have a natural definition of a "local patch", namely the 1-ring of the vertex. This led Desbrun et al. [2] to define the mean curvature normal simply as the gradient of the area of the 1-ring normalized by the area of the 1-ring.

Definition 8.2 The mean curvature normal for a mesh vertex is

$$\mathbf{H}(\mathbf{p}_i) = \frac{1}{2} \frac{\nabla A_i^{1\text{-ring}}}{A_i^{1\text{-ring}}}$$

$$= \frac{1}{4A_i^{1\text{-ring}}} \sum_{\mathbf{p}_j \in \mathcal{N}_i} (\cot \alpha_{ij} + \cot \beta_{ij})(\mathbf{p}_i - \mathbf{p}_j), \tag{8.4}$$

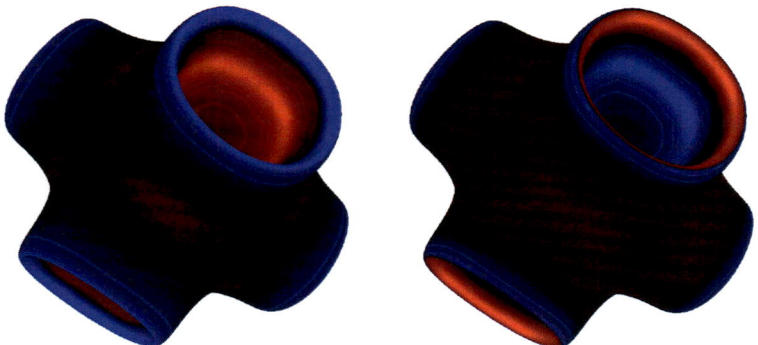

Fig. 8.2 A comparison of mean curvature (*left*) to Gaußian curvature (*right*). *Blue* is positive and *red* is negative, intensity indicates *magnitude* and the *white* stripes are spaced at equal intervals to indicate the rate of change

Fig. 8.3 The angles α and β are the angles opposite to edge ij

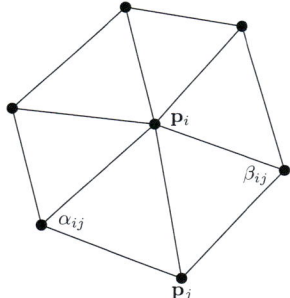

where $A_i^{1\text{-ring}}$ is the area of the 1-ring. In this equation we sum over all the edges in the 1-ring and weight the edge by the cotangent of the angles at the vertices opposite the edge as illustrated in Fig. 8.3. The details of how to derive this formula are found in the Appendix to this chapter.

Figure 8.2 shows the difference between mean and Gaußian curvature.

A desirable property of the mean curvature normal as defined above is that if the 1-ring is flat, the mean curvature normal is **0**. In practice this means that the overall size and shape of the triangles are unchanged if the mean curvature normal is used for smoothing as discussed in Sect. 8.7.

It is important to note that the mean curvature normal is really the Laplace–Beltrami operator (cf. (3.53)) applied to the vertex coordinates independently. Thus, (8.4) can be seen as a discrete Laplace–Beltrami operator applied to the vertex coordinate **p**.

8.3   Estimating Gaußian Curvature using Angle Defects

While mean curvature is the average of the principal curvatures, Gaußian curvature is their product, but we recall from Chap. 3 that there is also a more intuitive way to define Gaußian curvature.

We can construct a closed curve $\mathbf{c} \in S$ around $\mathbf{p}$ in the surface S. This curve will have a corresponding curve $\mathbf{c}'$ in the image of S on the Gauß map. Let the area enclosed by $\mathbf{c}$ be denoted A_S and the area enclosed by the corresponding curve $\mathbf{c}'$ be denoted A_G. The Gaußian curvature is defined as the limit of the ratio of these two areas

$$K = \lim_{A_S \to 0} \frac{A_G}{A_S}, \tag{8.5}$$

with the further condition that the greatest distance between two points on the curve $\mathbf{c}$ also tends to zero (cf. Sect. 3.7).

Gaußian curvature is clearly zero almost anywhere on a triangle mesh. The neighborhood around a point on a *triangle face* is completely flat and maps to a single point on the sphere, and the neighborhood around a point on an edge can be unfolded into a flat surface. However, for a vertex it is rarely so. If the 1-ring around a vertex is to be "flattened", the angles which the incident triangles make at the vertex must sum to 2π. Otherwise, the vertex has a non-zero Gaußian curvature.

To compute the Gaußian curvature, the limit of the areas in (8.5) can be approximated using finite areas. We begin by finding the area on the unit sphere of a curve around the vertex v. Clearly there is a normal for each incident face, and these normals become the vertices of a spherical polygon[1] whose area is

$$A_G(\mathbf{p}_i) = 2\pi - \sum_j \theta_j, \tag{8.6}$$

where θ_j is the angle of face j at the vertex $\mathbf{p}_i$, and j is an index running over all faces in the 1-ring. Equation (8.6) is sometimes called the angle deficit.

The next step is to find the corresponding area on the triangle mesh. Here it is reasonable to divide the area of each triangle among its three vertices; hence

$$A_S(\mathbf{p}_i) = \frac{1}{3} \sum_j A_j,$$

again, index j runs over all triangles in the 1-ring of $\mathbf{p}_i$, and A_j is the area of triangle j in the 1-ring. This leads to the following expression for the Gaußian curvature of a vertex.

---

[1]A spherical polygon is a polygon on a unit sphere: a polygon whose vertices are points on a unit sphere and whose edges are segments of great circles connecting these vertices.

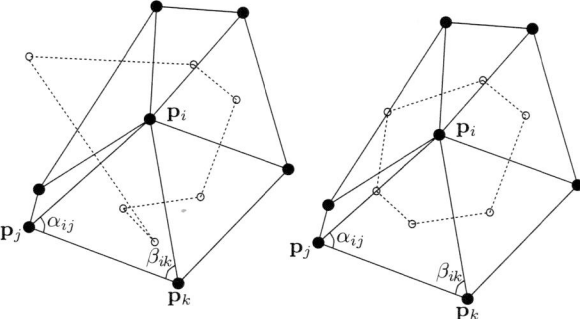

Fig. 8.4 The Voronoi region (*left*) and the region used for computing the mixed area (*right*). The region used for the Voronoi area is not included in the 1-ring in the case of obtuse triangles which is the motivation for the mixed area. For each obtuse triangle, the mixed-area region is obtained by connecting a corner of the region to the middle of the edge opposite the obtuse angle

Definition 8.3 The Gaußian curvature at a vertex of a triangle mesh is

$$K(\mathbf{p}_i) = \frac{A_G(\mathbf{p}_i)}{A_S(\mathbf{p}_i)} = \frac{2\pi - \sum_j \theta_j}{\frac{1}{3}\sum_j A_j}. \tag{8.7}$$

The Gauß–Bonnet theorem (cf. Sect. 3.8) can also be used to justify this formula, see [6] for a very lucid explanation.

Assigning a third of the triangle areas to each triangle is not ideal. We would like to assign to a given vertex the area of the region of the mesh closest to that vertex. In other words, the area of the Voronoi neighborhood [6]. Unfortunately, as observed by Meyer et al. [6], the Voronoi neighborhood is not contained in the 1-ring in the presence of obtuse triangles (see Fig. 8.4). Following [6] the solution is embodied in Algorithm 8.1 for the mixed area where the symbols are shown in Fig. 8.4. This algorithm uses the Voronoi area where possible but restricts the region used for area computation to the 1-ring area. The regions are also non-overlapping and tile the mesh [6].

Algorithm 8.1 Mixed Area of vertex i

---

1. Let $A_{\text{mix}} = 0$
2. For each face f incident on vertex i:
3.   If f is not obtuse then
4. $A_{\text{mix}} = A_{\text{mix}} + \frac{1}{8}(\cot(\alpha_{ij})\|\mathbf{p}_i - \mathbf{p}_j\|^2 + \cot(\beta_{ik})\|\mathbf{p}_i - \mathbf{p}_k\|^2)$
5.   Else If the obtuse angle is at vertex i: $A_{\text{mix}} = A_{\text{mix}} + \frac{1}{2}A(f)$
6.   Else $A_{\text{mix}} = A_{\text{mix}} + \frac{1}{4}A(f)$

Fig. 8.5 The dihedral angle
of an edge is the angle
between the two normals
sharing that edge

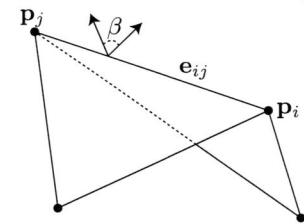

Fig. 8.6 An edge replaced
by a cylindrical blend

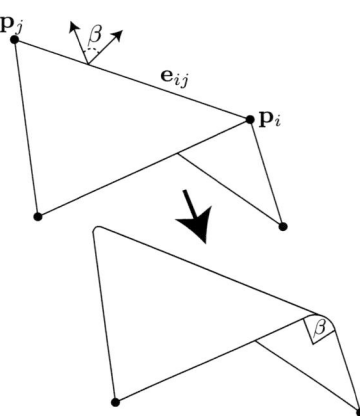

8.4   Curvature Estimation based on Dihedral Angles

Estimation of both the mean curvature and the shape operator (cf., Sect. 3.3) may
be based on the fact that for triangle meshes the surface only bends across edges.
If we imagine that a sharp edge is replaced by a cylindrical blend of radius r, the
curvature is well-defined on this blend. The curvature is zero in the direction of the
edge and $\frac{1}{r}$ in the direction perpendicular to the edge (see Fig. 8.6). At any point on
the cylinder, the mean curvature is $H = \frac{1}{2r}$.

The angle β between the face normals on either side of the edge is denoted the
dihedral angle, see Fig. 8.5. Since the cylindrical blend should meet the faces on
either side smoothly (i.e., the cylinder normal should be the same as the face normal
where they meet) the slice of the cylinder should also have an angle of β—regardless
of the radius. Let $\mathbf{e}$ be a vector in the direction of the edge and having the same
length. The area of the cylindrical blend is $\beta r \|\mathbf{e}\|$, and thus the mean curvature
integrated over the entire cylindrical blend, B, is

$$\int_B H = \frac{1}{2}\beta \|\mathbf{e}\| \, dA,$$

regardless of r. Consequently, the formula also holds in the limit as r tends to zero,
and it is reasonable to define the integral of the mean curvature as follows.

Definition 8.4 The integral of the mean curvature over the edge is

$$H(\mathbf{e}) = \frac{1}{2}\beta\|\mathbf{e}\|. \tag{8.8}$$

The edge may be concave as well as convex. Assign a positive sign to β if the edge is concave and a negative sign if it is convex.

To integrate the mean curvature over the surface we only need to sum H_e over all edges leading to the following formula for the integral absolute mean curvature:

$$\int_S H \, \mathrm{d}A = \sum_{i=1}^{|\mathcal{E}|} H(\mathbf{e}_i) = \frac{1}{2}\sum_{i=1}^{|\mathcal{E}|} \beta_i \|\mathbf{e}_i\|. \tag{8.9}$$

In practice one often needs the *integral absolute mean curvature* which is defined as follows:

$$\int_S |H| \, \mathrm{d}A = \frac{1}{2}\sum_{i=1}^{|\mathcal{E}|} |\beta_i| \|\mathbf{e}_i\|, \tag{8.10}$$

where the sign of β is ignored. For instance, this curvature measure is sometimes minimized when one wishes to optimize a triangulation. See Sect. 8.7.

It is also possible to define a shape operator integral for edges using the notions above. Instead of the ordinary 2×2 matrix, we will define the shape operator as a 3×3 matrix. This matrix should have an eigenvalue of zero and a corresponding eigenvector pointing in the normal direction. Its two non-zero eigenvectors correspond to the min and max curvature, but its eigenvectors are swapped. In other words, the eigenvector corresponding to the maximum eigenvalue is in the direction of minimum curvature.

For an edge, the 3×3 shape operator is defined as follows:

$$\mathbf{S}(\mathbf{e}) = \frac{\beta \mathbf{e}\mathbf{e}^T}{\|\mathbf{e}\|^2}, \tag{8.11}$$

where β is defined in the same way as above (also with regard to sign). Note that $\mathbf{S}(\mathbf{e})\mathbf{e} = \beta\mathbf{e}$, and $\mathbf{S}(\mathbf{e})\mathbf{v} = \mathbf{0}$ for any vector, $\mathbf{v}$, that is orthogonal to $\mathbf{e}$.

To obtain the shape operator at a given vertex (or, generally, point) on the mesh, a good strategy is to select a region, R, around the vertex. One then sums the shape operators for each edge wholly or partially within R multiplied by the length of the intersection of R and the edge segment. Finally, divide by the area of R to obtain the average for the region. We have the following definition.

Definition 8.5 The shape operator for a point on a triangle mesh is

$$\mathbf{S}(\mathbf{p}) = \frac{\sum_{\mathbf{e}_i \cap R \neq \emptyset}[\mathrm{length}(\mathbf{e}_i \cap R)\beta_i \mathbf{e}_i \mathbf{e}_i^T \frac{1}{\|\mathbf{e}_i\|^2}]}{\mathrm{area}(R)}. \tag{8.12}$$

See [7] for a more rigorous derivation of these formulas based on the theory of normal cycles. Also, a very similar shape operator has been defined by Polthier et al. [8]. If we make the simplifying assumption that all edges are split at their midpoint we obtain the somewhat more readable formula

$$\mathbf{S}(\mathbf{p}_i) = \frac{\sum_{j \in \mathcal{N}_i} [\frac{1}{\|\mathbf{e}_{ij}\|} \beta_{ij} \mathbf{e}_{ij} \mathbf{e}_{ij}^T]}{2A_S(\mathbf{p}_i)}. \tag{8.13}$$

8.5 Fitting a Smooth Surface

In this section, we discuss an alternative way of obtaining the shape operator, this time as a 2×2 matrix.

An obvious way of estimating the curvature of a triangle mesh is based on fitting a smooth surface to the mesh. Finding a global surface that fits the mesh is a difficult problem. Instead, one simply approximates the surface in a local neighborhood of a vertex, $\mathbf{p}_i$. Fortunately, any smooth surface can be represented locally as a height function $f(u, v)$ where u and v are coordinates in an estimated tangent plane with normal $\mathbf{n}$. We require that $f(0, 0) = \mathbf{p}_i$ and that the surface normal of f at $\mathbf{p}_i$ is $\mathbf{n}$. A good surface to use is the paraboloid given by

$$f(u, v) = \frac{1}{2}(au^2 + 2buv + cv^2). \tag{8.14}$$

If, in fact, $b = 0$ then a and c *are* the principal curvatures. However, this requires that the two axes spanning the tangent plane are aligned with the principal directions which is not the case in general.

The first step is to find the surface normal. Clearly, the mean curvature normal discussed in Sect. 8.2 could be used.

The next obvious question is: how many points to use? The surface has only three degrees of freedom, so only three points are needed. If too many points are used, the result is an overdetermined system, but this system can be solved in the *least squares sense* [9]. In practice, it is probably a good idea to simply use all the neighbors of a vertex. If the triangle mesh is not degenerate there will be at least three neighbors.

Based on the normal, $\mathbf{n}$, we need to compute a pair of vectors $\mathbf{a}$ and $\mathbf{b}$ spanning the tangent plane. We can find the first one by picking a random vector $\tilde{\mathbf{a}}$ (not parallel to $\mathbf{n}$) and by subtracting its projection onto $\mathbf{n}$:

$$\mathbf{a} = \frac{\tilde{\mathbf{a}} - \mathbf{n}(\tilde{\mathbf{a}} \cdot \mathbf{n})}{\|\tilde{\mathbf{a}} - \mathbf{n}(\tilde{\mathbf{a}} \cdot \mathbf{n})\|},$$

and $\mathbf{b} = \mathbf{n} \times \mathbf{a}$. $\mathbf{a}$, $\mathbf{b}$, and $\mathbf{n}$ define a frame in space. We choose to let this frame attach to the point $\mathbf{p}_i$ for which we wish to fit the surface. By construction, this means that the uv-coordinates of $\mathbf{p}_i$ are at $(0, 0)$. For all other points, we find the uv-coordinates in the following way: let $\mathbf{T} = [\mathbf{ab}]$ be a 3×2 matrix whose columns are $\mathbf{a}$ and $\mathbf{b}$.

Fig. 8.7 The projection of a vertex into the tangent plane associated with **n**

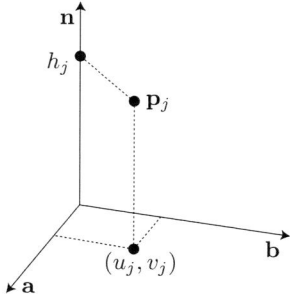

The uv coordinates are

$$(u_j, v_j) = \mathbf{T}^T(\mathbf{p}_j - \mathbf{p}_i),$$

and the height values are

$$h_j = \mathbf{n} \cdot (\mathbf{p}_j - \mathbf{p}_i).$$

See Fig. 8.7. For all points $\mathbf{p}_j$, we can form the equation

$$h_j = f(u_j, v_j),$$

where the coefficients are unknown. These equations can be rewritten in matrix form $\mathbf{UX} = \mathbf{F}$ where $\mathbf{U}$ is the matrix of parameter values, $\mathbf{X}$ is the vector of coefficients for the polynomial surface, and $\mathbf{F}$ is the vector of height values

$$
\begin{bmatrix}
\cdot & \cdot & \cdot \\
\frac{u_j^2}{2} & u_j v_j & \frac{v_j^2}{2} \\
\cdot & \cdot & \cdot
\end{bmatrix}
\begin{bmatrix}
a \\ b \\ c
\end{bmatrix}
=
\begin{bmatrix}
\cdot \\ h_j \\ \cdot
\end{bmatrix}.
\tag{8.15}
$$

We will assume that we have at least three points. In this case, we can solve (8.15) in the least squares sense. The least squares solution is

$$\mathbf{X} = \mathbf{DF} = \left((\mathbf{U}^T\mathbf{U})^{-1}\mathbf{U}^T\right)\mathbf{F}.
\tag{8.16}$$

The final step is to estimate the curvature values form the coefficient vector. The shape operator (as a 2×2 matrix) can be computed directly from the derivatives of f.

Definition 8.6 The Shape operator for a mesh vertex (2×2 matrix) is

$$
\mathbf{S} = -\frac{1}{\sqrt{1 + f_u^2 + f_v^2}}
\begin{bmatrix}
1 + f_u^2 & f_u f_v \\
f_u f_v & 1 + f_v^2
\end{bmatrix}^{-1}
\begin{bmatrix}
f_{uu} & f_{uv} \\
f_{uv} & f_{vv}
\end{bmatrix},
$$

where it is understood that all derivatives are evaluated at $(u, v) = (0, 0)$. Fortunately, $f_u(0, 0) = f_v(0, 0) = 0$ so the first matrix is the identity; we have $f_{uu} = a$, $f_{uv} = b$, and $f_{vv} = c$. Consequently,

$$\mathbf{S} = - \begin{bmatrix} a & b \\ b & c \end{bmatrix}. \tag{8.17}$$

From $\mathbf{S}$, the principal curvatures and directions at the point $\mathbf{p}_i$ can be computed as discussed in the next section. It is also straightforward to transform $\mathbf{S}$ into a 3×3 matrix using the matrix $\mathbf{T}$:

$$\mathbf{S}_{3D} = \mathbf{T} \mathbf{S} \mathbf{T}^T.$$

For a more detailed discussion of the method above, the reader is referred to Hamann [10].

8.6   Estimating Principal Curvatures and Directions

Given a shape operator, $\mathbf{S}$, represented as either a 2×2 or 3×3 matrix, the problem of finding the principal curvatures and the corresponding directions (an example of minimum curvature directions is shown in Fig. 8.8) is reduced to finding the eigenvalues and eigenvectors of $\mathbf{S}$. Note that $\mathbf{S}$ is not necessarily symmetric—this depends on the parametrization. However, the methods which we have discussed above both produce symmetric shape operators. This may be important since simple methods for computing eigensolutions rely on the matrix being symmetric.

If $\mathbf{S}$ is represented as a 3×3 matrix, the eigenvector corresponding to the numerically smallest eigenvalue (it should be zero, but it is possible that it is not exactly zero due to numerical issues) is the normal direction. The two remaining eigensolutions correspond to the principal curvatures and directions. If one of the principal curvatures is zero, it is clearly not possible to distinguish between the principal direction and the normal. However, if we know the normal (which can easily be estimated) the cross product of the normal and the first principal direction (eigenvector) will give the second principal direction. If all eigenvalues are 0, the surface is clearly planar and the principal directions are not defined.

If $\mathbf{S}$ is a 2×2 matrix, the eigensolutions simply correspond to the principal curvatures and directions.

Note, though, that if the dihedral angle method from Sect. 8.4 was used to estimate the shape operator, the directions are flipped so that the eigenvector corresponding to the greatest eigenvalue is the direction of minimal curvature.

Once the principal curvatures have been computed, it is easy to find the Gaußian and mean curvatures using the relations

$$H = \frac{1}{2}(\kappa_{\min} + \kappa_{\max}) = \frac{1}{2} \text{trace}(\mathbf{S}),$$

Fig. 8.8 The min curvature directions of the bunny. The directions were estimated using the dihedral angle method from Sect. 8.4. The shape operators at each vertex were averaged with their neighbors to smoothen the field, and then the eigensolutions were found as discussed in Sect. 8.6

and

$$K = \kappa_{\min}\kappa_{\max} = \det(\mathbf{S}).$$

8.7 Discussion

For an interesting comparison of curvature estimation methods see [11]. Surazhky et al. construct meshes from known smooth surfaces and compare the estimated curvature to the analytic.

Clearly, not all techniques for estimating curvature have been discussed. Probably a good starting point for a more in-depth look at the literature is [6] where Meyer et al. propose a systematic framework for estimating curvatures from triangle meshes. Two methods (which were not discussed here) for estimating the shape operator were proposed by Taubin [12] and by Hildebrandt and Polthier [8]. The latter is similar but not identical to the method discussed in Sect. 8.4. Note also that this is an active area of research, and new methods for computing differential geometric properties of meshes still appear.

The most obvious application of curvature measures on triangle meshes is simply the enhancement of visualization. Small curvature variations may not be obvious from a shaded rendition of 3D model but are easy to detect if the curvature is mapped onto the surface.

A related area is non-photorealistic rendering which is concerned with the rendering of 3D models in an artistic fashion. One artistic style that has been simulated is "hatching" where the artist draws strokes in order to enhance shape and simulate shading. Very often these strokes are aligned with the directions of min or max curvature as discussed in [13].

In some cases, we wish to improve a triangulation while maintaining the vertices unchanged. The most popular way of doing this is by edge flipping. The edge shared by two adjacent triangles can be replaced by a transverse edge. One approach is to loop over all edges and flip if it reduces some energy. This means that some energy function is required, and an effective choice is the integral absolute mean curvature (8.10). See Sect. 11.2 for more details on this method.

There is a large body of literature on this topic of smoothing and removing noise from triangle meshes. A popular method is to move vertices in the opposite direction of the mean curvature normal, i.e. in the direction of $-\mathbf{H}$. This will minimize the surface area leading to a smoother surface. In fact, (8.4) was (re)introduced in the context of surface smoothing by Desbrun et al. [2]. It is worth noting, though, that volume is also reduced which is often a bit of a problem. For more on smoothing see Chap. 9.

In [14] Pierre Alliez et al. proposed a technique for "remeshing" a polygonal mesh to produce a new (quad-dominant) mesh whose edges would be aligned with the min and max curvature directions. This method was based on the shape operator discussed in Sect. 8.4. Since this paper a number of other solutions to the same problem have appeared, e.g. [15].

8.8 Exercises

Exercise 8.1 Write a program which computes the Gaußian of all vertices. Visualize the values as a scalar field defined on the surface.

[**GEL Users**] GEL provides a tool for visualization of scalar fields on meshes.

Exercise 8.2 Write a program which computes the mean curvature normal at all vertices. Visualize the values as a scalar field defined on the surface. Since the mean curvature normal is a vector, it must be converted to a scalar. Take the dot product of the angle weighted normal and the mean curvature normal at each vertex. The mean curvature is the sign of this dot product times the length of the mean curvature normal.

Fig. 8.9 A triangle with a vertex $\mathbf{p}_i$

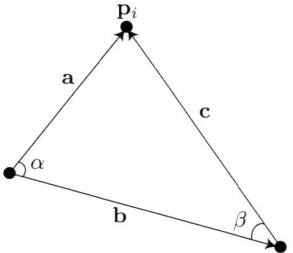

Exercise 8.3 Compute the integral of the Gaußian curvature over a mesh of known Euler characteristic, χ. Verify that the integral corresponds almost exactly to $2\pi\chi$.

Exercise 8.4 Compute the shape operator for each vertex of a mesh. Obtain the principal curvature directions and visualize those.

[**GEL Users**] GEL provides a line field visualization tool.

Appendix

In this appendix, we derive the Cotan formula for the gradient of the area of a triangle. Given a triangle (shown in Fig. 8.9) whose vertex $\mathbf{p}_i$ is movable, compute the gradient of the area of the triangle as a function of $\mathbf{p}_i$.

It is clear that the gradient is perpendicular to the plane of the triangle since moving $\mathbf{p}_i$ either in the positive or negative direction along the triangle normal will increase the area. Hence, the present position is a minimum. Moving $\mathbf{p}_i$ parallel to the base line $\mathbf{b}$ will not change the area. It follows that the gradient is in the plane of the triangle and orthogonal to $\mathbf{b}$. It is trivial to find the length of the gradient, and this leads to the first line of the equation below. After a number of steps, we reach the expression used in (8.4).

Some of the steps may be a little tricky. The bottom line is that we need to use the fact that the cotangent of an angle between two vectors $\mathbf{a}$ and $\mathbf{b}$ is equal to $\frac{\mathbf{a}^T\mathbf{b}}{\|\mathbf{a}\times\mathbf{b}\|}$:

$$
\begin{aligned}
\nabla A(\mathbf{p}_i) &= \frac{(\mathbf{b}\times\mathbf{a})\times\mathbf{b}}{2\|\mathbf{b}\times\mathbf{a}\|} \\
&= \frac{(\mathbf{b}^t\mathbf{b})\mathbf{a} - (\mathbf{b}^t\mathbf{a})\mathbf{b}}{2\|\mathbf{b}\times\mathbf{a}\|} \\
&= \frac{(\mathbf{b}^t\mathbf{b})\mathbf{a} - (\mathbf{b}^t\mathbf{a})\mathbf{a} + (\mathbf{b}^t\mathbf{a})\mathbf{a} - (\mathbf{b}^t\mathbf{a})\mathbf{b}}{2\|\mathbf{b}\times\mathbf{a}\|} \\
&= \frac{-(\mathbf{c}^t\mathbf{b})\mathbf{a}}{2\|\mathbf{c}\times-\mathbf{b}\|} + \frac{(\mathbf{b}^t\mathbf{a})\mathbf{c}}{2\|\mathbf{b}\times\mathbf{a}\|}
\end{aligned}
$$

$$= \frac{(\mathbf{c}^t - \mathbf{b})\mathbf{a}}{2\|\mathbf{c} \times -\mathbf{b}\|} + \frac{(\mathbf{b}^t \mathbf{a})\mathbf{c}}{2\|\mathbf{b} \times \mathbf{a}\|}$$

$$= \frac{1}{2}(\mathbf{a}\cot\beta + \mathbf{c}\cot\alpha).$$

References

1. Pinkall, U., Polthier, K.: Computing discrete minimal surfaces and their conjugates. Exp. Math. **2**(1), 15–36 (1993)
2. Desbrun, M., Meyer, M., Schröder, P., Barr, A.H.: Implicit fairing of irregular meshes using diffusion and curvature flow. In: Proceedings of the 26th Annual Conference on Computer Graphics and Interactive Techniques, SIGGRAPH '99, pp. 317–324. ACM Press, New York (1999). doi:10.1145/311535.311576
3. Thürmer, G., Wüthrich, C.A.: Computing vertex normals from polygonal facets. J. Graph. Tools **3**(1), 43–46 (1998)
4. Bærentzen, J., Aanæs, H.: Signed distance computation using the angle weighted pseudo-normal. IEEE Trans. Vis. Comput. Graph. **11**(3), 243–253 (2005)
5. do Carmo, M.P.: Differential Geometry of Curves and Surfaces. Prentice-Hall, Englewood Cliffs (1976)
6. Meyer, M., Desbrun, M., Schröder, P., Barr, A.H.: Discrete differential-geometry operators for triangulated 2-manifolds. In: Hege, H.-C., Polthier, K. (eds.) Visualization and Mathematics III, pp. 35–57. Springer, Heidelberg (2003)
7. Cohen-Steiner, D., Morvan, J.-M.: Restricted Delaunay triangulations and normal cycle. In: Proceedings of the Nineteenth Annual Symposium on Computational Geometry, SCG '03, pp. 312–321. ACM Press, New York (2003). doi:10.1145/777792.777839
8. Hildebrandt, K., Polthier, K.: Anisotropic filtering of non-linear surface features. Comput. Graph. Forum **23**(3), 391–400 (2004)
9. Golub, G.H., van Loan, C.F.: Matrix Computations, 3rd edn. John Hopkins, Baltimore (1996)
10. Hamann, B.: Curvature approximation for triangulated surfaces. In: Geometric Modelling, pp. 139–153. Springer, London (1993)
11. Surazhsky, T., Magid, E., Soldea, O., Elber, G., Rivlin, E.: A comparison of Gaussian and mean curvatures triangular meshes. In: Proceedings of IEEE International Automation (ICRA2003), Taipei, Taiwan, 14–19 September, pp. 1021–1026 (2003)
12. Taubin, G.: Estimating the tensor of curvature of a surface from a polyhedral approximation. In: Proceedings of the Fifth International Conference on Computer Vision, ICCV '95, p. 902. IEEE Comput. Soc., Washington (1995)
13. Hertzmann, A., Zorin, D.: Illustrating smooth surfaces. In: Proceedings of the 27th Annual Conference on Computer Graphics and Interactive Techniques, SIGGRAPH '00, pp. 517–526. ACM, New York (2000). doi:10.1145/344779.345074
14. Alliez, P., Cohen-Steiner, D., Devillers, O., Lévy, B., Desbrun, M.: Anisotropic polygonal remeshing. ACM Trans. Graph. **22**(3), 485–493 (2003). doi:10.1145/882262.882296
15. Kälberer, F., Nieser, M., Polthier, K.: QuadCover-Surface Parameterization using Branched Coverings. Comput. Graph. Forum **26**(3), 375–384 (2007)

Mesh Smoothing and Variational Subdivision

A big part of the motivation for this book is the need to deal with *acquired* geometry: for instance, triangle meshes produced from optically scanned point clouds. Scanning is measurement, and any measurement is subject to noise. Apart from detracting from the visual quality of the model, noise can also be a problem for other geometric algorithms. Therefore, removing noise from the "signal" in a triangle mesh is an important concern, and, in fact, the analogy is excellent since we can construe the vertex positions of a triangle mesh as a discrete signal on an irregular grid. The signal analysis viewpoint is explored in Sect. 9.1 where we revisit some of the basic notions which will be used later.

In Sect. 9.2, we introduce the simplest type of mesh smoothing, Laplacian smoothing, where a vertex is replaced by the average of its neighbors. Laplacian smoothing is effective but results in a great deal of shrinkage. Taubin smoothing, which is described next, does a better job of preserving the coarse features, but both Taubin and Laplacian smoothing also have a tangential component [1], causing the shape of the triangles to be smoothed as well as the geometry of the shape they represent.

In Sect. 9.3 we replace the Laplacian with the *Laplace–Beltrami* operator (cf., Sect. 3.9) and arrive at mean curvature flow, which does a much better job of preserving the shape of the triangles while smoothing the geometry represented by the mesh. The Laplace–Beltrami operator for a triangle mesh can be written as a matrix as we shall discuss in Sect. 9.4, and we can use the eigenvectors of this matrix to perform a spectral analysis of the shape completely analogous to the discrete Fourier transform. This can be used for smoothing although for large meshes it is quite challenging to implement efficiently.

In Sect. 9.5 we discuss some smoothing methods that preserve sharp edges and corners. Such features are treated as noise by other smoothing algorithms.

Finally, smoothing combined with an up-sampling of the triangle mesh leads to a simple technique for generating very smooth surfaces. This technique is known as variational subdivision, which is the topic of Sect. 9.6.

J.A. Bærentzen et al., *Guide to Computational Geometry Processing*,
DOI 10.1007/978-1-4471-4075-7_9, © Springer-Verlag London 2012

Fig. 9.1 Simple smoothing of a 1D signal: the signal on top contains high frequencies which have been removed by convolving the signal with a function that serves as a low-pass filter. In the discrete setting this is simply equivalent to replacing a value by the weighted average of its neighbors

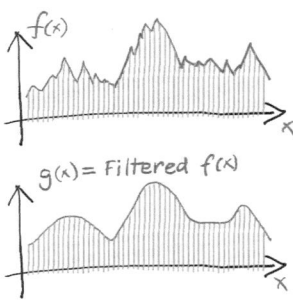

9.1   Signal Processing

The immediate problem with smoothing a triangle mesh is that it is not quite clear what smoothing means. We can approach the problem by viewing a triangle mesh as a signal. In many ways, a mesh can be seen as a 2D signal on an irregular grid, just like an image is a 2D signal on a regular grid of pixels. On a regular grid, smoothing is performed simply by averaging samples. A simple example of a 1D signal and its smoothed counterpart are shown in Fig. 9.1.

For nD signals (smooth or discrete on a regular grid) smoothing can be understood as low-pass filtering. Recall that both continuous and discrete signals can be transformed into the frequency domain via the Fourier transform and the discrete Fourier transform, respectively. In the following, we will consider just 1D discrete signal, f, consisting of N samples. We can express f as the sum

$$f(k) = \sum_{n=0}^{N-1} F(n) \exp(2\pi i k n / N),$$

where $F(n)$ is a coefficient which controls how much of the pure wave $\exp(2\pi i k n / N)$ at frequency n that contributes to f.

In the frequency domain, a low-pass filter is simply a function which attenuates the coefficients, producing new coefficients $F_{\text{low}}(n) = L(n)F(n)$. The ideal low-pass filter removes high frequencies by simply having $L(n) = 0$ for frequencies above some threshold and $L(n) = 1$ elsewhere.

This removes high frequency detail in the signal, and since noise is frequently more "high frequency" than the real signal content, this is often desirable. Of course, there is no guarantee that the high frequencies contain only noise. In fact, it is a fundamental problem that smoothing tends to remove not only noise but also parts of the signal.

It is not only possible to perform low-pass filtering in the frequency domain. If $L(n)$ and $F(n)$ are, respectively, the Fourier transform of the filter and of a function, the low-pass filtering is simply $F_{\text{low}}(n) = F(n)L(n)$. In the time domain, the same operation can be expressed in terms of convolution, $f_{\text{low}}(k) = f(k) * l(k) = \sum_m f(m)l(k - m)$, where $*$ denotes convolution. Assuming l is symmetric and of *compact support*, we can write the filtering

$$f_{\text{low}}(k) = \sum_{m=-w}^{w} f(m)l(m-k), \tag{9.1}$$

which is simply a weighted sum of the values of f in the vicinity of k where the support w tells us over how wide a neighborhood this average is taken. l is just the weighting function. Unfortunately, the ideal low-pass filter is not of compact support, but often we would smooth a signal simply using a symmetric filter with a small support. In other words, we would locally average the signal values. As we shall see, this carries over into the mesh domain.

9.2   Laplacian and Taubin Smoothing

In analogy to (9.1), we can smooth the vertex positions, $\mathbf{p}_i$, of our mesh using

$$\mathbf{p}_i \leftarrow \frac{1}{|\mathcal{N}_i|} \sum_{\mathbf{p}_j \in \mathcal{N}_i} \mathbf{p}_j,$$

where $\mathcal{N}_i$ is the set of neighbors of i. In other words, a vertex is replaced by the average of its neighbors. This is known as Laplacian smoothing [2]. A weight λ can be used to control the degree of smoothing:

$$\mathbf{p}_i \leftarrow (1-\lambda)\mathbf{p}_i + \frac{\lambda}{|\mathcal{N}_i|} \sum_{\mathbf{p}_j \in \mathcal{N}_i} \mathbf{p}_j.$$

These formulas can be written in terms of a *mesh Laplacian*. The Laplacian of a 2D function is $\Delta\mathbf{f} = \mathbf{f}_{uu} + \mathbf{f}_{vv}$. It is not obvious what that means on a mesh. The *umbrella operator* [2] is based on the observation that we can least squares fit a smooth second order surface $\mathbf{f}: \mathbb{R}^2 \to \mathbb{R}^3$ to the vertices in the 1-ring neighborhood of a given vertex. With a particular choice of parameter domain,

$$\mathcal{L}(\mathbf{p}_i) = \frac{1}{|\mathcal{N}_i|} \sum_{\mathbf{p}_j \in \mathcal{N}_i} (\mathbf{p}_j - \mathbf{p}_i) \tag{9.2}$$

is the value of $\Delta\mathbf{f}$ at $\mathbf{p}_i$. For a complete derivation, see the Appendix. In terms of $\mathcal{L}$, the weighted averaging can now be rewritten

$$\mathbf{p}_i \leftarrow \mathbf{p}_i + \lambda\mathcal{L}(\mathbf{p}_i). \tag{9.3}$$

For a practical implementation, we should avoid overwriting the positions of some vertices before we have computed the Laplacian at other vertices. Thus, a good strategy is to first compute the Laplacian, using (9.2), at all vertices and store these in an array. Subsequently, the positions are updated at all vertices using (9.3). The result of Laplacian smoothing implemented in this fashion is shown in Fig. 9.2(a).

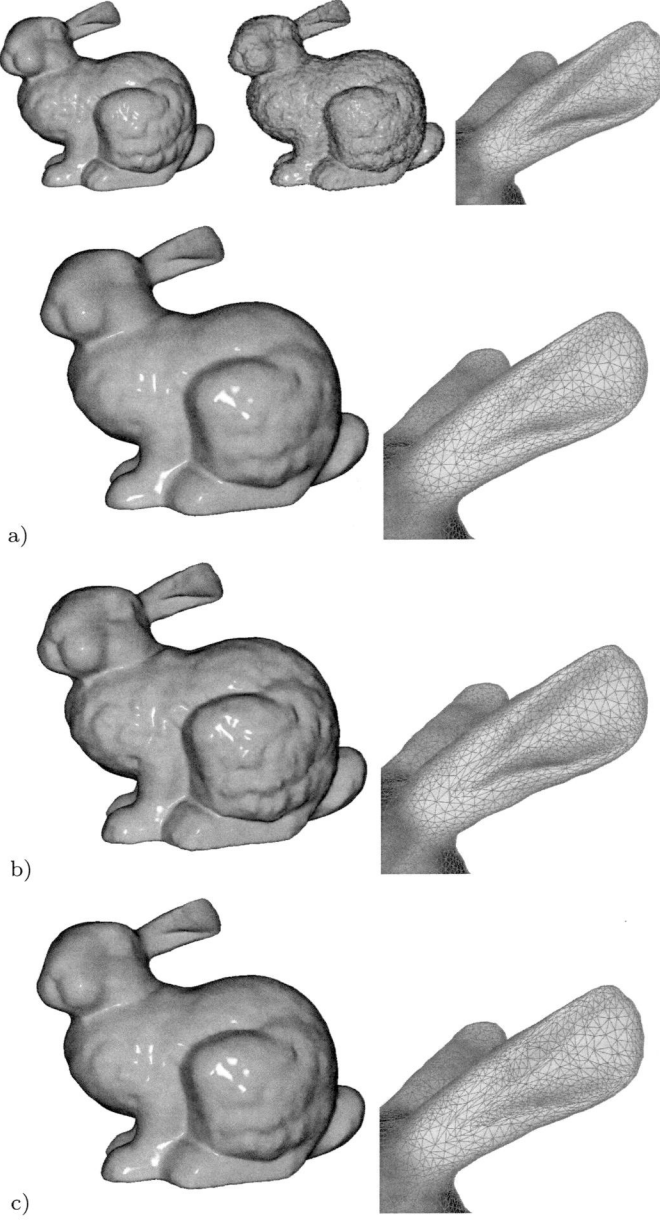

Fig. 9.2 *Top*: the reconstructed bunny used for these experiments consisting of 35341 vertices, a version where noise has been added to the vertices (normal direction displacement) and a closeup of the ear. (**a**) The bunny after smoothing with 10 iterations of Laplacian smoothing, weight $\lambda = 0.5$. (**b**) Smoothing with 50 iterations of Taubin smoothing. (**c**) Mean curvature smoothing with 10 iterations, weight $\lambda = 0.5$

Fig. 9.3 The result of over smoothing. A comparison of 4000 iterations of Laplacian smoothing (*left*) and Taubin smoothing (*right*)

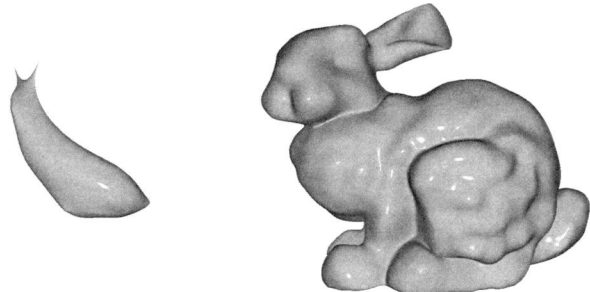

Returning to the low-pass filter outlook, Taubin observed that it is possible to define notions of frequency and vibration modes for a triangle mesh, and these concepts are strongly linked to the Laplacian. Laplacian smoothing (9.3) has the effect that it attenuates all frequencies (except 0). See [2] for details. This indiscriminate attenuation leads to severe shrinkage (as shown in Fig. 9.3), and Taubin designed a filtering process with less shrinkage. The result is the $\lambda|\mu$ algorithm,[1] which is identical to Laplacian smoothing except that for every other iteration

$$\mathbf{p}_i \leftarrow \mathbf{p}_i + \lambda \mathcal{L}(\mathbf{p}_i),$$

and for every remaining iteration

$$\mathbf{p}_i \leftarrow \mathbf{p}_i - \mu \mathcal{L}(\mathbf{p}_i).$$

If the constants λ and μ are chosen according to certain guidelines, this procedure will attenuate large frequencies and actually enhance low frequencies slightly. In the experiments in this chapter, we chose $\lambda = 0.5$ and $\mu = -0.52$.

The $\lambda|\mu$ algorithm is effective in that it does not shrink nearly as much as Laplacian smoothing while it does remove high frequency noise. The theoretical explanation for this is that it is a better approximation of a low-pass filter: unlike in Laplacian smoothing, the low frequencies are not attenuated. In fact they are boosted slightly. A more intuitive explanation is that it alternately shrinks and expands the model. Thus, the total shrinkage is far less than in plain Laplacian smoothing. Row (b) of Fig. 9.2 shows the effect of Taubin smoothing and Fig. 9.3 shows the difference between 4000 iterations of Laplacian smoothing versus 4000 iterations of Taubin smoothing. While the former shrinks the mesh to a no longer recognizable version, the latter turns it into a bit of a caricature by enhancing some fairly coarse features (due to the aforementioned boosting of low frequencies).

---

[1]Pronounced "Lambda-Mu", although generally referred to as Taubin smoothing.

9.3   Mean Curvature Flow

While $\lambda|\mu$ solved the shrinkage problem to a large degree it still did not solve all the problems. The $\lambda|\mu$ algorithm is an excellent choice for removing a little noise, but if too many iterations of $\lambda|\mu$ are used the result can be surprising since some frequencies are actually enhanced as discussed above.

Moreover, the particular choice of the Laplacian,

$$\mathcal{L}(\mathbf{p}_i) = \frac{1}{|\mathcal{N}_i|} \sum_{\mathbf{p}_j \in \mathcal{N}_i} (\mathbf{p}_j - \mathbf{p}_i),$$

does not take geometry into account. This has the practical implication that if the triangle mesh has edges of very irregular length both $\lambda|\mu$ and plain Laplacian smoothing will tend to distort the mesh to equalize the edge lengths. While this may be desired it can also lead to an unwanted distortion.

In fact, the umbrella operator is a poor choice as observed by Desbrun et al. [1] because it does not take the shape metric into account. What we should use instead is the Laplace–Beltrami operator, which is defined as the divergence of the gradient on a surface. For more details, see Sect. 3.9. As noted in Sect. 8.2, the Laplace–Beltrami operator applied to the vertex coordinates yields the mean curvature normal. This led to the Cotan formula [3] (8.4) which has the desirable property that it is zero if the 1-ring is flat. This property is not shared by the umbrella operator which will move a vertex to the centroid of its neighbors regardless of local geometry. This indicates that there is less distortion if we simply plug the mean curvature normal into (9.3). Put differently, Desbrun's use of the mean curvature normal rather than the umbrella operator is useful in order to smooth the shape rather than both the shape and the mesh. On the other hand, the mesh still shrinks with this approach. This problem was solved simply by rescaling the mesh to make up for lost volume [1].

Another contribution of [1] was that the authors introduced an implicit smoothing method which allowed for bigger time step albeit the algorithm requires the solving of a linear system. For a simple explicit solver, we should use the following operator:

$$\mathcal{L}^{\mathrm{LBO}}(\mathbf{p}_i) = \frac{1}{\sum_{\mathbf{p}_j \in \mathcal{N}_i} \cot \alpha_{ij} + \cot \beta_{ij}}$$
$$\times \sum_{\mathbf{p}_j \in \mathcal{N}_i} (\cot \alpha_{ij} + \cot \beta_{ij})(\mathbf{p}_j - \mathbf{p}_i), \qquad (9.4)$$

in lieu of $\mathcal{L}$ in (9.3). With reference to Fig. 9.4 the angles are computed as follows:

$$\alpha_{ij} = \mathrm{acos}\left(\frac{(\mathbf{p}_i - \mathbf{p}_k) \cdot (\mathbf{p}_j - \mathbf{p}_k)}{\|\mathbf{p}_i - \mathbf{p}_k\|\|\mathbf{p}_j - \mathbf{p}_k\|}\right),$$
$$\beta_{ij} = \mathrm{acos}\left(\frac{(\mathbf{p}_i - \mathbf{p}_l) \cdot (\mathbf{p}_j - \mathbf{p}_l)}{\|\mathbf{p}_i - \mathbf{p}_l\|\|\mathbf{p}_j - \mathbf{p}_l\|}\right). \qquad (9.5)$$

Fig. 9.4 The vertices and angles needed for computing the weights used in mean curvature flow

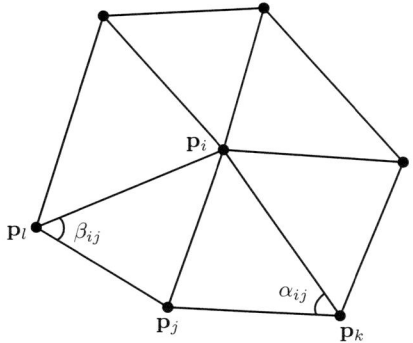

Some care should be taken to avoid numerical problems. On most computer systems known to us, acos will return an invalid (NaN) floating point value for input outside the range $[-1, 1]$, and due to numerical imprecision, the argument could be outside the range. Thus, it makes sense to clamp the argument to $[-1, 1]$ before applying acos. Likewise, if the mesh is degenerate and, say, $\mathbf{p}_i$ and $\mathbf{p}_k$ are at the same position, the normalization above results in a division by zero. Typically, we fix this by normalizing a vector only if its length is non-zero. Finally, if α_{ij} or β_{ij} are zero, the result will also be an invalid floating point value. Our solution here is to compute $\cot(\alpha) = \frac{1}{\tan(\epsilon + \alpha)}$ where ϵ is a small number around the machine precision.

With these precautions, mean curvature flow is an effective technique for smoothing which is recommended if it is important that the geometry and not the structure of the triangle mesh is smoothed. Figure 9.2(c) illustrates mean curvature flow.

9.4 Spectral Smoothing

As discussed, smoothing attenuates the high frequency details in the mesh, but it is difficult to filter out some frequencies while leaving other frequencies unscathed using methods such as Laplacian or Taubin smoothing. Thus, methods which allow us to deal more precisely with the frequency content of meshes are of great interest.

As observed by Taubin, the basis functions of the Fourier transform are the eigenfunctions of the Laplace operator [2]. On a finite discrete grid, we can represent the Laplace operator as a matrix, and the eigenvectors of this matrix form a basis in which we can represent functions on the grid.

This generalizes to triangle meshes where we would use the eigenvectors of the Laplace–Beltrami operator as our basis. Thus, we need to express the Laplace–Beltrami operator, $\mathcal{L}$, as a matrix $\mathbf{L}$ whose entries are defined as follows [4]:

$$\mathbf{L}_{ij} = \frac{1}{\sqrt{A_i A_j}}(\cot \alpha_{ij} + \cot \beta_{ij}) \quad \text{if } i, j \text{ connected},$$

$$\mathbf{L}_{ii} = -\sum_{j \in \mathcal{N}_i} \frac{1}{\sqrt{A_i A_j}},$$

(9.6)

where A_i is the area belonging to vertex i and the angles α and β are defined as shown in Fig. 8.3. Ideally, the area used should be the Voronoi area, i.e., the area of the mesh closest to vertex i. Unfortunately, there is no guarantee that this area is contained in the set of faces incident on vertex i and then the formula for computing the Voronoi area does not yield the correct result. In practice, we, therefore, suggest using the mixed area proposed by [5] where the Voronoi area is used except at obtuse triangles. The normalization in (9.6) is a little different from the definition in (8.4). That is because for a spectral decomposition we need the eigenvectors to be orthogonal, which requires $\mathbf{L}$ to be symmetric, i.e., $\mathbf{L}_{ij} = \mathbf{L}_{ji}$. Therefore, instead of dividing all entries in row i by the area associated with vertex i, we divide each entry, $\mathbf{L}_{ij}$ by the square root of the product of the areas associated with vertices i and j.

For a mesh of N vertices, $\mathbf{L}$ has dimensions $N \times N$, and we can compute the Laplacian of a function whose values are defined per vertex by constructing a vector of these values, $\mathbf{v}$, and multiplying it on to the matrix, i.e., $\mathbf{Lv}$.

To perform spectral smoothing, we compute the N eigenvectors, $\mathbf{e}_i$, of the mesh. Given all the $\mathbf{X}$, $\mathbf{Y}$, and $\mathbf{Z}$ coordinates of the vertices in three separate vectors, we can now do a spectral decomposition of the mesh by computing the projection of the coordinate vectors onto each eigenvector as follows:

$$
\begin{aligned}
k_i^x &= \mathbf{e}_i \cdot \mathbf{X}, \\
k_i^y &= \mathbf{e}_i \cdot \mathbf{Y}, \\
k_i^z &= \mathbf{e}_i \cdot \mathbf{Z}.
\end{aligned}
\tag{9.7}
$$

To get back our coordinate vectors, we sum up the contributions

$$
\begin{aligned}
\mathbf{X} &= \sum_{i=1}^{N} k_i^x \mathbf{e}_i, \\
\mathbf{Y} &= \sum_{i=1}^{N} k_i^y \mathbf{e}_i, \\
\mathbf{Z} &= \sum_{i=1}^{N} k_i^z \mathbf{e}_i,
\end{aligned}
\tag{9.8}
$$

and if choose N smaller than the total number of vertices we get a smoothed result. Note that the eigenvectors should be sorted in ascending order of eigenvalue since the eigenvalue of an eigenvector corresponds to its frequency.

Figure 9.5 illustrates what happens if we reconstruct using 2, 4, 50, and 500 eigenvectors. The mesh shown has 3581 vertices.

This is an effective method for smoothing, but it is perhaps not the most practical method. Using a standard numerics library, running on a modest computer, it can easily take minutes to compute the eigenvectors of the Laplace–Beltrami operator for even a small mesh. Better results can be obtained if we use a sparse matrix library and restrict ourselves to computing a subset of the eigenvectors. For details

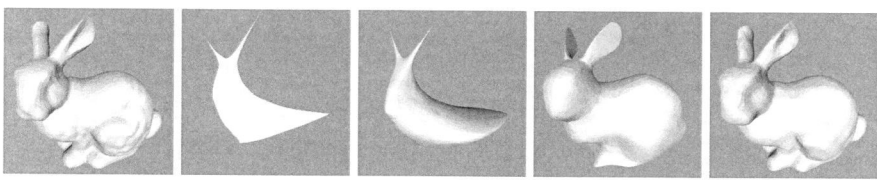

Fig. 9.5   The vertex positions of the original (*simplified*) bunny mesh (*far left*) were projected onto the eigenvectors of the Laplace–Beltrami operator. From this representation, we reconstructed the mesh using 2, 4, 50, and 500 eigenvectors as shown from *left* to *right*

on such numerical issues and other practical as well as theoretical aspects of spectral mesh processing, the interested reader is referred to the course notes by Levy and Zhang [6].

9.5   Feature-Preserving Smoothing

All of the methods discussed so far indiscriminately attenuate high frequency content whether it corresponds to noise or to actual fine details such as corners and edges.

An effective method for feature-preserving smoothing, known as FVM (*Fuzzy Vector Median*) filtering, is due to Shen and Barner [7]. Their approach is to perform first a filtering of the face normals and to then fit the mesh to this new set of normals. Naively taking the average of a face normal and the normals of adjacent faces would not preserve features. Instead, the authors propose to use a median filter. Initially, for each face, f, we form a set of faces which are incident on f. The median normal is the face normal with the smallest angle to the other normals in that set. We obtain the smoothed normal of face f by computing the weighted average of all the normals in the incident face set, where the weight is computed using a Gaußian function of the angular distance to the median normal.

In the second step, the vertices are moved so as to minimize the difference between the filtered normal and the actual face normal [7]. One iteration of the update algorithm consists of applying

$$\mathbf{p}_i \leftarrow \mathbf{p}_i + k \sum_{j \in \mathcal{N}_i} \sum_{f \in \mathcal{F}_{ij}} \mathbf{n}_f \big(\mathbf{n}_f^T (\mathbf{p}_i - \mathbf{p}_i) \big), \qquad (9.9)$$

where $\mathcal{F}_{ij}$ is the set of (one or two) faces adjacent to edge ij, and k is a small constant. We suggest to use around 20 iterations and $k = 0.05$.

The method is fairly simple to implement, and the results are convincing. Figure 9.6 shows how the fuzzy median filter can be used to reconstruct the geometry of a cylinder with uneven sized faces from a slightly noisy version. Clearly, in this case, Taubin smoothing only makes matters worse. The method is not very efficient, however, and, recently, Zheng et al. proposed a somewhat similar algorithm based on bilateral filtering of normals [8]. This method appears to be more efficient since

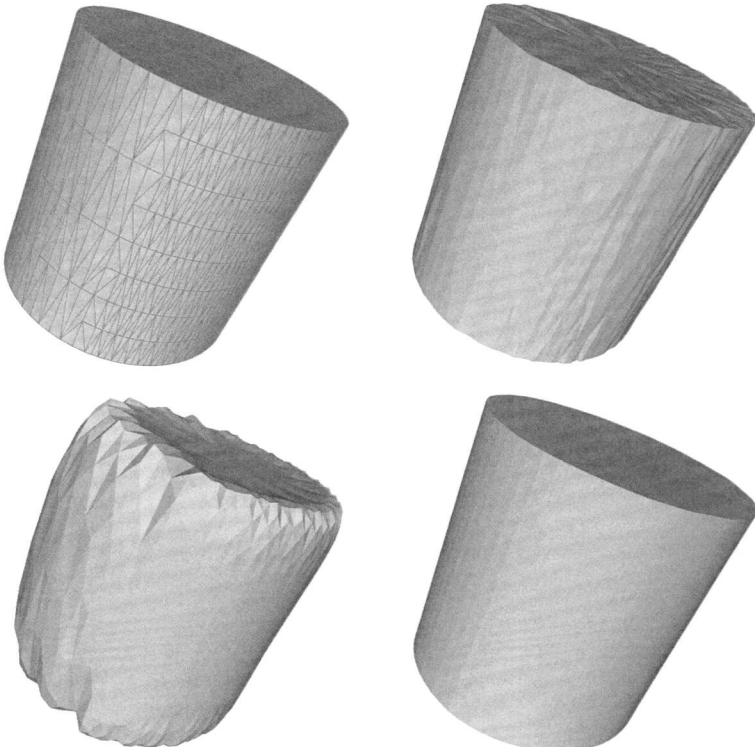

Fig. 9.6 A cylinder triangulated with triangles of uneven size (*top left*) has had some noise added (*top right*). Below, a comparison of Taubin smoothing (*bottom left*) and FVM smoothing [7] (*bottom right*)

the bilateral filter only compares the normal of f to its adjacent faces where the median filter also compares the normals of adjacent faces to each other.

9.6   Variational Subdivision

In the preceding sections, we have thought of smoothing mostly as a form of low-pass filtering, but we can also see it as a diffusion process or as energy minimization which is the outlook we will adopt in the rest of this chapter: the surface is changed to reduce some energy, and if we combine this type of energy minimizing smoothing with refinement of the mesh, we obtain what has been called *variational subdivision* [9].

Using variational subdivision, we iteratively place new vertices on the midpoints of edges to introduce more detail and then move these new vertices to minimize an energy. Since we want the mesh to interpolate the original points, these are never moved—only inserted points are moved. This means that the method is relatively indifferent to the size of the input, but is very sensitive to the size of the output since the amount of work done is proportional to the number of output vertices in the final triangle mesh.

9.6.1 Energy Functionals

Briefly put, calculus of variations allows us to find optimal functions by minimizing the values of *functionals* [10]. A functional, E, is a mapping from a space of functions to a real number which we can think of as an energy. The optimal function according to this energy measure is the one that minimizes E. We will consider two functionals: the *membrane energy* and the *bending energy*. The former is a function of the values of the first derivatives, i.e., it penalizes stretch, whereas the latter is a function of the second derivatives, hence penalizing bending.

The membrane energy functional is

$$E_M[f] = \frac{1}{2} \int f_u^2 + f_v^2. \tag{9.10}$$

In order to find the function f which minimizes this energy, we must find the so called variational derivative, $\frac{\delta E_M}{\delta f}$ and then the point where the variational derivative is 0. This is completely analogous to normal analysis where the first derivative is 0 at an extremal point. Again, it is also just a necessary and not a sufficient condition for finding the global minimum that the variational derivative is 0. However, we shall assume that we do find the global minimum.

For a generic functional, J, the equation $\frac{\delta J}{\delta f} = 0$ is called the Euler–Lagrange equation. Let J have the form

$$J[f] = \int F(u, v, f, f_u, f_v) \, du \, dv,$$

where F is a function which depends on u, v, f and its partial derivatives with respect to u and v. The Euler–Lagrange equation is then as follows:

$$F_f - \frac{\partial}{\partial u} F_{f_u} - \frac{\partial}{\partial v} F_{f_v} = 0,$$

where, e.g., F_{f_u} is the partial derivative of F w.r.t. f_u [10].

In our particular case, the functional E_M does not directly depend on f, hence we obtain

$$0 = -\frac{\partial}{\partial u} F_{f_u} - \frac{\partial}{\partial v} F_{f_v}$$

$$= -\frac{\partial}{\partial u} f_u - \frac{\partial}{\partial v} f_v$$

$$= -f_{uu} - f_{vv}.$$

In other words, if the membrane energy is minimal, we have the Laplace equation

$$\Delta f = f_{uu} + f_{vv} = 0. \tag{9.11}$$

The thin plate or bending energy is

$$E_B[f] = \frac{1}{2} \int f_{uu}^2 + 2f_{uv}^2 + f_{vv}^2. \tag{9.12}$$

Again, the minimum is found by finding the solution to the Euler–Lagrange equation which turns out to be the biharmonic equation

$$0 = f_{uuuu} + 2f_{uuvv} + f_{vvvv}, \tag{9.13}$$

and note that the right hand side is simply the squared Laplace operator, i.e.,

$$f_{uuuu} + 2f_{uuvv} + f_{vvvv} = \Delta^2 f.$$

To summarize, we have two simple schemes. If we wish to minimize the membrane energy, we need to find the solution to

$$\Delta f = 0,$$

whereas the bending energy is minimized by

$$\Delta^2 f = 0.$$

The strategy is to find a discrete version of the two expressions. Here the umbrella operator which is discussed above and derived in the appendix is the most obvious choice of Laplace operator. As noted in the sections on smoothing, the umbrella operator will smooth not only the geometry but also the triangulation. However, for variational subdivision that is likely to be an advantage. To solve for the bending energy, the Laplacian can be applied twice to obtain the square Laplacian.

9.6.2   Minimizing the Energies

At this point, we have a Laplacian operator for a mesh. This operator can be expressed as a matrix, $\mathbf{L}$ (applying the Laplacian to all vertices at once).

$$\mathbf{LP} = -K\mathbf{P}, \tag{9.14}$$

where $\mathbf{P}$ is the vector of vertex positions, I is the identity, $K = I - W$, W_{ij} is $1/|\mathcal{N}_i|$ if vertex i and j are connected.

To minimize the membrane energy for a mesh, we need to solve the following equation:

$$\mathbf{LP} = -K\mathbf{P} = (W - I)\mathbf{P} = 0. \tag{9.15}$$

In other words, the Laplacian must be 0 for the entire mesh, and we can solve this system of equations using Gauß–Seidel iteration [11]. The idea behind Gauß–Seidel iteration is to iteratively compute the solution to a system of equations. For each row, we solve for the value of the diagonal element using the current value of all the other columns. When we move on to the next row, the value from the previous row is used. Provided we have a vertex i with valence three, row i of the equation above looks like this

$$-\mathbf{p}_i + \frac{1}{3}\mathbf{p}_j + \frac{1}{3}\mathbf{p}_k + \frac{1}{3}\mathbf{p}_l = 0,$$

where j, k, l are the indices of the neighbors of vertex i. In a Gauß–Seidel scheme we would update as follows:

$$\mathbf{p}_i \leftarrow \frac{1}{3}\mathbf{p}_j + \frac{1}{3}\mathbf{p}_k + \frac{1}{3}\mathbf{p}_l.$$

From this example, we can distill the general rule

$$\mathbf{p}_i \leftarrow \frac{1}{|\mathcal{N}_i|} \sum_{\mathbf{p}_j \in \mathcal{N}_i} \mathbf{p}_j, \tag{9.16}$$

which can also be expressed in terms of the Laplacian

$$\mathbf{p}_i \leftarrow \mathcal{L}(\mathbf{p}_i) + \mathbf{p}_i. \tag{9.17}$$

The squared Laplacian is similar. We simply apply the Laplacian to the values of the Laplacian at each vertex. Solving the equations is a tiny bit more complicated, however, since we do not have ones down the diagonal. The squared Laplacian can be written in full

$$\mathcal{L}^2(\mathbf{p}_i) = \frac{1}{|\mathcal{N}_i|} \sum_{\mathbf{p}_j \in \mathcal{N}_i} \left(\mathcal{L}(\mathbf{p}_j) - \mathcal{L}(\mathbf{p}_i) \right)$$

$$= \frac{1}{|\mathcal{N}_i|} \sum_{\mathbf{p}_j \in \mathcal{N}_i} \left[\frac{1}{|\mathcal{N}_j|} \sum_{\mathbf{p}_k \in \mathcal{N}_j} (\mathbf{p}_k - \mathbf{p}_j) - \frac{1}{|\mathcal{N}_i|} \sum_{\mathbf{p}_k \in \mathcal{N}_i} (\mathbf{p}_k - \mathbf{p}_i) \right].$$

From this equation, we can simply sum up the weights for $\mathbf{p}_i$ to compute the update rule. The weights are

$$w = \frac{1}{|\mathcal{N}_i|} \sum_{\mathbf{p}_j \in \mathcal{N}_i} \left[\frac{1}{|\mathcal{N}_j|} + 1 \right]. \tag{9.18}$$

Thus, $\mathcal{L}^2(\mathbf{p}_i) = 0$ is the same as

$$\mathcal{L}^2(\mathbf{p}_i) - w\mathbf{p}_i + w\mathbf{p}_i = 0$$
$$\frac{1}{w}\mathcal{L}^2(\mathbf{p}_i) - \mathbf{p}_i + \mathbf{p}_i = 0$$
$$\mathbf{p}_i - \frac{1}{w}\mathcal{L}^2(\mathbf{p}_i) = \mathbf{p}_i,$$

leading to the following update rule:

$$\mathbf{p}_i \leftarrow \mathbf{p}_i - \frac{1}{w}\mathcal{L}^2(\mathbf{p}_i). \tag{9.19}$$

Kobbelt notes a few important details [12]. In particular, we cannot use Gauß–Seidel iteration if we precompute the Laplacian for each vertex. Hence, we need to first compute all the Laplacians and then (having computed the squared Laplacians) we compute temporary vertex positions for each vertex. Finally, we assign these temporary positions to the vertices which completes one iteration. This corresponds to Jacobi iterations rather than Gauß–Seidel [11]. We generally use a damping factor of 0.4 which leads to the following (final) update rules for the squared Laplacian:

$$\mathbf{p}_i \leftarrow \mathbf{p}_i - \frac{0.4}{w}\mathcal{L}^2(\mathbf{p}_i). \tag{9.20}$$

Kobbelt provides some more precise guidelines for choosing the damping parameter [12].

9.6.3 Implementation

We now have a vertex update rule for both the membrane energy and the bending energy. We also know that the membrane energy can be minimized using a simple Gauß–Seidel scheme while the bending energy can be minimized using Jacobi iteration. A remaining question is how to handle boundary conditions. Vertices $\mathbf{p}_i$ which are part of the input should not be moved, and vertices on the boundary should also be treated differently.

- In our implementation, boundary vertices are simply never moved. This is also true if they have been introduced during subdivision (which will be discussed later). The value of the Laplacian is simply 0 for a boundary vertex.
- Vertices which were part of the input but do not lie on the boundary are simply not moved but they are used when computing $\mathcal{L}$ and $\mathcal{L}^2$ of their neighbors.

The goal of the algorithm is to introduce new vertices and to move these to positions so as to minimize either the bending energy or the membrane energy.

New detail is introduced by splitting edges. An edge split introduces a new vertex which is located on the midpoint of the split edge. Consequently, the (one or) two triangles sharing this edge are also split.

The aim is to create a mesh which minimizes the desired energy, but, in general, we would also like the mesh to be as regular as possible. To attain these goals the following three steps are repeated.

- To make the edge lengths more even, edges are split if they are longer than 1.5 times the average edge length.
- Edges are flipped if it maximizes the minimum angle and the dihedral angle between the two triangles is very low. This step also improves regularity.
- The energy minimization procedure discussed in the preceding section is run until the error is below a set threshold (or a maximum number of iterations has been reached.

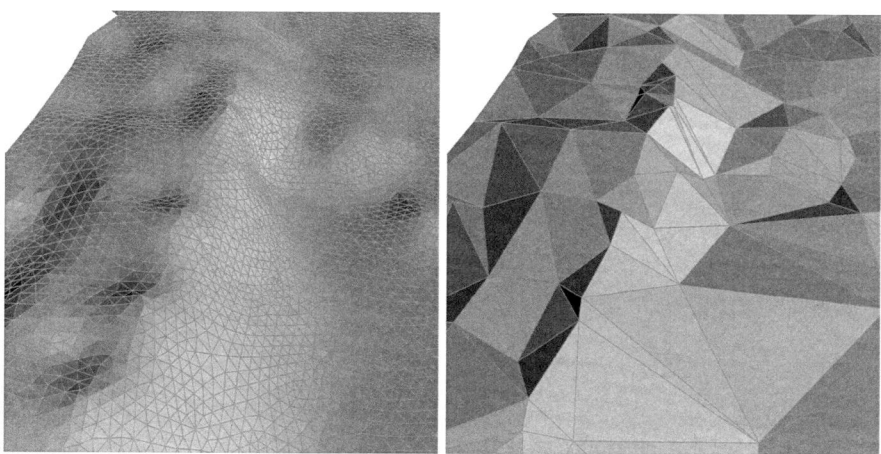

Fig. 9.7 From *top* to *bottom*: a detail of a smooth terrain obtained via $\mathcal{L}^2$ variational subdivision and a corresponding detail of the original mesh

This is iterated for a number of iterations (say 20) until no edge is longer than the average.

The next step is to introduce more detail. For a fixed number of iterations (typically 2), the precise same procedure as described above is performed. The only difference is that edges are split if they are longer than 0.75 times the average length.

The result of variational subdivision applied to the Delaunay triangulation of a terrain data set is shown in Fig. 9.7. As seen, a relatively smooth surface is produced. This is because the bending rather than the membrane energy has been used. If the membrane energy is minimized, the terrain looks more like stretched fabric.

9.7 Exercises

Exercise 9.1 Implement Laplacian smoothing, i.e., perform two passes over a triangles mesh: in the first pass computing the Laplacians using (9.2) and in the second pass updating the vertex positions using (9.3).

Exercise 9.2 Extend the previous exercise to Taubin smoothing.

Exercise 9.3 Extend the previous exercise to smoothing based on mean curvature flow.

Exercise 9.4 Implement variational subdivision using the guidelines from Sect. 9.6.3.

[GEL Users] Note that GEL supplies functions for splitting edges and triangulating the resulting mesh.

Appendix: A Laplace Operator for a Triangle Mesh

There are a number of techniques for finding a Δ operator on a triangle mesh. In many cases, we prefer a discrete Laplace–Beltrami operator, but the umbrella operator [2] has also been used extensively in this chapter and will be derived in the following.

To start with, we consider how to find the value of Δ locally at a single vertex. Notice that in the following, we will move from a scalar function $f : \mathbb{R}^2 \to \mathbb{R}$ to a vector function $\mathbf{f} : \mathbb{R}^2 \to \mathbb{R}^3$. This simply means that we minimize three functions f^x, f^y, and f^z independently, where $\mathbf{f}(u, v) = [f^x(u, v) \ f^y(u, v) \ f^z(u, v)]^T$.

In the above, we assumed a parametric surface representation. However, this parametric representation need not be global. All we need is a local function $\mathbf{f}$ which approximates the surface in the neighborhood of that vertex. Given such a function $\mathbf{f}$, we can simply compute its

$$\Delta \mathbf{f} = \mathbf{f}_{uu} + \mathbf{f}_{vv},$$

and consider that to be the Laplacian of the mesh at the vertex in question. Typically, we find a local surface approximation by least squares fitting a second order polynomial surface to the 1-ring neighborhood of a vertex

$$\mathbf{f}(u, v) = \mathbf{f}_0 + u\mathbf{f_u} + v\mathbf{f_v} + \frac{u^2}{2}\mathbf{f_{uu}} + uv\mathbf{f_{uv}} + \frac{v^2}{2}\mathbf{f_{vv}}. \tag{9.21}$$

Note that the coefficient $\mathbf{f}$ corresponds precisely to the surface point $\mathbf{f}(0, 0)$ and the other coefficients correspond to the first and second derivatives of the surface at $(u, v) = (0, 0)$.

In order to find a suitable function, $\mathbf{f}$, at a given mesh vertex we must assign a pair of parameter coordinates (u, v) to each vertex. Let the centre vertex have parameter coordinates $(0,0)$ and let the other (u, v) coordinates be numbered from 1 to n where n is the valency (number of neighbors) of the vertex. This is illustrated in Fig. 9.8.

Let j be the index of a vertex in the 1-ring. We reserve the index 0 for the centre vertex, and the neighbors are numbered from 1 to n. Let $\mathbf{p}_j$ be the 3D space position of vertex j. Since $(u, v) = (0, 0)$ has been assigned to the centre vertex,

$$\mathbf{f}(0, 0) = \mathbf{f}_0 = \mathbf{p}_0.$$

We now subtract the equation above from (9.21) for all the neighboring vertices. We can then write the resulting system of equations (n equations, one for each neighboring vertex) in matrix form $UF = \mathbf{P}$ where U is the matrix of parameter values, F is the vector of coefficients for the polynomial surface, and $\mathbf{P}$ is the vector of geometric positions:

$$\begin{bmatrix} \cdot & \cdot & \cdot & \cdot & \cdot \\ \cdot & \cdot & \cdot & \cdot & \cdot \\ u_j & v_j & \frac{u_j^2}{2} & u_j v_j & \frac{v_j^2}{2} \\ \cdot & \cdot & \cdot & \cdot & \cdot \\ \cdot & \cdot & \cdot & \cdot & \cdot \end{bmatrix} \begin{bmatrix} \mathbf{f_u} \\ \mathbf{f_v} \\ \mathbf{f_{uu}} \\ \mathbf{f_{uv}} \\ \mathbf{f_{vv}} \end{bmatrix} = \begin{bmatrix} \cdot \\ \mathbf{p_j} - \mathbf{p_0} \\ \cdot \end{bmatrix} . \tag{9.22}$$

Fig. 9.8 This figure
illustrates the 1-ring of a
vertex and shows the
parameter coordinates for the
centre vertex and its
neighbors

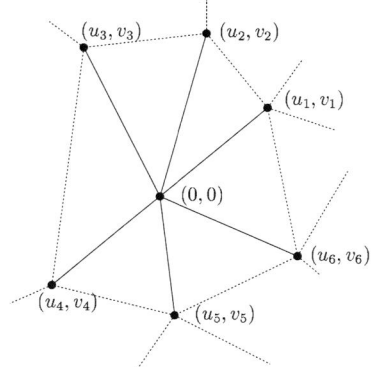

Note that each entry in F and $\mathbf{P}$ is really a 3D vector. This is not important for the analysis, since we can consider each component separately.

Clearly, the matrix above has a solution only in the case where $n = 5$ since the matrix U is not square otherwise. We will assume that we have at least five neighbors. In this case, we can solve (9.22) in the least squares sense

$$\mathbf{F} = \mathbf{DP} = \left((\mathbf{U}^T\mathbf{U})^{-1}\mathbf{U}^T\right)\mathbf{P}. \tag{9.23}$$

To sum up, F is the vector of coefficients to our polynomial surface, and the third and fifth entry of F correspond to the second order derivatives of $\mathbf{f}(u, v)$. We can compute the Laplacian simply by

$$\Delta\mathbf{f} = \mathbf{F}_3 + \mathbf{F}_5 = (\mathbf{D}_{3\_} + \mathbf{D}_{5\_})\mathbf{P}. \tag{9.24}$$

Till now, we have ignored the fact that we do not really have a parametrization. Now we need to find a parametrization since the $\mathbf{U}$ and, hence, $\mathbf{D}$ requires it. We now parametrize in a way that is completely independent of geometry. A very simple method is to just distribute the parameter coordinates evenly on the unit circle as shown in Fig. 9.9.

With this distribution, we can precompute parameter coordinates for a particular valence as follows:

$$(u_j, v_j) = \left(\cos(2\pi j/n), \sin(2\pi j/n)\right).$$

If we plug the above values into the U matrix and compute (9.24), we obtain an unexpectedly simple result. It turns out that

$$\mathbf{D}_{3\_} + \mathbf{D}_{5\_} = [4/n \quad 4/n \quad \dots \quad 4/n],$$

Fig. 9.9 This figure illustrates the 1-ring of a vertex and shows the parameter coordinates for the centre vertex and its neighbors. There are eight neighbors, and hence the angle between them is $2\pi/8$

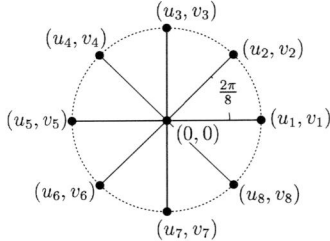

(apparently) for all valencies greater than 5. For $n < 5$ the system is singular. This result leads to the following very simple formula:

$$\Delta\mathbf{f} = \frac{4}{n} \sum_{j=1}^{n} (\mathbf{p}_j - \mathbf{p}_0).   \tag{9.25}$$

In other words, the Laplacian is simply the average of the vectors from the centre vertex to its neighbors times 4. This factor is of no importance since we want to solve $\Delta f = 0$.

Let $\mathcal{N}_i$ be the set of indices of neighbors to the vertex of index i and let $n = |\mathcal{N}_i|$ be the valency of vertex i. Dropping the 4 factor, we can now write a discrete triangle mesh Laplacian, $\mathcal{L}$, operating on a vertex $\mathbf{p}_i$

$$\mathcal{L}(\mathbf{p}_i) = \frac{1}{|\mathcal{N}_i|} \sum_{\mathbf{p}_j \in \mathcal{N}_i} (\mathbf{p}_j - \mathbf{p}_i)$$

$$= \left(\frac{1}{|\mathcal{N}_i|} \sum_{\mathbf{p}_j \in \mathcal{N}_i} \mathbf{p}_j \right) - \mathbf{p}_i.   \tag{9.26}$$

This operator is usually referred to as the umbrella operator, and we can express it as a matrix (applying the Laplacian to all vertices at once).

$$\mathbf{LP} = -\mathbf{KP},   \tag{9.27}$$

where $\mathbf{P}$ is the vector of vertex positions, I is the identity, $\mathbf{K} = I - W$, $\mathbf{W}_{ij}$ is $1/|\mathcal{N}_i|$ if vertex i and j are connected. This Laplacian is also used for vertices of valence less than 5, but note that these are rare—unless the mesh is very irregular.

References

1. Desbrun, M., Meyer, M., Schröder, P., Barr, A.H.: Implicit fairing of irregular meshes using diffusion and curvature flow. In: SIGGRAPH'99: Proceedings of the 26th Annual Conference on Computer Graphics and Interactive Techniques, pp. 317–324. ACM Press, New York (1999). doi:10.1145/311535.311576
2. Taubin, G.: A signal processing approach to fair surface design. In: ACM SIGGRAPH'95 Proceedings (1995)

3. Pinkall, U., Polthier, K.: Computing discrete minimal surfaces and their conjugates. Exp. Math. **2**(1), 15–36 (1993)
4. Vallet, B., Lévy, B.: Spectral geometry processing with manifold harmonics. Comput. Graph. Forum **27**(2), 251–260 (2008)
5. Meyer, M., Desbrun, M., Schröder, P., Barr, A.H.: Discrete differential-geometry operators for triangulated 2-manifolds. In: Hege, H.-C., Polthier, K. (eds.) Visualization and Mathematics III, pp. 35–57. Springer, Heidelberg (2003)
6. Levy, B., Zhang, R.H.: Spectral geometry processing. In: ACM SIGGRAPH Course Notes (2010)
7. Shen, Y., Barner, K.E.: Fuzzy vector median-based surface smoothing. IEEE Trans. Vis. Comput. Graph. **10**(3), 252–265 (2004). doi:10.1109/TVCG.2004.1272725
8. Zheng, Y., Fu, H., Au, O.K.-C., Tai, C.-L.: Bilateral normal filtering for mesh denoising. IEEE Trans. Vis. Comput. Graph. **17**(10), 1521–1530 (2011). doi:10.1109/TVCG.2010.264
9. Kobbelt, L.P.: Discrete fairing and variational subdivision for freeform surface design. Vis. Comput. **16**, 142–158 (2000)
10. Gelfand, I.M., Fomin, S.V.: Calculus of Variations. Dover, New York (2000)
11. Golub, G.H., van Loan, C.F.: Matrix Computations, 3rd edn. John Hopkins, Baltimore (1996)
12. Kobbelt, L., Campagna, S., Vorsatz, J., Seidel, H.-P.: Interactive multi-resolution modeling on arbitrary meshes. In: ACM SIGGRAPH'98 Proceedings, pp. 105–114 (1998)

Parametrization of Meshes 10

Previously in this book, we have discussed parametric surfaces which are mappings from a 2D domain into 3D. However, many, if not most, 3D models in computer graphics are made of triangles with texture. Texture in this context means an image which is glued on to the 3D model as illustrated in Fig. 10.1. Unfortunately, since the images are flat and the 3D model is generally curved (albeit piecewise planar) the image needs to be deformed in order to precisely fit the 3D model or conversely the 3D model needs to be flattened to map it onto the image. Typically, we perform the following steps, illustrated in Fig. 10.2, in order to define the mapping from texture onto 3D model [1]:

1. the 3D mesh is cut into smaller pieces (each with disc topology) meaning that the piece can be smoothly deformed to a disc, cf. Sect. 2.3,
2. each piece is made planar, and
3. the pieces are packed into a 2D texture image.

For each vertex, we now have its texture coordinates, i.e., its 2D position in the texture image. Thus, when we need a mapping from a point inside a triangle and into the texture image, we simply interpolate the texture coordinates at the corners of the triangle.

Neither the cutting nor the packing are trivial steps, and in some cases an effort is made to avoid these steps. For instance, many objects have sphere topology and the mesh may, in this case, be deformed into a sphere instead of a disc [2]. Also, some recent methods produce periodic parametrizations covering the entire shape [3]. In the present chapter, we shall focus on methods for flattening 3D surfaces which are already of disc topology. Bruno Levy et al. give a fairly complete overview of the entire process of cutting, flattening and packing the maps in the paper, which introduced *least squares conformal maps* [1].

Flattening of 3D surfaces, or parametrization, is not useful exclusively for texture mapping. Many algorithms also work in the parametrization domain rather than directly on the 3D shape since this can sometimes simplify algorithms for, e.g., remeshing [4]. Moreover, this is clearly not a new topic. In fact, map projections, see Fig. 10.3, are probably the best known examples of parametrizations. The task is

J.A. Bærentzen et al., *Guide to Computational Geometry Processing*,
DOI 10.1007/978-1-4471-4075-7_10, © Springer-Verlag London 2012

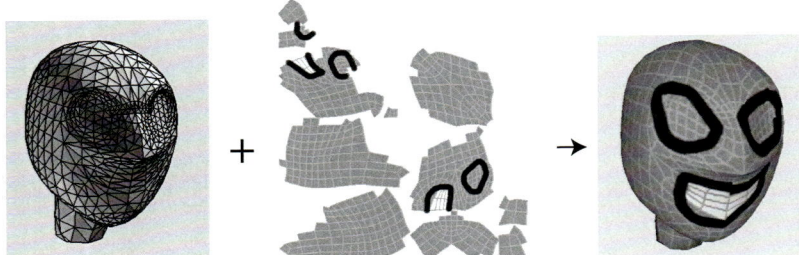

Fig. 10.1 This figure illustrates the principle of texture mapping. Texture stored in a planar image is mapped onto the mesh

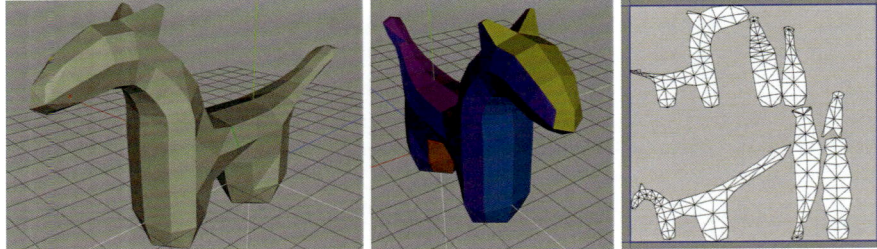

Fig. 10.2 This figure illustrates the steps in parametrization (carried out) in the Wings 3D program. The simple geometric shape (*left*) is cut into pieces as indicated by the colored patches (*center*) and each pieces is then flattened and packed (poorly) into a 2D image (*right*)

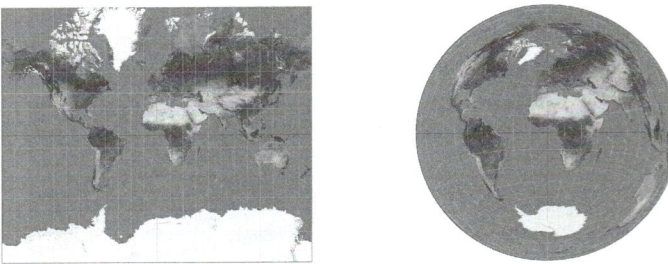

Fig. 10.3 To the *left* Mercator projection (1569). An example of a conformal parametrization. To the right Lambert Azimuthal projection (1772). An example of an area preserving parametrization

to find a one-to-one mapping between a region on a surface, in this case the sphere, and a region in the plane.

10.1 Properties of Parametrizations

A parametrization is a mapping from a curved surface to a planar surface (and vice versa, cf. Sect. 3.1). The ideal situation would of course be if such a mapping could be to scale, i.e., the mapping preserves lengths and by implication angles. Such a mapping is known as an *isometric* map, and, in general, we cannot get such maps, so one has to settle for less.

A weaker condition on the parametrization is preservation of angles, in which case the parametrization is called a *conformal* map. The Mercator projection is an example, see Fig. 10.3. Another is preservation of area in which case the parametrization is called a *equiarea* map. The Lambert Azimuthal projection is an example, see Fig. 10.3. Let $\mathbf{x} : U \subset \mathbb{R}^2 \to S \subset \mathbb{R}^3$ be a parametrization of a surface and let I be the first fundamental form. By Theorem 3.1 we have the following.

1. The parametrization $\mathbf{x}$ is an isometry if and only if $I = \begin{pmatrix} 1 & 0 \\ 0 & 1 \end{pmatrix}$.

2. The parametrization $\mathbf{x}$ is conformal (cf. Fig. 10.3) if and only if $I = \begin{pmatrix} \lambda & 0 \\ 0 & \lambda \end{pmatrix}$, where λ is a function on U.

3. The parametrization $\mathbf{x}$ is equiarea (cf. Fig. 10.3) if and only if $\det I = 1$.

It is, in general, impossible to have an isometric parametrization. It is only developable surfaces, cylinders, cones, and tangent developables that can be unrolled in the plane. They are characterized by having zero Gaußian curvature.

A parametrization is by definition invertible so we can consider the inverse map $\mathbf{u} = \mathbf{x}^{-1} : S \to U$. Any of the properties, 1–3, above obviously holds for $\mathbf{x}$ if and only if it holds for $\mathbf{u}$. In our applications, the smooth surface S is replaced by a triangular mesh and in the parametrization algorithms we present in this chapter it is an invertible, and piecewise linear, map, $\mathbf{u} : S \to U \subseteq \mathbb{R}^2$, we construct.

We can identify the plane with the field of complex numbers and a mapping between two planar domains is conformal if and only if it is holomorphic, which means that it is complex differentiable. So the *uniformization theorem* [5, 6], a famous result in complex function theory, tells us that any simply connected surface, i.e., with disk topology, can be parametrized conformally and the parameter domain can be any simply connected domain in the plane.

As we shall soon see, this is in contrast to the piecewise linear situation where the conformal factor of a conformal map has to be constant, so the map is, essentially, an isometry.

Going back to planar domains $U_1, U_2 \subseteq \mathbb{R}^2$ and a mapping $f : U_1 \to U_2 :$ $(u, v) \mapsto (x, y)$ the first fundamental form is

$$I = \begin{pmatrix} \frac{\partial x}{\partial u} & \frac{\partial y}{\partial u} \\ \frac{\partial x}{\partial v} & \frac{\partial y}{\partial v} \end{pmatrix} \begin{pmatrix} \frac{\partial x}{\partial u} & \frac{\partial x}{\partial v} \\ \frac{\partial y}{\partial u} & \frac{\partial y}{\partial v} \end{pmatrix} = \begin{pmatrix} (\frac{\partial x}{\partial u})^2 + (\frac{\partial y}{\partial u})^2 & \frac{\partial x}{\partial u}\frac{\partial x}{\partial v} + \frac{\partial y}{\partial u}\frac{\partial y}{\partial v} \\ \frac{\partial x}{\partial u}\frac{\partial x}{\partial v} + \frac{\partial y}{\partial u}\frac{\partial y}{\partial v} & (\frac{\partial x}{\partial v})^2 + (\frac{\partial y}{\partial v})^2 \end{pmatrix},$$

so f is conformal if and only if the Cauchy–Riemann equations hold:

$$\frac{\partial x}{\partial u} = \frac{\partial y}{\partial v} \quad \text{and} \quad \frac{\partial x}{\partial v} = -\frac{\partial y}{\partial u}. \tag{10.1}$$

If we differentiate the first equation with respect to u and the second with respect to v and add the result we see that

$$\triangle x = \frac{\partial^2 x}{\partial u^2} + \frac{\partial^2 x}{\partial v^2} = 0, \quad \text{likewise } \triangle y = \frac{\partial^2 y}{\partial u^2} + \frac{\partial^2 y}{\partial v^2} = 0. \tag{10.2}$$

The definition of a *harmonic function* $f : \mathbb{R}^N \to \mathbb{R}$ is precisely that $\triangle f = 0$, and we see that the coordinate functions $x(u, v)$ and $y(u, v)$ of a conformal map are harmonic functions.

Definition 10.1 A harmonic map $\mathbf{x} : (u, v) \mapsto (x(u, v), y(u, v))$ from one planar domain to another is a map where the two coordinate functions, x and y, are harmonic.

Harmonic functions minimize the *Dirichlet energy*,

$$\mathcal{E}(f) = \frac{1}{2} \int \|\nabla f\|^2 \, du \, dv. \tag{10.3}$$

This turns out to be attractive: the energy minimization property implies that harmonic functions and maps are smooth. So while the latter are neither angle nor area preserving, they are often used in practice.

   Any continuous map of the boundary of two regions, i.e., $\partial U_1 \to \partial U_2$, extends uniquely to a harmonic map between the interiors $U_1 \to U_2$. Furthermore, if the *target* U_2 is convex then the Rado–Kneser–Choquet Theorem, [7–9] tells us that this harmonic map is a *diffeomorphism*. A diffeomorphism is a homeomorphism (cf. Sect. 2.3), where both the map and its inverse is differentiable.

   A map $f : S \to U$ from a surface $S \subseteq \mathbb{R}^3$ to a domain $U \subseteq \mathbb{R}^2$ is called harmonic if, in (10.2), we replace the Laplace operator $\triangle$ in $\mathbb{R}^2$ with the Laplace–Beltrami operator on the surface S, see Sect. 3.9. Again the Rado–Kneser–Choquet Theorem holds, i.e., if U is convex then any homeomorphism $\partial S \to \partial U$ of the boundaries extends uniquely to a harmonic map $S \to U$ which is a diffeomorphism on the interior.

10.1.1 Goals of Triangle Mesh Parametrization

In the context of triangle meshes, we often strive for a conformal mapping—because local preservation of the angles in the triangles corresponds well with our intuitive notion of shape preservation. Unfortunately, it is, generally, not possible to make a truly conformal mapping of a triangle mesh. A conformal map would mean that all triangles are mapped from 3D and into the plane without changing their angles (only their sizes). Unfortunately, this is not possible in general: Two triangles sharing an edge would have to be scaled by the same factor. Since the mesh is connected this means that all triangles must be scaled by the same factor.

While conformality is not attainable, a number of useful and much used techniques for parametrization of triangle meshes have been developed. In Sect. 10.2 the basic algorithm is introduced in the context of convex combination mappings: The method corresponds to fixing the boundary of a disk shaped mesh to a 2D polygon and then smoothing the interior of the mesh until convergence. In Sect. 10.3, mean value (or Floater) coordinates are used. This method employs different weights for the smoothing procedure which have the special property that for an already planar mesh the algorithm does not change anything. In Sect. 10.4, harmonic maps are introduced. These maps tend to give smooth parametrizations, and in Sect. 10.5 the boundary is allowed to move which enables us to create parametrizations, which are least squares conformal.

For a more in-depth treatment, the reader is referred to the survey by Floater and Hormann [10] or the recent book by Botsch et al. [11] which covers parametrization and many other aspects of mesh processing.

10.2   Convex Combination Mappings

A practical way of computing the flattening of a 3D mesh is as follows. Since the mesh must have disc topology, it has a boundary. The boundary vertices are placed in such a way that they form a convex shape in the 2D domain. All other vertices are now computed in such a way that

$$\mathbf{u}_i = \sum_{j \in \mathcal{N}_i} \lambda_{ij} \mathbf{u}_j, \tag{10.4}$$

or, equivalently,

$$\sum_{j \in \mathcal{N}_i} \lambda_{ij} (\mathbf{u}_j - \mathbf{u}_i) = 0, \tag{10.5}$$

where $\mathbf{u}_i = [u_i, v_i]$ is the vector of 2D parameter space coordinates for vertex i and $\mathcal{N}_i$ its set of neighbors. The weights, λ_{ij}, must have the property that $\sum_{j \in \mathcal{N}_i} \lambda_{ij} = 1$, and $\lambda_{ij} > 0$. If this is fulfilled, we know according to Tutte's theorem [10] that the mapping is one to one. In other words, the mesh does not fold as it is flattened. In the case where $\lambda_{ij} = 1/|\mathcal{N}_i|$ we call the resulting weights *barycentric weights*.

Algorithm 10.1 shows a simple pseudocode listing of the algorithm for flattening a mesh of disk topology. Having run the algorithm, the mesh is turned into the unit disk. Of course, the algorithm works only if the mesh has precisely one simple boundary loop.

As illustrated in Fig. 10.5 (top) barycentric weights do lead to a lot of distortion and a less than fair parametrization. The figure compares different techniques for flattening the mesh shown in Fig. 10.4.

Algorithm 10.1 Flatten Mesh

---

1. Let B be the set of boundary vertices
2. Let $n = |B|, k = 0$
3. Visit vertices $i \in B$ in counter clockwise order:
4. $\mathbf{u}_i = [\cos(2\pi k/n)\ \sin(2\pi k/n)]$
5. $k = k + 1$
6. Visit vertices $i \notin B$: $\mathbf{u}_i = [0\,0]$
7. While vertices move significantly:
8.   For each vertex $i \notin B$:
9. $\mathbf{u}_i = \sum_{j \in \mathcal{N}_i} \lambda_{ij} \mathbf{u}_j$

Fig. 10.4 A 3D model of a head. Note that this triangulated manifold is in fact homeomorphic to a disc since it is cut off at the neck. Consequently, it can be flattened

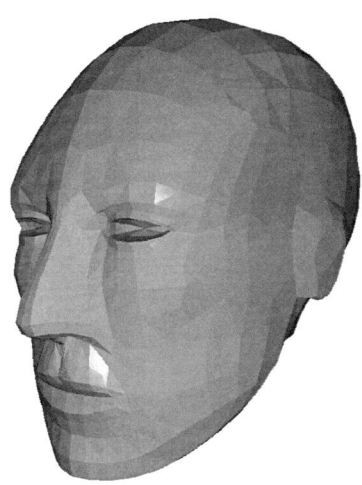

10.3 Mean Value Coordinates

Michael Floater proposed a different set of weights known as *mean value coordinates* [12] and defined as follows:

$$\lambda_{ij} = \frac{w_{ij}}{\sum_{k \in \mathcal{N}_i} w_{ik}}, \quad w_{ij} = \frac{\tan(\frac{\theta_{i-1}}{2}) + \tan(\frac{\theta_i}{2})}{\|\mathbf{p}_i - \mathbf{p}_j\|}, \tag{10.6}$$

where the angles are illustrated in Fig. 10.6. Mean value coordinates have some concrete advantages. They are always positive since $\theta_i < \pi$ and they also seem to do a much better job of preserving the structure of the mesh in a 2D parametrization than barycentric coordinates as illustrated in Fig. 10.5. Perhaps their power of reproduction is not surprising since these coordinates are designed to fulfill

$$\sum_{j \in \mathcal{N}_i} \lambda_{ij} \mathbf{p}_j = \mathbf{p}_i, \tag{10.7}$$

Fig. 10.5 Comparison of
mesh flattening by solving the
Laplace equation with the
boundary vertices placed on a
circle. Various edge weights
were used for the discrete
Laplace operator: From *top* to
bottom, we used barycentric,
harmonic, and floater (mean
value) weights. Images on the
left show a grid mapped from
the parameter domain back
onto the mesh. Images on the
right show parts of the
parameter domain

when the points $\mathbf{p}_i$ all lie in a planar configuration which does not fold. Thus, a planar triangle mesh, which does not fold, is its own parametrization with respect to mean value coordinates. Or put more simply, running the parametrization algorithm does not change such a planar mesh if we use mean value coordinates.

Since mean value coordinates depend on geometry (and do not change), it is a good idea to compute and store λ_{ij} before the main loop of Algorithm 10.1. Otherwise, the procedure is the same.

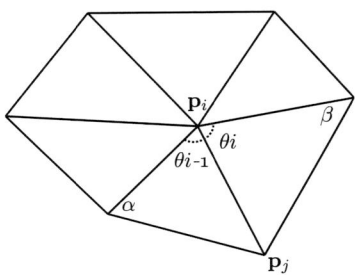

Fig. 10.6 A one ring with the angles used in the cotangent formula and the vertices of the corresponding edge

10.4   Harmonic Mappings

As discussed above, harmonic maps are smooth thanks to the property that they minimize the Dirichlet energy. When it comes to mesh parametrization, we seek a harmonic map **u** from the mesh to a planar domain. Thus, it is important to note that we are now operating on a 3D surface, and the Laplace operator does not take the surface metric into account. Instead, we should use the *Laplace–Beltrami* operator which is defined as the divergence of the gradient on the surface. Stating this precisely in a smooth setting requires more formalism, than we wish to introduce here. However, the Laplace–Beltrami operator applied to vertex positions is precisely two times the mean curvature, thus, for the discrete version, we can use (8.4) simply replacing the vertex positions with an arbitrary function f:

$$\Delta_S f = \frac{1}{4 A_i^{\text{one ring}}} \sum_{\mathbf{p}_j \in \mathcal{N}_i} (\cot \alpha + \cot \beta)(f_i - f_j),$$

where $A_i^{\text{one ring}}$ is the area of the triangles incident on vertex i. Typically we do not normalize with one ring area but instead with the sum of the weights, i.e., we solve (10.4) with

$$\lambda_{ij} = \frac{w_{ij}}{\sum_{k \in \mathcal{N}_i} w_{ik}}, \quad w_{ij} = \frac{1}{2}(\cot \alpha + \cot \beta), \tag{10.8}$$

for every vertex i, where N_i is the set of indices of neighbors to vertex i as illustrated in Fig. 10.6.

Algorithm 10.1 remains unchanged for harmonic mappings, but a problematic issue is that the cots can be negative if one or both of α and β are obtuse. A simple fix is to clamp the edge weights to a small positive number. Subdivision is another possibility as mentioned in [11].

10.5   Least Squares Conformal Mappings

In the methods described so far, we do not allow the boundary of the parametrization to change. This limits how much we can reduce distortion. It is possible to come closer to conformal maps if we allow the boundary to change. First of all, we can view harmonic maps in a different way: The Dirichlet energy

$$E_D(\mathbf{u}) = \frac{1}{2} \int_S \|\nabla \mathbf{u}\| \, \mathrm{d}A \qquad (10.9)$$

is minimized by solving $\Delta_S \mathbf{u} = 0$. The lower bound of the Dirichlet energy is the area of the map [13] and if the minimum is reached then the map is indeed conformal. This has motivated the introduction of the conformal energy

$$E_C(\mathbf{u}) = E_D(\mathbf{u}) - A(\mathbf{u}) \qquad (10.10)$$

where $A(\mathbf{u})$ is simply the area of the parametrization domain.

In a discrete setting, we observe that moving vertices in the interior of the 2D parametrization domain does not change the area. Thus, the area gradient is zero. This, however, is not true on the boundary. Thus, to go from a harmonic mapping which minimizes the Dirichlet energy to a mapping which minimizes the conformal energy, we should impose as boundary conditions that the discrete Laplace–Beltrami operator applied to the coordinates in the parameter domain is identical to the area gradient.

Again, we can use Algorithm 10.1 for the practical implementation. The main difference is that while we still initialize the boundary vertices to lie on a unit disk, we only fix two boundary vertices. The remaining boundary vertices are updated during the loop, but the new position is set to

$$
\begin{aligned}
\mathbf{u}_i &= \frac{1}{\sum_{j \in \mathcal{N}_i} w_{ij}} \sum_{j \in \mathcal{N}_i} w_{ij} \mathbf{u}_j & i \notin B \\
\mathbf{u}_i &= \frac{1}{\sum_{j \in \mathcal{N}_i} w_{ij}} \left(\sum_{j \in \mathcal{N}_i} w_{ij} \mathbf{u}_j - \mathbf{g}_i \right) & i \in B
\end{aligned}
\qquad (10.11)
$$

where $\mathbf{g}_i$ is the gradient of the area of triangles incident on vertex i (as a function of the position of vertex i) and B is the set of boundary vertices. When computing the weights w_{ij} we now have to deal with boundary edges that have only one incident triangle. In this case, the weight is computed using (10.8) with only one of the cot terms.

10.5.1 Natural Boundary Conditions

This method was reached in different ways by Desbrun et al. [14] and Levy et al. [1] and later these two approaches were shown equivalent in a brief unpublished note by Cohen–Steiner and Desbrun. The approach by Levy et al. was to find a discrete expression for the conformal energy and minimize it in the least squares sense. Hence, they called the result *least squares conformal maps*.

The approach in [14] is, essentially, that it is "natural" that the discrete Laplacian equals the area gradient on the boundary because in the planar case, the two are, in fact, the same. For this reason, Desbrun et al. called these boundary conditions

Fig. 10.7 A single triangle
and the angles used in the
cotangent formula

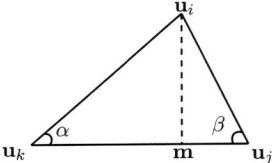

the *natural boundary conditions*. As we see in (10.13) below, the area gradient of
a triangle in 2D can be computed simply by rotating the baseline $\mathbf{u}_k\mathbf{u}_j$ 90 degrees
so that it points in the direction of the height. Thus, for all triangles incident on a
boundary vertex i, we rotate by 90 degrees and sum all edges $\mathbf{u}_k\mathbf{u}_j$ where k, j are
the indices of an edge opposite vertex i and belonging to its one ring. In the case of
a boundary vertex, this sum replaces zero on the right hand side of (10.5), where we
assume that harmonic weights are used.

Referring to Fig. 10.7, the per triangle contribution to (10.8) is

$$\frac{1}{2}\left(\cot\alpha(\mathbf{u}_i - \mathbf{u}_j) + \cot\beta(\mathbf{u}_i - \mathbf{u}_k)\right), \tag{10.12}$$

where we now assume that the angles are also computed from the flattened mesh.
We can rewrite (10.12) to show that it is the same as the gradient of the area of the
triangle, $A(\mathbf{u}_i)$, as a function of the position of $\mathbf{u}_i$. For simplicity, we will simply
assume a planar configuration below. We introduce a new point $\mathbf{m}$ where the height
of the triangle intersects the baseline. Now,

$$\frac{1}{2}\left(\cot\alpha(\mathbf{u}_i - \mathbf{u}_j) + \cot\beta(\mathbf{u}_i - \mathbf{u}_k)\right)$$

$$= \frac{1}{2}\left(\frac{\|\mathbf{m} - \mathbf{u}_k\|}{\|\mathbf{u}_i - \mathbf{m}\|}(\mathbf{u}_i - \mathbf{u}_j) + \frac{\|\mathbf{m} - \mathbf{u}_j\|}{\|\mathbf{u}_i - \mathbf{m}\|}(\mathbf{u}_i - \mathbf{u}_k)\right)$$

$$= \frac{1}{2}\left(\frac{\|\mathbf{m} - \mathbf{u}_k\|}{\|\mathbf{u}_i - \mathbf{m}\|}(\mathbf{u}_i - \mathbf{m} + \mathbf{m} - \mathbf{u}_j) + \frac{\|\mathbf{m} - \mathbf{u}_j\|}{\|\mathbf{u}_i - \mathbf{m}\|}(\mathbf{u}_i - \mathbf{m} + \mathbf{m} - \mathbf{u}_k)\right)$$

$$= \frac{1}{2}\left(\frac{\|\mathbf{u}_j - \mathbf{u}_k\|(\mathbf{u}_i - \mathbf{m})}{\|\mathbf{u}_i - \mathbf{m}\|}\right)$$

$$= \nabla A(\mathbf{u}_i). \tag{10.13}$$

Why is the last equality true? We will make a very simple argument. If we move
$\mathbf{u}_i$ parallel to the baseline $\mathbf{u}_k\mathbf{u}_j$, its area does not change and the derivative in that
direction is zero. As for the magnitude, the area is a linear function of distance to
the baseline $\mathbf{u}_k\mathbf{u}_j$ and proportional to half its length.

In Fig. 10.8 we compare harmonic parametrization with natural boundary conditions to regular harmonic parametrization where the boundary vertices are pinned
to the unit circle. The difference is not enormous, but clearly pinning the boundary vertices to the circle creates a significant distortion precisely near the boundary.
However, natural boundary conditions requires us to specify two vertices in the

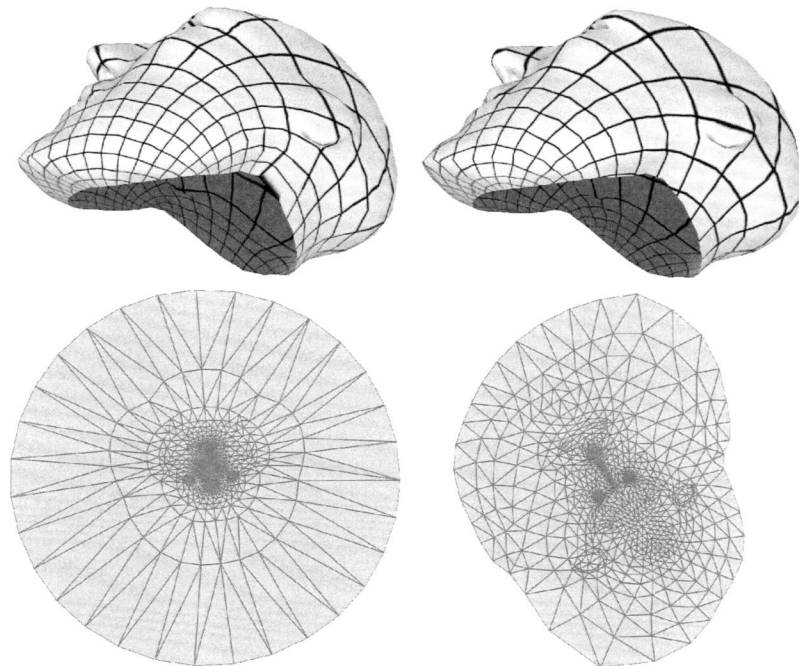

Fig. 10.8 Comparison of harmonic parametrization with natural boundary conditions (*top right*, *bottom right*) and with boundary vertices pinned to a circle (*top left, bottom left*)

boundary to lock scaling, rotation and translation. In [15] it is argued that this is a problem because the solution depends on what vertices we choose. This is probably true, but the dependency can only be due to numerical issues. Indeed, implementing this method we found that it was much more sensitive to the quality of the mesh, and our simple iterative solver would fail for meshes with poor triangulations near the boundary.

10.6   Exercises

Exercise 10.1 Implement Algorithm 10.1 and try it using barycentric, harmonic, and mean value coordinates. The mesh used in this chapter is provided as an example on the book homepage.

Exercise 10.2 Implement least squares conformal maps by adding natural boundary conditions to the harmonic maps implementation from the first exercise.

References

1. Lévy, B., Petitjean, S., Ray, N., Maillot, J.: Least squares conformal maps for automatic texture atlas generation. ACM Trans. Graph. **21**(3), 362–371 (2002)
2. Praun, E., Hoppe, H.: Spherical parametrization and remeshing. ACM Trans. Graph. **22**(3), 340–349 (2003)
3. Kälberer, F., Nieser, M., Polthier, K.: QuadCover-surface parameterization using Branched Coverings. Comput. Graph. Forum **26**(3), 375–384 (2007). Wiley Online Library
4. Alliez, P., Meyer, M., Desbrun, M.: Interactive geometry remeshing. ACM Trans. Graph. **21**(3), 347–354 (2002)
5. Farkas, H.M., Kra, I.: Riemann Surfaces. Graduate Texts in Mathematics, vol. 71, p. 337. Springer, New York (1980)
6. Forster, O.: Lectures on Riemann Surfaces. Graduate Texts in Mathematics, vol. 81, p. 254. Springer, New York (1991). Translated from the 1977 German original by Bruce Gilligan, Reprint of the 1981 English translation
7. Radó, T.: Aufgabe 41. Dtsch. Math.-Ver. **35**, 49 (1926)
8. Kneser, H.: Lösung der Aufgabe 41. Dtsch. Math.-Ver. **35**, 123–124 (1926)
9. Choquet, G.: Sur un type de transformation analytique généralisant la représentation conforme et définie au moyen de fonctions harmoniques. Bull. Sci. Math. **69**, 156–165 (1945)
10. Floater, M.S., Hormann, K.: Surface parameterization: a tutorial and survey. In: Advances in Multiresolution for Geometric Modelling, pp. 157–186 (2005)
11. Botsch, M., Kobbelt, L., Pauly, M., Alliez, P., Levy, B.: Polygon Mesh Processing. AK Peters, Wellesley (2010)
12. Floater, M.S.: Mean value coordinates. Comput. Aided Geom. Des. **20**(1), 19–27 (2003)
13. Pinkall, U., Polthier, K.: Computing discrete minimal surfaces and their conjugates. Exp. Math. **2**(1), 15–36 (1993)
14. Desbrun, M., Meyer, M., Alliez, P.: Intrinsic parameterizations of surface meshes. Comput. Graph. Forum **21**(3), 209–218 (2002)
15. Mullen, P., Tong, Y., Alliez, P., Desbrun, M.: Spectral conformal parameterization. Comput. Graph. Forum **27**(5), 1487–1494 (2008)

Simplifying and Optimizing Triangle Meshes 11

In Chap. 9 we considered methods for improving triangle quality by smoothing and thus moving the mesh vertices. In this chapter, we will consider methods which modify triangle meshes (mostly) without moving vertices. This can be quite important when the vertices are known to have little noise as, for instance, terrain points measured by a surveyor. It could also be that smoothing is simply done in a different part of the pipeline as we shall discuss in Sect. 11.3.

The two main tools for mesh manipulation, which we will use in this chapter, are edge collapse and edge flip. The former tool is very effective when it comes to reduction of the complexity of a triangle mesh which is the topic of Sect. 11.1. Triangle meshes acquired from real objects or from tessellated iso-surfaces (cf. Chap. 18) are almost invariably too detailed for rendering or further processing, and reducing the number of triangles can significantly speed up any such downstream processing.

In Sect. 11.2 we discuss how edge flips can be used to improve a triangle mesh. This is useful when we have a mesh without much redundancy in the set of vertices but whose geometry can still be improved. These improvements could be made both to improve the quality of the triangles (removing triangles with very small or big angles) and also to improve the overall geometry of the shape. Figure 11.1 illustrates how a scanned 3D model of a Greek bust is significantly improved by edge flipping; the most obvious improvement being that the dark spots in the rendering, which are due to nearly degenerate triangles, are removed by optimization. However, as Fig. 11.10 shows, mesh optimization can sometimes find structure in the data which the original triangulation does not show.

Edge flips alone do not suffice if our goal is to create a nearly regular mesh where most vertices have valency six and triangles are nearly equilateral. In Sect. 11.3 we consider how to combine optimization by edge flipping with smoothing as well as both coarsening and refinement in order to produce a nearly regular mesh.

A general theme of this chapter is optimization. In the case of simplification, we look for the mesh which best approximates an original mesh but with a smaller number of vertices. In the case of mesh optimization, we try to find a mesh which minimizes a geometric measure (e.g., curvature) or which maximizes triangle quality (e.g., making them as equilateral as possible). In most cases, we use simple

J.A. Bærentzen et al., *Guide to Computational Geometry Processing*,
DOI 10.1007/978-1-4471-4075-7_11, © Springer-Verlag London 2012

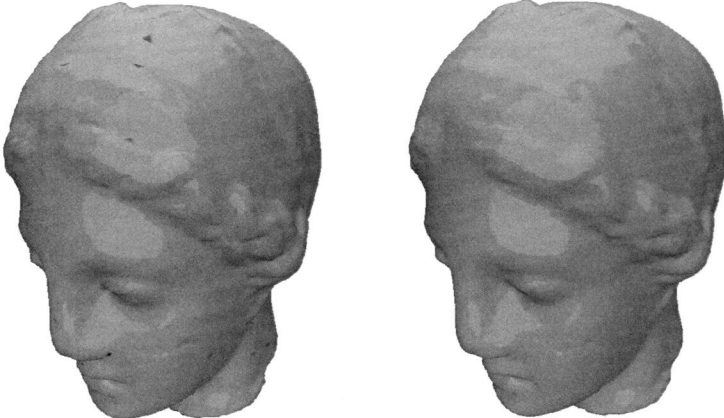

Fig. 11.1 The Egea statue on the *left* was optimized by edge flipping producing the result on the *right*. Initially, the integral absolute mean curvature was greedily optimized using edge flips. Next, the minimum angle was maximized using the same method, but with a threshold on the dihedral angle disallowing flips across sharp edges

greedy strategies where we always choose the best next step. Such strategies often work well but might be trapped in local minima. If the results are not satisfactory a "bigger hammer" that often improves things significantly is simulated annealing which is discussed more in Sect. 11.2.3

11.1   Simplification of Triangle Meshes

Acquired triangle meshes have a tendency to contain a lot of redundancy. For instance, an optical scanner measures the 3D position of points on a surface and it is not generally possible to tune the number of measured points to how detailed the surface is. Consequently, we have to deal with meshes containing, say, millions of polygons, and in many cases it is desirable to remove this redundancy by some sort of simplification.

Perhaps the method that first comes to mind first is clustering of vertices. For instance, we could divide space into a regular grid of rectangular boxes, and all vertices with a given box are replaced with a single new vertex at the average position of the old vertices. Clearly, some triangles have all vertices inside the box and are reduced to points (and are discarded), other triangles will be reduced to line segments, while some remain triangular.

This procedure has the advantage that it can simplify topology, i.e., we can simplify a sponge object effectively using this method. On the other hand, we often want to disallow topology changes. Another issue with the method is that the geometry of the object will change a lot: sharp edges and corners are not preserved when we simply compute average positions. In general, clustering is too simplistic and produces results far worse both with regard to triangle quality and to approximation

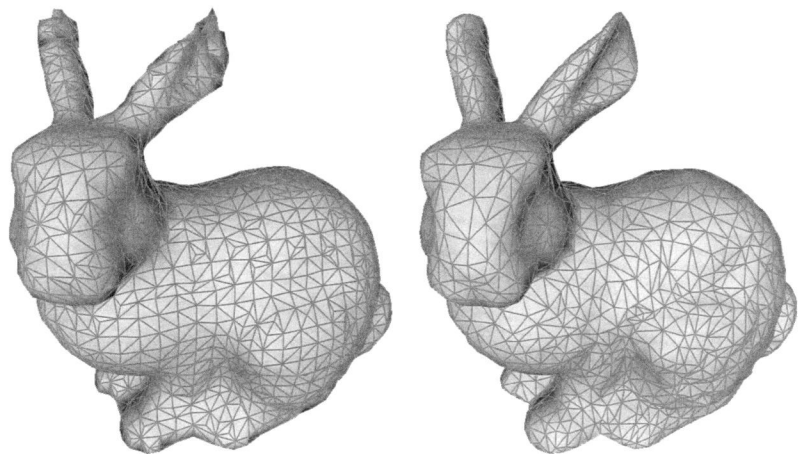

Fig. 11.2 A simple comparison of simplification by clustering on the *left* to Garland–Heckbert simplification on the *right*. The meshes contain the same number of triangles, 3682

of the original object than the method which we will discuss later in this chapter. For a comparison see Fig. 11.2.

Another fairly early method consists of removing a vertex along with all incident triangles and then triangulating the hole. If the hole has a fairly regular shape and the triangles lie in what is reasonable approximation of a plane [1]. This procedure known as *decimation* works well, but it has been superseded by a slightly simpler alternative which we will describe shortly. Decimation can also be seen as a precursor of more recent work where triangle clusters are found and replaced by individual polygons [2]. These methods, called variational shape approximation, directly compute simplified polygonal meshes from very detailed triangle meshes. The central idea is to find a set of clusters which minimizes an energy function that describes how well these clusters fit a smooth patch such as a plane or more generally a quadric surface [3]. Thus, instead of clustering vertices in space with no regard for connectivity, we cluster faces on the surface according to how well they approximate a smooth surface.

More recently, attention has been turned to other methods, which are based on creating periodic parametrizations of the triangles meshes based on vector fields such as those produced by the principal curvature directions. A good example of such a method is QuadCover by Kälberer et al. [4]. The advantage of these methods is that they produce quad meshes that align well with the symmetry directions of the object and generally have an appearance similar to meshes that have been created by a human designer.

However, when it comes triangle mesh simplification, a relatively early algorithm due to Garland and Heckbert is still much used [5]. Instead of removing a vertex along with its incident faces, the vertex is removed by merging it with a neighboring vertex thus collapsing their shared edge (cf. Sect. 5.4). In fact, this operation can more generally be seen as merging any two vertices in which case topological

---

Algorithm 11.1 Mesh simplification

1. For each edge we store the cost of a collapse in a priority queue.
2. Extract the collapse which is cheapest and perform it. This removes an edge along with its adjacent triangles. Its two vertices are merged into a single vertex, possibly with a new position.
3. Recompute the collapse cost for all edges affected by the collapse and update their position in the priority queue.
4. Until the stop condition is met, we go to 2.

---

changes are still possible, but in most cases we only merge connected vertices and often explicitly disallow a merge if the topology would change.

11.1.1 Simplification by Edge Collapses

When performing mesh simplification we generally wish to reduce the complexity of the model while preserving the geometry as much as possible. In the context of a method based on edge collapse, a *greedy strategy* would be to always choose the edge collapse which has the least impact on geometry and to go on until the simplification goal has been reached. Thus, we need a cost function which tells us how much the geometry changes for a given edge collapse. With such a cost function, we can perform mesh simplification using Algorithm 11.1. This algorithm requires a priority queue, which is an off-the-shelf data structure. However, a banal but tricky issue is that we need to update the entries in the priority queue in Step 3. A simple alternative is to put a time stamp on the edges. When an edge cost is updated, we update the time stamp and put an entry with the new time stamp in the priority queue. When an element is extracted from the queue, we discard it if the time stamps do not match.

The next important concern is the stop condition. The easiest condition is to simply stop when a given mesh complexity has been reached, e.g., how many of the original vertices or faces are left. Ideally, we would want the algorithm to stop when the changes to the mesh are too great. This is essentially what the cost function typically measures and an obvious function is one which measures the distance from the simplified shape to the original. This can be stated more precisely as the *Hausdorff distance*. If we regard the original mesh as a point set A and the simplified mesh as a point set B, we find, for all points in B, the closest point in A. The directed Hausdorff distance is then the greatest of these:

$$h(A, B) = \max_{\mathbf{a} \in A}\left(\min_{\mathbf{b} \in B}\left(\|\mathbf{a} - \mathbf{b}\|\right)\right). \tag{11.1}$$

This is not a symmetric measure, since the tip of a small feature in A could be far from B while all points on B are quite close to some point of A as shown in Fig. 11.3. To make it symmetric, we simply take the maximum of the directed

Fig. 11.3 The directed
Haussdorf distance from A to
B is greater than that from B
to A since the closest point on
B is relatively far from the tip
of the protrusion on A

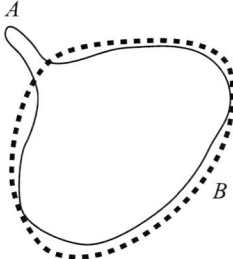

Hausdorff distance from A to B and B to A:

$$H(A, B) = \max\big(h(A, B), h(B, A)\big). \tag{11.2}$$

To compute the symmetric Hausdorff distance in practice, we will have to sample points on all of the triangles to create a dense set of points (capturing all features) for both A and B.

Thus, we could implement an algorithm which iteratively collapsed edges, always picking the edge whose collapse would cause the smallest increase in Hausdorff distance, until the Hausdorff distance were above a given threshold or a given number of triangle had been reached. In practice this algorithm will be quite costly from a computational point of view. A feasible alternative is to precompute a *distance field* (cf. Chap. 17) for the original mesh. If we denote the original B, the Hausdorff distance from the decimated mesh A to B is computed by sampling a set of points on A and for each point looking up the distance in B's distance field. The maximum of the distances found is an approximation of the directed Hausdorff distance. Although the grids used in [6] are binary in the sense that the grid points have binary values, this approach is similar to that adopted by Zelinka and Garland.

11.1.2 Quadric Error Metrics

However, previously Garland and Heckbert proposed another strategy [5]. Given a plane in 3D space which we will define through a normal, $\mathbf{n}$, and the distance of the plane from the origin, d, we can compute the square distance to a point, $\mathbf{p}$,

$$\begin{aligned}
Q(\mathbf{p}) &= (\mathbf{p} \cdot \mathbf{n} - d)^2 \\
&= \big(\mathbf{p}^T \mathbf{n} - d\big)\big(\mathbf{n}^T \mathbf{p} - d\big) \\
&= \mathbf{p}^T \mathbf{A} \mathbf{p} + \mathbf{b}^T \mathbf{p} + d^2,
\end{aligned} \tag{11.3}$$

where $\mathbf{A} = \mathbf{n}\mathbf{n}^T$ and $\mathbf{b} = -2d\mathbf{n}$. Normally, we define the plane from a triangle where a point of the triangle is used to compute the distance to the origin. If we call that point $\mathbf{p}_0$, then $d = \mathbf{n} \cdot \mathbf{p}_0$.

Thus, Q, which we will denote a *quadric error metric* (QEM), is a quadratic function which we can represent by the triple $\langle \mathbf{A}, \mathbf{b}, d \rangle$. Alternatively, using homogeneous coordinates, we could also represent Q as a 4×4 matrix, however, we will use the triple notation. Assuming we have two planes, measuring the sum of square distance is straightforward:

$$Q_1(\mathbf{p}) + Q_2(\mathbf{p}) = \sum_{i \in \{1,2\}} \mathbf{p}^T \mathbf{A}_i \mathbf{p} + \mathbf{b}_i^T \mathbf{p} + d_i^2$$

$$= \mathbf{p}^T (\mathbf{A}_1 + \mathbf{A}_2)\mathbf{p} + (\mathbf{b}_1^T + \mathbf{b}_2^T)\mathbf{p} + (d_1^2 + d_2^2). \quad (11.4)$$

Or we could simply sum the two QEMs

$$Q = Q_1 + Q_2 = \langle \mathbf{A}_1 + \mathbf{A}_2, \mathbf{b}_1 + \mathbf{b}_2, d_1 + d_2 \rangle. \quad (11.5)$$

If Q represents the sum of two or more QEMs, $Q(\mathbf{p})$ represents the sum of squared distances to the original planes. Computationally, there is no difference between computing the squared distance to a single plane and computing the sum of squared distances to a great number of planes.

Informally, a QEM is a function which measures how close a point is to a bunch of planes (containing a subset of the triangles in a 3D model). The value may be high either because the point is far from the planes or because the planes are not well aligned. No matter how many points are involved, the computational cost is constant. If $\mathbf{p}$ is a vertex, and we find Q, the QEM which represents the planes of all triangles incident on $\mathbf{p}$, then $Q(\mathbf{p}) = 0$. Conversely, as $\mathbf{p}$ moves away from this point, the value increases.

For the purpose of simplification, we initially compute a QEM for each vertex by summing the QEMs for the planes of all incident triangles. We then compute Q, a QEM for each edge, by summing the end-point QEMs. The error of performing an edge collapse is then simply $Q(\mathbf{p})$ where $\mathbf{p}$ is the position of the collapsed vertex. That position could, for instance, be the position of one of the end-points $\mathbf{q}_1$ or $\mathbf{q}_2$, depending on whether $Q(\mathbf{p}_1) < Q(\mathbf{p}_2)$. Alternatively, we could pick the optimal position. Since Q is quadratic, we know that its minimum is where $\nabla Q = 0$ and

$$\nabla Q(\mathbf{p}) = 2\mathbf{A}\mathbf{p} + \mathbf{b}, \quad (11.6)$$

so we can easily find the optimal position, $\mathbf{p}_{\mathrm{opt}}$,

$$\nabla Q(\mathbf{p}_{\mathrm{opt}}) = 2\mathbf{A}\mathbf{p}_{\mathrm{opt}} + \mathbf{b} = 0 \implies \mathbf{p}_{\mathrm{opt}} = -\frac{1}{2}\mathbf{A}^{-1}\mathbf{b}. \quad (11.7)$$

Unfortunately, $\mathbf{A}$ may not be full rank which corresponds to the optimum occurring at any point in a plane or on a line. For instance, if all the faces that contributed to Q lie in the same plane, $Q = 0$ for any point in that plane. To solve that problem, one generally computes the singular value decomposition of $\mathbf{A} = \mathbf{U}\mathbf{\Sigma}\mathbf{V}^T$ to find the pseudo inverse, $\mathbf{A}^+ = \mathbf{V}\mathbf{\Sigma}^+\mathbf{U}^T$ where small singular values of $\mathbf{\Sigma}$ have been set to zero to obtain a stable solution. An advantage of using the optimal position is that we generally get a better simplified mesh. However, we pay the price that the set of vertices of the simplified mesh is not the subset of the vertices of the original mesh. If we want to compute a progressive mesh [7] such that we can dynamically change

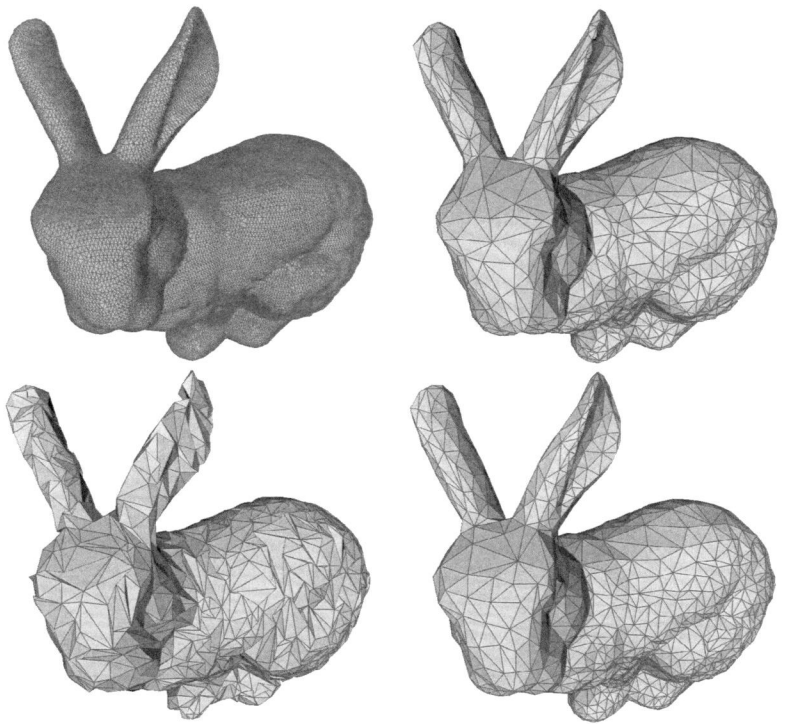

Fig. 11.4 Four bunnies. The *top left* image shows the Stanford Bunny. To the *right* it has been simplified using the Garland–Heckbert method with 33000 edge collapses. On the *bottom left*, 33000 random collapses were used, and on the *bottom right* 33000 edge collapses with optimal point placement

the number of vertices we use for a given 3D model, it is a great advantage that the vertices of the simplified mesh are a subset. However, for simply computing a simplified mesh, it is preferable to use the optimal positions.

Whatever strategy is chosen, we find the new vertex position $\mathbf{p}_{new}$ and compute its cost $Q(\mathbf{p}_{new})$ which we store in a priority queue. We then extract the cheapest collapse and perform it by removing the edge and its to adjacent triangles. The new merged vertex is then placed according to the chosen strategy. The QEM of the merged vertex is simply the sum of the QEMs of the vertices which were merged. Finally, all edges incident on the new, merged vertex are updated and reinserted in the priority queue.

The results of the presented algorithm are shown in Fig. 11.4. It is very reassuring that the random collapses in the bottom left image are clearly inferior to the QEM-based collapses in the right images. If we look closely, and especially at the ears of the Stanford Bunny, it is also clear that optimal placement produces better meshes with more equilateral triangles than simply choosing one of the end-points.

11.1.3 Some Implementation Details

In Garland and Heckbert's original description of the algorithm, any pair of vertices can be contracted, but it is often implemented on top of a halfedge data structure which does not support non-manifold surfaces. This means that vertices belonging to different components will not be selected for merging, but also that we must take care not to perform simplifications that might result in non-manifold meshes. These cases were discussed in Sect. 5.4.

A more subtle issue is related to the fact that if $\mathbf{A}$ is singular (or so close that we set some of the singular values to zero) $\mathbf{p}_{opt} = -\frac{1}{2}\mathbf{A}^+\mathbf{b}$ yields the optimal point closest to the origin. That is usually not desirable since the origin could be placed quite arbitrarily. We would much prefer, say, the optimal point closest to the center of the edge being collapsed. Fortunately, we can remedy the situation since columns of $\mathbf{V}$ which correspond to zeroes in $\mathbf{\Sigma}$ form a basis for the null space. Thus, we can add any linear combination of those columns to $\mathbf{p}_{opt}$. So, in practice, if $\mathbf{v}_1$ and $\mathbf{v}_2$ are columns whose corresponding singular values are zero, we compute the new position

$$\mathbf{p}_{new} = \mathbf{p}_{opt} + \mathbf{v}_1(\mathbf{v}_1 \cdot \mathbf{p}_0) + \mathbf{v}_2(\mathbf{v}_2 \cdot \mathbf{p}_0), \tag{11.8}$$

where $\mathbf{p}_0$ is the point we wish should serve as the origin. In practice we use the edge midpoint.

Garland and Heckbert adopted a slightly different strategy which is explained in Michael Garland's thesis [8]. If the matrix $\mathbf{A}$ is regular, they compute the inverse. If it is singular they search along the line segment for an optimal position with the end-points of the contracted edge as their final fall back option.

The precautions discussed above do not entirely guarantee that the new vertex position is a sound choice. The new vertex position could introduce flipped triangles with a very big dihedral angle along some edge. To prevent this, we check whether the new vertex position lies inside the region defined by the triangles that will form its new 1-ring [8].

A final issue is that of mesh boundaries. A vertex on the boundary can certainly be collapsed. If two vertices that are collapsed both lie on the boundary we will normally require that the edge between them is also a boundary edge, since we otherwise introduce a vertex which lies on two boundary loops. From a geometric point of view, boundaries also cause trouble. Boundary vertices which result from a collapse could easily be pulled away from the boundary which, effectively, could cause holes in the mesh to grow. This problem is easily fixed by adding extra QEMs to the QEMs of boundary vertices. For a boundary vertex, we compute two planes both of which contain its normal and one of its two incident boundary edges. These planes which are perpendicular to the surface and contain the boundary edges suffice to prevent the boundary loops from degenerating.

Algorithm 11.2 Mesh optimization

1. Initially, ΔF is computed for all edges, and for each edge e a pair $\langle \Delta F(e), e \rangle$ is inserted into a priority queue if $\Delta F(e) < 0$.
2. The next step is a loop where we iteratively extract and remove the record with the ΔF corresponding to the greatest decrease in energy from the heap and flip the corresponding edge.
3. After an edge flip, $\Delta F(e')$ must be recomputed for any edge e' if its $\Delta F(e')$ has changed as the result of e being flipped.
4. The loop continues until the priority queue is empty.

11.2   Triangle Mesh Optimization by Edge Flips

As we have seen above, simplification can be performed through simple greedy optimization where we select the cheapest edge to collapse until the stop criterion is reached. A very similar algorithm can be used for a wide range of tasks if—instead of edge collapse—we use edge flips (cf. Fig. 11.6). This method simply reconnects a set of points making no changes to the vertex positions, but it does reconnect vertices. This algorithm applies whenever we wish to improve the geometry of the mesh without changing the number or position of the vertices.

An energy function and a priority queue are needed and the basic scheme is to always perform the edge flip which leads to the greatest reduction in energy F. Given an edge, e, we denote by $\Delta F(e)$ the energy after minus the energy before an edge flip:

$$\Delta F(e) = F_{\text{after}}(e) - F_{\text{before}}(e). \qquad (11.9)$$

Thus, to optimize greedily we need to pick always the edge e with the most negative $\Delta F(e)$. The details are in Algorithm 11.2. When updating the priority queue, one can use the time stamp method discussed above in Sect. 11.1.1.

This algorithm can be used to turn any planar triangulation into a Delaunay triangulation. A Delaunay triangulation has the property that it minimizes the maximum angle. If a flip which changes a configuration of two triangles $\mathbf{p}_0\mathbf{p}_1\mathbf{p}_2$ and $\mathbf{p}_0\mathbf{p}_2\mathbf{p}_3$ into $\mathbf{p}_0\mathbf{p}_1\mathbf{p}_3$ and $\mathbf{p}_3\mathbf{p}_1\mathbf{p}_0$ (cf. Fig. 11.7) increases the smallest of the six angles in the configuration it should be made. This corresponds to locally making the edge Delaunay, and when all such edges have been processed, the mesh is Delaunay (cf. Sect. 14.2.3). Thus in this case

$$\Delta F = -(\text{smallest angle after flip} - \text{smallest angle before flip}). \qquad (11.10)$$

Note that in this case ΔF is only the local change in energy. Unless the smallest of the six angles happens to be the globally smallest angle, the energy of the entire mesh is unaffected. Nevertheless, if the algorithm is run until no more local changes can be made, the mesh is Delaunay [9].

Having performed the flip, we need to update $\Delta F(e')$ for any edge e' belonging to the two triangles which share the flipped edge e.

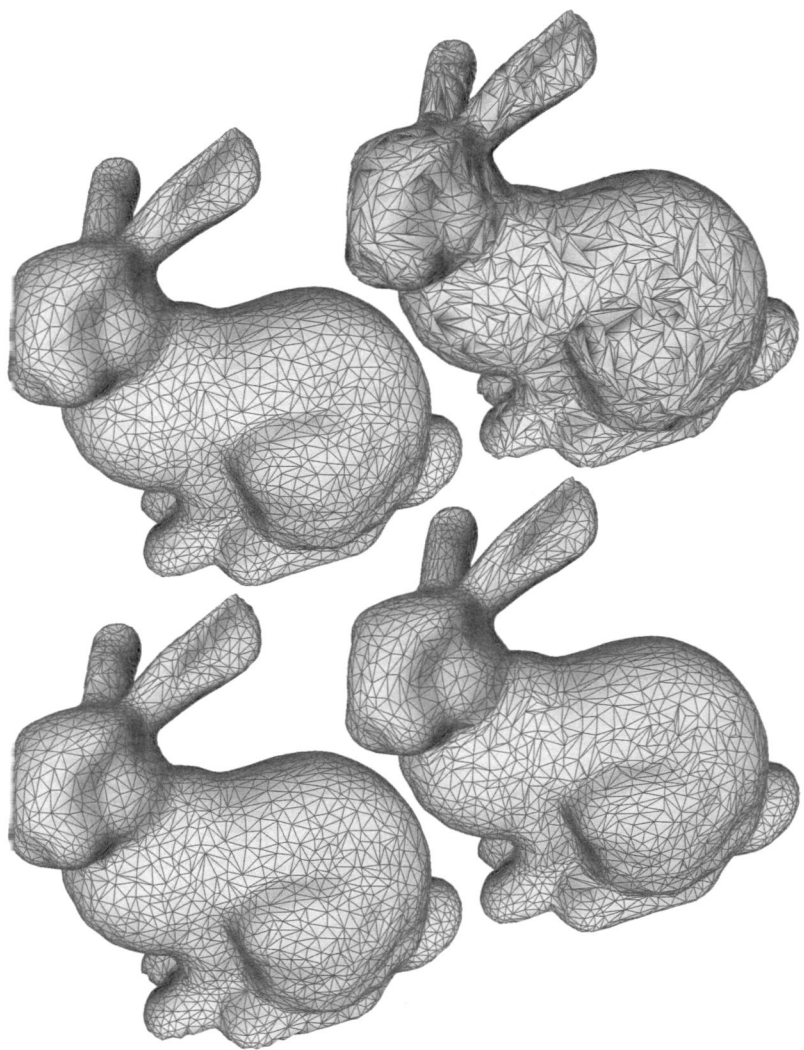

Fig. 11.5 The starting point (*top left*) is a simplified version of the Stanford bunny (simplified using the Garland–Heckbert method) which has been corrupted by random edge flips (*top right*). We then use the greedy optimization method just described to maximize the minimum angle (*bottom left*) and to minimize the integral absolute mean curvature (*bottom right*)

An important aspect of this algorithm is that it also works quite well on meshes which are not planar, and in Fig. 11.5 we see the result (bottom left) of applying the method to a simplified bunny where the connectivity has been corrupted by random edge flips (shown top right). The result is, in fact, very similar to the original model (top left) but near the bottom of the model, we notice that the silhouette has become a

Fig. 11.6 Edge Configurations: A triangle mesh approximates a surface with a sharp bend. On the *left* an edge is transverse to the bend. On the *right* an edge flip has been performed, and the new edge follows the bend

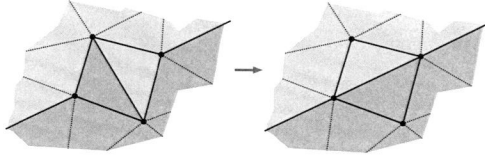

Fig. 11.7 This illustration shows an edge e, the normals of its adjacent faces, its end-points and the dihedral angle

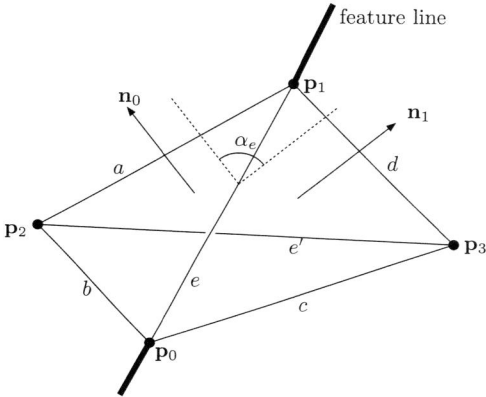

bit jagged. This is unsurprising because in a sense our energy only takes the triangle quality and not the mesh quality into account.

11.2.1 Energy Functions Based on the Dihedral Angles

Maximizing the minimum angle improves triangle quality but not necessarily the mesh geometry. Consider Fig. 11.6 where points are sampled near a sharp feature line (or bend). Possibly the best configuration considering only the triangle quality is the one on the left, but clearly the configuration on the right is a better representation of what we expect the geometry should look like.

Dyn et al. considered various curvature-based energy measures which could be used to improve triangle meshes using edge flips [10]. One of their findings was that the integral absolute mean curvature works well and is fairly cheap to compute. There are several ways of computing the integral absolute mean curvature. One of these is to sum the dihedral angle times edge length for all edges (8.10). The approach used by Dyn et al. is similar, but they compute a spatial average for each vertex and then sum the per-vertex measures of curvature. We find that directly using (8.10) leads to meshes of just as high quality. If we denote the dihedral angle by β

Fig. 11.8 Plot of F^c for a
single edge e as a function of
dihedral angle β_e with $\gamma = 4$
and $l_e = 1$

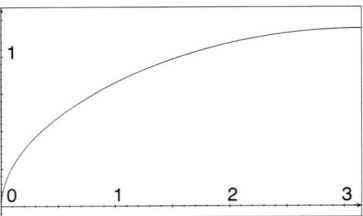

(cf. Fig. 11.7), the energy is simply

$$F^\beta = \frac{1}{2} \sum_{i=1}^{|\mathcal{E}|} |\beta_i| \|\mathbf{e}_i\|. \qquad (11.11)$$

This functional is almost the same as F_2 in [10], except that we have energy per
edge rather than per vertex.

When an edge flip is performed, the difference in energy,

$$\Delta F^\beta = F^\beta_{\text{after}} - F^\beta_{\text{before}},$$

is easily computed since only the five edges belonging to the two triangles adjacent
to e are affected. This energy measure works quite well and was used in Fig. 11.5
(bottom right). It is clear that minimizing the integral absolute mean curvature sometimes produces triangles that have a smaller minimum angle but a better overall
mesh geometry. In particular, we notice that the jagged silhouette from the bottom
left image is straight in the bottom right image.

For some tasks, we have found that variations F^β can be quite useful. These
tasks, mostly related to optimization of triangle meshes representing terrain models,
were investigated in a technical report [11] from which we relate central findings in
this chapter.

One simplification we can make to F^β is due to the fact that the angle β is
generally computed from the normal vectors and we do not need to take the inverse
cosine to obtain the angle. More importantly, if we consider Fig. 11.6 it is clear that
the flipped edge in the configuration on the right could have a very sharp dihedral
angle such that ΔF^β does not decrease because a single big dihedral angle after the
flip could outweigh the sum of the moderately big dihedral angles before the flip.
We address this problem by introducing a parameter γ which is used to bias the
energy towards making flips. Putting these things together, we obtain

$$F^c = \sum_{i=1}^{|\mathcal{E}|} \left[l_e \big(1 - \cos(\beta_e)\big) \right]^{1/\gamma}, \qquad (11.12)$$

where we often choose $\gamma = 4$. A plot of F^c as a function of β_e is shown in Fig. 11.8.
An example of the use of this energy which we shall return to is shown in Fig. 11.10.

Fig. 11.9 This figure illustrates the problem of flipping an edge which is adjacent to a vertex of valence three

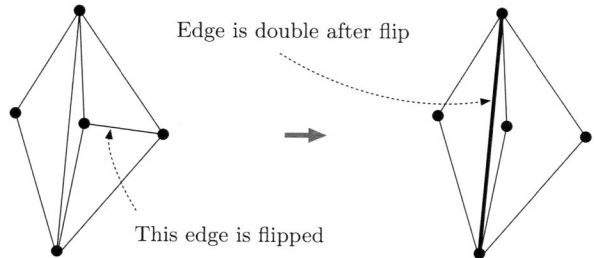

Edge is double after flip

This edge is flipped

11.2.2 Avoiding Degenerate Triangles

Not all edges should be flipped even if the flip will cause an energy reduction. Sometimes an edge flip can cause a very sharp edge to be introduced. Since the method, essentially, tries to concentrate dihedral angle in a few edges while keeping most edges smooth, the issue is not surprising, but it should be avoided by setting an upper threshold, $\tau \in [0, \pi]$, on the dihedral angle. If an edge flip results in a dihedral angle greater than this threshold, the flip is not allowed. Usually, this threshold is set to $2.09 \, \text{rad} \approx 120°$.

In the case of terrain data, we know that the triangles should always face up, consistently. If an edge flip results in a triangle facing down, the flip is not allowed.

If we are dealing with terrain data, it seems that either of these two rules suffices. For general triangle meshes, the first rule is necessary, and the second does not apply.

However, more rules are necessary: It is assumed that the triangle mesh represents a manifold surface [12], possibly with boundary. It is fairly obvious that we cannot flip boundary edges. However, some flips will also violate the manifold property of the mesh and these are not allowed. In practical terms, manifoldness means that every edge is adjacent to two triangles, and that the triangles sharing a vertex form a single cycle around that vertex (cf. Sect. 5.2). Moreover, the intersection of two triangles should be either empty or the shared edge.

To preserve manifoldness, an edge is flipped only on two conditions:
1. Both vertices at the end-points of the edge must have valence (i.e., the number of adjacent edges) greater than three.
2. The two vertices which will be connected by the flip must not be connected by a different edge.

The first rule ensures that we do not have vertices of valence two after the flip. If an interior vertex has valence two, it is connected to two edges which in turn are both adjacent to the same two triangles. Then these two triangles share all three vertices, which means that they have collapsed as shown in Fig. 11.9.

Since all edges are straight, a violation of the second rule means that two edges after the flip are geometrically identical. Since an interior edge is shared by two faces, we would have at least three and in general four faces meeting at the same geometric edge after the flip.

11.2.3 Simulated Annealing

A problem with the greedy strategy is that for many problems, including the present, all edges can be in a configuration that is locally optimal while the configuration is not globally optimal. Simulated annealing [13] is a general framework for optimization which is well suited to avoid these local optima, although the algorithm is often slow.

Larry Schumaker initially suggested using simulated annealing as a tool for computing optimal triangulations via edge flips [14]. Applied to the problem at hand, the method works as follows. We iteratively, pick a random edge from $\mathcal{E}$ and compute the ΔF associated with flipping this edge. If $\Delta F \leq 0$ the flip is performed since the energy decreases. If $\Delta F > 0$, the flip is performed with probability:

$$P_{\text{flip}}(e) = e^{\frac{-\Delta F}{T}}, \tag{11.13}$$

where T is the temperature. A random number, r is generated, and if $P_{\text{flip}}(e) > r$ then e is flipped. For a given T, a small ΔF means a high probability, and for a given ΔF a high temperature means high probability.

Initially, the temperature is very high, which means that all flips are probable. After some time, the temperature is lowered according to an *annealing schedule*, and as the temperature approaches zero, so does the probability of making flips which cause the energy to increase. The intuition behind the method is that the initial random flips help avoid being trapped in poor local optima.

Simulated annealing requires some initial parameters. For the experiments in this chapter, these have been selected experimentally based on advice given in [14]. The initial temperature is

$$T_0 = -2 \min_e \big(\Delta F(e)\big).$$

All edges are visited five times in random order (and are either flipped or not according the scheme above) before the temperature is lowered. The new temperature is then computed according to

$$T \leftarrow T \cdot 0.9.$$

When the temperature becomes very low, flips that increase the energy cease to occur in practice, and when all edges have been visited without any flips being made, the algorithm terminates. Pseudocode which sums up the steps discussed above is provided in Algorithm 11.3.

The method has been tested on several models. The row of images in Fig. 11.10 show a terrain model generated using Delaunay triangulation and after various types of optimization. The most important feature is a road which runs along the left edge of the terrain. Fortunately, it is very easy to verify whether a triangulation correctly captures this feature, and that is plainly not the case for the Delaunay triangulation shown on the left. In the image on the center left, F^c has been minimized using a greedy approach. As feared, it does get stuck in a local minimum, and the edges

---

Algorithm 11.3 Simulated annealing on meshes

---

1. Set $T = -2 \min_e(\Delta F(e))$
2. Perform five iterations where each edge e is visited in random order:
   - if $\Delta F(e) \leq 0$ flip e
   - else perform flip with probability $\exp(\frac{-\Delta F(e)}{T})$
3. $T \leftarrow T \cdot 0.9$
4. if any flips were made, go to 2

Fig. 11.10 From *left* to *right*, this figure shows the original Delaunay triangulation of the height points, the triangulation after a greedy optimization (with $\gamma = 4$), the triangulation after simulated annealing with $\gamma = 1$, and, finally, the triangulation after optimization using simulated annealing with $\gamma = 4$. Notice how the road running along the *left side* of the terrain is correctly reconstructed in the image on the *right*

are only partially aligned with the road. In the image on the center right, simulated annealing has been used with $\gamma = 1$, but again the result is not satisfactory. Only the combination of $\gamma = 4$ and simulated annealing gives a good reconstruction of the road as seen in the image on the right. The images in Fig. 11.11 show height curves from the terrain generated directly from the Delaunay triangulation and after optimization (the same as in the top right image).

The Venus model was used for a test which is shown in Fig. 11.12. The initial model was corrupted by performing random edge flips resulting in the model shown in the middle. Using simulated annealing in conjunction with F^β produced the result on the right.

Unfortunately, simulated annealing is not efficient, and for very large models, run times of hours could be required. Perhaps for this reason Dyn et al. [10] took a different approach to avoiding local minima. Instead of using simulated annealing they considered combinations of two flips which combined would give an energy reduction. To be more precise, a single flip might sometimes lead to an energy increase, but if we combine it with a second flip, the combined operation decreases the energy.

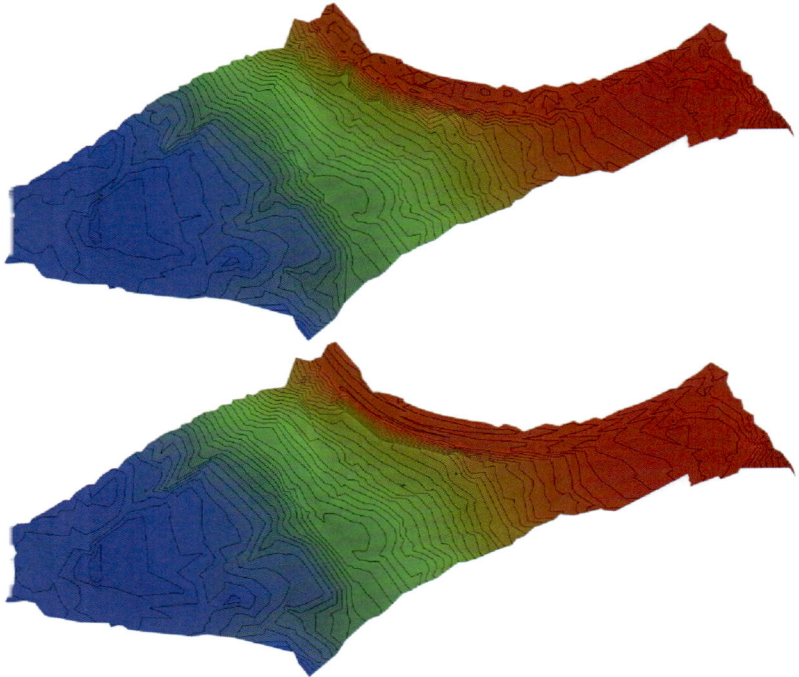

Fig. 11.11 The two images show the height curves of the model before and after minimization of F^c, $\gamma = 4$

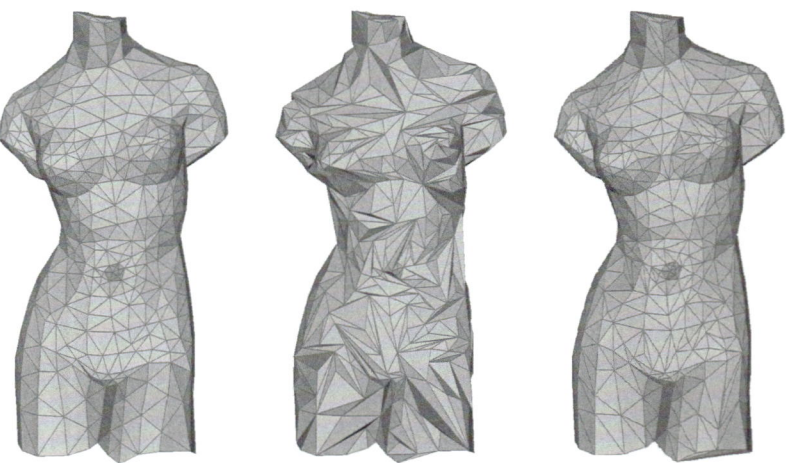

Fig. 11.12 The original Venus model (*left*), the Venus model corrupted by random edge flips (*middle*), the Venus model after minimization of F^β using simulated annealing (*right*)

Algorithm 11.4 Remeshing by local operations

1. Split all edges longer than $4/3 \, l$ at their midpoint.
2. Collapse all edges shorter than $4/5 \, l$ to their midpoint.
3. Optimize the mesh to improve regularity using Algorithm 11.2.
4. Smooth the mesh.
5. Project vertices onto the original surface.

Thus, a scheme similar to the greedy approach described above is used, but the priority queue now also contains these double flips. On the other hand, an advantage of simulated annealing is that it is simpler to implement and almost always gives good results provided the parameters have been set to reasonable values. Therefore, simulated annealing is often a good choice for applications where quality of the result is important and run time less so.

11.3   Remeshing by Local Operations

In the previous section, we discussed schemes that only reconnect vertices. While that is sufficient for many tasks, we sometimes need to do more. For instance, the meshes produced by isosurface polygonization algorithms as discussed in Chap. 18 often have vertices which are extremely close together, and, clearly, flipping will not improve on that. Edge collapses would help, and we can add edge splits to break edges that are too long. If we also add smoothing to the repertoire (cf. Chap. 9), we can achieve a semi-regular mesh of high quality. Unfortunately, when we smooth the mesh, we invariably pull it a bit away from the original surface. To ensure that we blur details as little as possible, we should also project the vertices back onto the original surface. If the mesh is a triangulated isosurface, projection onto the isosurface is straightforward (and the procedure is discussed below). If we only have the original triangle mesh, projection is a bit more difficult but still feasible. One approach is to create a distance field from the triangle mesh.

The algorithm we discuss in the following is the one used in [15] which was in turn adapted from [16]. The input is a triangle mesh and a representation of the original surface onto which we can project points that have moved. We initially compute the median edge length l of the original triangle mesh and then perform ten iterations of Algorithm 11.4. As explained in [17] the numbers 4/3 and 4/5 are not arbitrary, but judiciously chosen to ensure that collapses and splits result in edge lengths closer to the target length l than before. Unfortunately, collapsing an edge shorter than 4/5 of the target length could cause some of the edges connected to the old end-points to exceed 4/3 of the target length. It is important that we disallow a collapse if that situation arises [17]. Otherwise, the mesh does not approach a stable configuration.

The optimization in Step 3 is simply carried out using the greedy edge flip algorithm discussed in Sect. 11.2 but using deviation from regularity as energy. In a regular triangle mesh, all vertices have valency six. Thus, we can define a regularity energy

$$F^r = \frac{1}{2} \sum_{i=1}^{|\mathcal{V}|} (|\mathcal{N}_i| - 6)^2, \qquad (11.14)$$

which is clearly minimal for a completely regular mesh. However, for boundary vertices the number 6 should be replaced with 4 since that is the valency of regular boundary vertices.

In our experience, maximizing the minimum angle also works well as an alternative to directly optimizing the valency. To promote the regularity of the triangle areas, tangential, area weighted Laplacian smoothing is used in Step 4. This is similar to normal Laplacian smoothing except for two features.

- Each vertex is weighted with the area of its 1-ring. This has the effect that vertices with a big 1-ring area pull more than vertices with a small 1-ring area promoting equalization of the 1-ring areas.
- The Laplacian vector is projected into the tangent plane of the vertex. This prevents shrinkage and ensures that vertices stay on or at least close to the surface.

In our experience Taubin smoothing also works quite well: like tangential Laplacian smoothing it shrinks the surface very little. On the other hand, some shrinkage is unavoidable, and, in Step 5, we adjust the mesh to compensate for the shrinkage which occurs during smoothing. Say our surface is implicitly represented (cf. Sect. 3.10) via a function $\Phi : \mathbb{R}^3 \to \mathbb{R}$ where τ is the isovalue, i.e., the value such that we can define our surface $S = \{\mathbf{x}|\Phi(\mathbf{x}) = \tau\} = \Phi^{-1}(\tau)$. In that case, we can simply project the point using

$$\mathbf{p}_{\text{new}} = \mathbf{p} - \left(\Phi(\mathbf{x}) - \tau\right) \frac{\nabla\Phi(\mathbf{x})}{\|\nabla\Phi(\mathbf{x})\|^2}. \qquad (11.15)$$

We usually do not have a closed form expression for Φ which is rather represented as a regular 3D grid of samples. Thus, we would typically use central differences (cf. Sect. 4.1) to approximate the gradient $\nabla\Phi$ and (trilinear) interpolation to compute values away from grid points.

If Φ is not close to being a linear function of distance to the surface, we may need to iterate (11.15) a few times to get sufficiently close to the isosurface, but note that since a tessellation of the isosurface is a starting point, $\mathbf{p}$ is originally very close to the isosurface and (11.15) generally works well with very few iterations.

As also argued in [17] this algorithm is simpler to implement than many of the alternatives. Figure 11.13 shows the result of the application of the algorithm to a mesh produced by Marching Cubes (cf. Chap. 18). The difference is hard to see on the rendered image on top, but quite obvious in the reflection line and wireframe images. However, the real reason for performing this optimization is that if we are to use the mesh for numerics, a semi-regular triangle mesh is preferable to an irregular mesh [16].

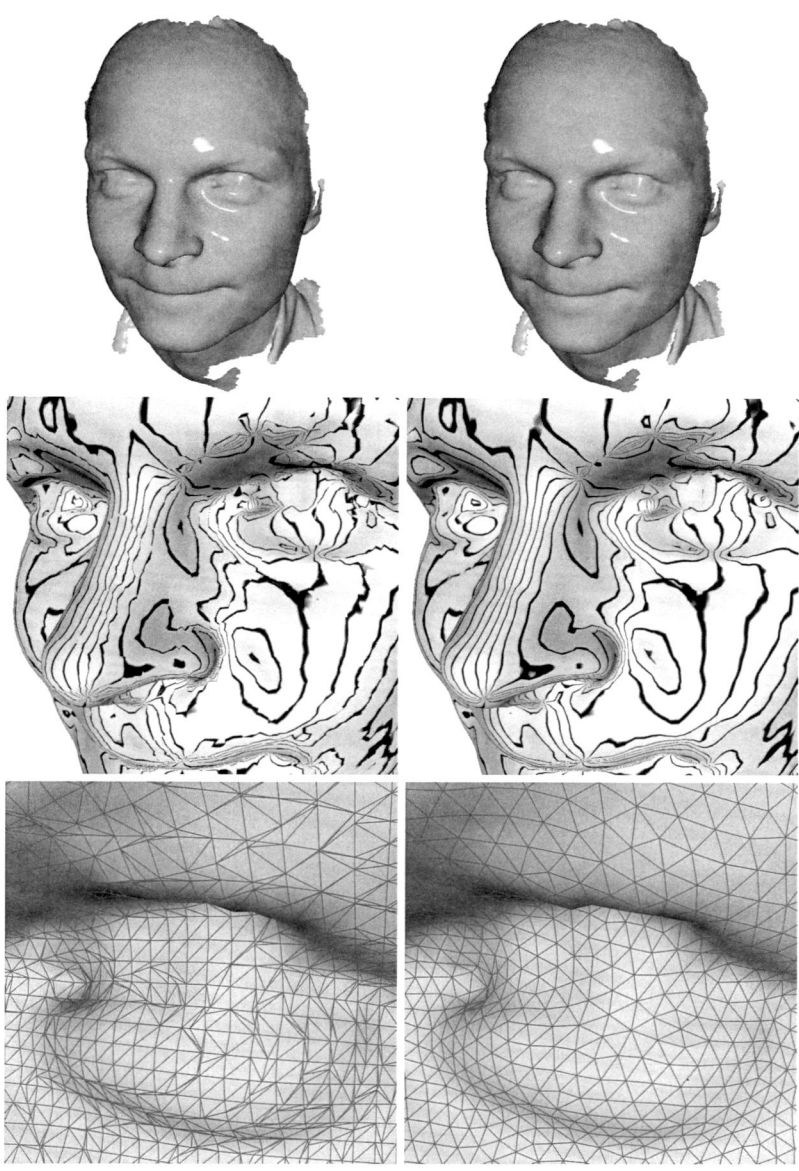

Fig. 11.13 This figure shows several images of a mesh reconstructed from a structured light scan of the first author of [15]. The *left column* of images show the result before the optimization method discussed in the present section has been applied. The *right column* shows the result after. From *top* to *bottom* we have normal rendering, rendering of reflection lines and wireframe

11.4 Discussion

The focus of this chapter has been on simple algorithms which can be used to improve triangle meshes. When it comes to optimization of triangles meshes without moving the vertices, our options are fairly limited, and the methods in Sect. 11.2 are probably the most reasonable methods to apply. On the other hand, simplification and remeshing (Sects. 11.1 and 11.3) can be done in many ways and much research has gone into these two topics. Incidentally, we can see simplification as being just one goal of remeshing. Other goals, which we did not consider here would be to obtain a quad mesh aligned with the principal curvature directions and out-of-core methods which allow for remeshing of meshes too large to be contained in a computer's main memory.

The book by Botsch et al. [17] is a good place to begin further investigation of this and other topics pertaining to remeshing. Hoppe's early work on optimization of meshes and subdivision surfaces so that they fit a point cloud is also recommended [18, 19]. In a broader perspective, the goal of mesh optimization is to produce a CAD model from a laser scan or otherwise acquired model of a real world object. If we go a bit further than just reconstructing triangle meshes, and reconstruct parametric (e.g., NURBS) patches, we arrive at *reverse engineering* for CAD which is an important topic in its own right.

11.5 Exercises

Exercise 11.5.1 [GEL Users] GEL supports functions for mesh optimization through edge flipping and also Garland–Heckbert simplification. Experiment with these methods. In particular, try to combine mesh simplification and mesh optimization in an iterative algorithm that both simplifies and improves triangle meshes.

Exercise 11.5.2 Based only on a mesh data structure that supports edge flips, implement the scheme for mesh optimization outlined in Algorithm 11.2. In particular, we suggest you test the implementation on the two problems of maximizing minimum angle and minimizing the integral absolute mean curvature.

References

1. Schroeder, W.J., Zarge, J.A., Lorensen, W.E.: Decimation of triangle meshes. Comput. Graph. **26**(2), 65–70 (1992)
2. Cohen-Steiner, D., Alliez, P., Desbrun, M.: Variational shape approximation. ACM Trans. Graph. **23**(3), 905–914 (2004)
3. Yan, D.M., Liu, Y., Wang, W.: Quadric surface extraction by variational shape approximation. In: Geometric Modeling and Processing-GMP 2006, pp. 73–86 (2006)
4. Kälberer, F., Nieser, M., Polthier, K.: QuadCover-surface parameterization using branched coverings. Comput. Graph. Forum **26**(3), 375–384 (2007)
5. Garland, M., Heckbert, P.S.: Surface simplification using quadric error metrics. In: Proceedings of the 24th Annual Conference on Computer Graphics and interactive Techniques, pp. 209–216 (1997)

6. Zelinka, S., Garland, M.: Permission grids: practical, error-bounded simplification. ACM Trans. Graph. **21**(2), 207–229 (2002)
7. Hoppe, H.: Progressive meshes. In: Proceedings of the 23rd Annual Conference on Computer Graphics and Interactive Techniques, pp. 99–108. ACM, New York (1996)
8. Garland, M.: Quadric-based polygonal surface simplification. Ph.D. thesis, Georgia Institute of Technology (1999)
9. Shewchuck, J.R.: Lecture notes on delaunay mesh generation. Technical report, UC Berkeley (1999). http://www.cs.berkeley.edu/~jrs/meshpapers/delnotes.ps.gz
10. Dyn, N., Hormann, K., Kim, S.-J., Levin, D.: Optimizing 3d triangulations using discrete curvature analysis. In: Mathematical Methods for Curves and Surfaces, Oslo, 2000, pp. 135–146. Vanderbilt University, Nashville (2001)
11. Bærentzen, J.A.: Optimizing 3d triangulations to recapture sharp edges. Technical report, Technical University of Denmark, Informatics and Mathematical Modelling, Image Analysis and Computer Graphics (2006). http://www2.imm.dtu.dk/pubdb/p.php?4689
12. Hoffmann, C.M.: Geometric and Solid Modeling. Morgan Kaufmann, San Mateo (1989)
13. Kirkpatrick, S., Gelatt, C.D., Vecchi, M.P.: Optimization by simulated annealing. Science **220**(4598), 671–680 (1983)
14. Schumaker, L.L.: Computing optimal triangulations using simulated annealing. In: Selected Papers of the International Symposium on Free-form Curves and Free-form Surfaces, pp. 329–345. Elsevier, Amsterdam (1993). doi:10.1016/0167-8396(93)90045-5
15. Paulsen, R.R., Baerentzen, J.A., Larsen, R.: Markov random field surface reconstruction. In: IEEE Transactions on Visualization and Computer Graphics, pp. 636–646 (2009)
16. Botsch, M., Kobbelt, L.: A remeshing approach to multiresolution modeling. In: Proceedings of the 2004 Eurographics/ACM SIGGRAPH Symposium on Geometry Processing, SGP'04, pp. 185–192. ACM, New York (2004)
17. Botsch, M., Kobbelt, L., Pauly, M., Alliez, P., Levy, B.: Polygon Mesh Processing. AK Peters, Wellesley (2010)
18. Hoppe, H., DeRose, T., Duchamp, T., McDonald, J., Stuetzle, W.: Mesh optimization. In: Proc. ACM SIGGRAPH 93 Conf Comput Graphics and Proceedings of the ACM SIGGRAPH'93 Conference on Computer Graphics, pp. 19–25 (1993)
19. Hoppe, H., DeRose, T., Duchamp, T., Halstead, M., Jin, H., McDonald, J., Schweitzer, J., Stuetzle, W.: Piecewise smooth surface reconstruction. In: Proceedings of the 21st Annual Conference on Computer Graphics and Interactive Techniques, pp. 295–302. ACM, New York (1994)

Spatial Data Indexing and Point Location

This chapter is devoted to the spatial data structures that can be used in order to optimize the access to spatial objects, needed in many of the other chapters of this part, and in particular, the Iterative Closest Point algorithm. Indeed, in many geometric algorithms, one needs to get fast access to the neighbors of a given object in order to traverse them. This is also true in some application fields like robotics and computer vision. However, many geometric data sets are too large to be stored entirely in the Random Access Memory (RAM). The purpose of spatial data indexing is to decompose the space into regions placed along a space filling curve, where each region can be stored within a disk block, which is the smallest unit of memory that can be transferred between the Random Access Memory and the mass storage. The Z-order curve shown in Fig. 12.1 is a space filling curve. We call these curves "space filling curves", because the sequence of tiles given by the curves fills the space of the data set. The resulting order of the regions is given by the linear order of the regions along the space filling curve. An easy adversary argument implies that any linear order will fail to always have locations that are close in two- or higher-dimensional space close in this linear order. Thus, no spatial indexing in two- or higher-dimensional space is perfect. Some compromises must be made.

This chapter is organized as follows. Section 12.1 will present the context of spatial data indexing: databases, spatial data handling and spatial data models. Section 12.2 will present space-driven methods: the kD tree, the adaptive kD tree, the binary space partitioning tree and the different kinds of quadtree. Some of these methods are also driven by objects. Section 12.3 will present the R trees, which are object-driven spatial access methods. Finally, Sect. 12.4 concludes the chapter.

12.1 Databases, Spatial Data Handling and Spatial Data Models

For a more in-depth coverage of databases, the reader can refer to [1]. Spatial databases have been covered very extensively in [2, 3]. Spatial data sets tend to be large, thus whole data sets cannot always be stored in the RAM (Random Access Memory) or main memory. Therefore, data have to be organized physically on

J.A. Bærentzen et al., *Guide to Computational Geometry Processing*,
DOI 10.1007/978-1-4471-4075-7_12, © Springer-Verlag London 2012

the mass storage in order to be retrieved efficiently (and transferred to the RAM or main memory). Spatial data indexing can be used for both point location and range searches (search for all objects in a given region).

12.1.1 Databases

In usual non-spatial databases, the solution for fast retrieval of information is typically to index the tables of the database using an *index* file. Information in databases is stored in *tables*, which are collections of *records*. Each record is a *n*-tuple where each element of the tuple is a value that belongs to the same set, which is called a *domain*. The value type is called a *field*. Each table must have a minimal (in the set inclusion sense) set of fields, which identifies uniquely and unambiguously any record in the table. This minimal set of fields is called a *key*. An example of a key for a table of passengers is a passport number together with the country code. An index file stores pairs of keys and record numbers in the total ordering of the keys. Using binary search, any entry in an index file can be searched for by key in logarithmic time. Once the key–record number pair has been located, it is possible to retrieve in constant time the full record in the file that has been indexed. However, non-spatial information does not have intrinsically a topology (a consistent definition of neighborhoods), and it can be sorted using some total ordering.

12.1.2 Spatial Data Handling

Spatial data could be ordered using an order on any coordinate, but that would not make sense, because such an order would project an *n*-dimensional space into a one-dimensional space filling curve, which would not store spatial data according to its topology. Indeed, there is no spatial ordering that can guarantee that points that are close in the spatial sense (topologically or geometrically) are stored in adjacent memory locations or even adjacent memory pages. A simple adversary argument is easy to design whatever method you use to group points in memory pages. It will be easy for an adversary, given a decomposition of the space into buckets or regions (see Fig. 12.1), to take two points near the boundary of two different buckets or regions and placed on both sides of the boundary. For example, take two points close to but on opposite sides of the common boundary of tiles "13" and "31". Those two points are neighbors, but they are stored in memory pages corresponding to the different buckets or regions that are far away.

However, different methods have been designed in order to speed-up point or object retrieval in the mass memory of the computer. The general idea is to cluster spatial data stored within a spatial database. Each cluster will be stored in a mass memory page. Different ways of clustering spatial data, and therefore different corresponding spatial indices, have been proposed in order to facilitate the efficient retrieval of spatial information in a computer. One can classify spatial data indexing

Fig. 12.1 The Z-order

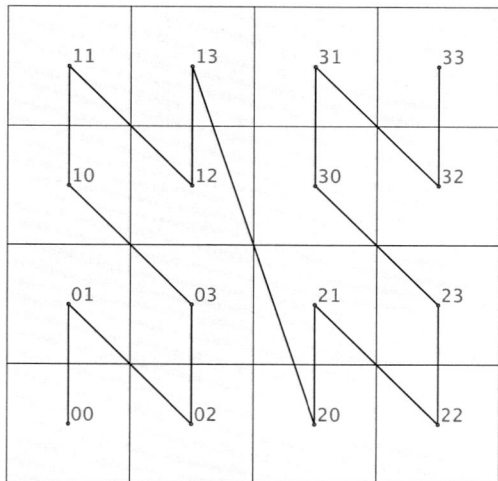

methods according to the mathematical constructs that are used to define the clusters. Spatial indexing methods have used similar approaches as *spatial data models*, which are models of spatial data.

12.1.3 Spatial Data Models

Spatial data models have been thought of in terms of the basic idea leading to the spatial data model: either the underlying space, or the objects that are in that space, or both. *Space-driven* spatial data models are models of spatial data, where space is partitioned into cells. In spatial databases and GIS, such a spatial data model with no gaps in the stored space is called "*continuous*". These space-driven spatial data models include the *raster spatial data model*, which is a regular *tessellation* of the space in rectangular cells (resulting from applying a grid onto the space). Space-driven spatial data models include also any regular or irregular tessellations. An important class of such tessellations are *space tilings*, among which we will see the Voronoi diagram in Chap. 14. The space-driven spatial data models are characterized by the fact that the neighborhood of the objects is given by the tiling, which is called an "*implicit topology*" in spatial databases or GIS jargon.

Object-driven spatial data models are structured by objects. Unless they are a partition of the space (like Voronoi diagrams), they are discontinuous because space is not actually stored: the only stored features are object boundaries. For example, a polygon is stored as the sequence of lines that compose the boundary of the polygon. The main example of the object-driven spatial data model is the vector spatial data model. The vector spatial data model does not store any neighborhood relationships between objects. In spatial databases and GIS, the vector spatial data model with no topology whatsoever is called the "*spaghetti*" spatial data model. Extensions of the vector spatial data models have been proposed for storing the connectivity relationships between objects (the *network spatial data model*) or both the connectivity and the spatial adjacency between objects (the *topological spatial data model*). In the spaghetti spatial data model, the connectivity and spatial adjacency relationships must be computed by using tolerances.

Hybrid space and object driven spatial data models include the Voronoi diagram, because it is an irregular tessellation or tiling (partition of the space into tiles) of the underlying space that adapts to objects. It is characterized by both an implicit topology, an explicit topology and a continuous space.

12.2   Space-Driven Spatial Access Methods

The general idea of spatial data indexing and point location is to locate spatial data in regions or clusters that are stored in the physical memory on a single page or disk block. Some methods form clusters by subdividing the space while others form clusters by grouping objects. This leads to the classification into space-based spatial indexing methods and object-based spatial indexing methods. However, some of the methods presented in this section are hybrid, in a sense. Another classification deals with the kind of memory that is dealt with by the spatial data indexing: either the main memory (e.g. RAM) or the mass memory (e.g. hard disks). This leads to the classification into main memory spatial indexing methods and spatial indexing aids point location, because it allows one to retrieve and, therefore, locate points efficiently in their cluster. Spatial indexing is a mechanism that aids the access and location of spatial objects. For this reason, the methods used in spatial data indexing are called spatial access methods. The main purpose of spatial access methods is to create spatial indices that allow to access and locate spatial data as fast as possible. A spatial index is the 2D equivalent of a book (binary tree) index (which is one-dimensional).

Space-driven spatial access methods include:

- the grid (regular tessellation)
- the kD tree (*n*-dimensional)
- the BSP tree (Binary Space Partitioning Tree, *n*-dimensional)
- the Quadtree (two-dimensional)
- the Octree (three-dimensional)

The grid or regular tessellation is very well adapted to handling image data, and corresponds to the raster spatial data model presented above. The kD tree is a special case of the more general BSP tree. The BSP tree is based on a decomposition of the space, which is an irregular tessellation. The BSP tree is at the origin of the constrained spatial data model, which is very well adapted to handling straight line segments, polygons and points. The quadtrees and the octrees are also based on irregular tessellations.

Fig. 12.2 A kD tree ($K = 2$)

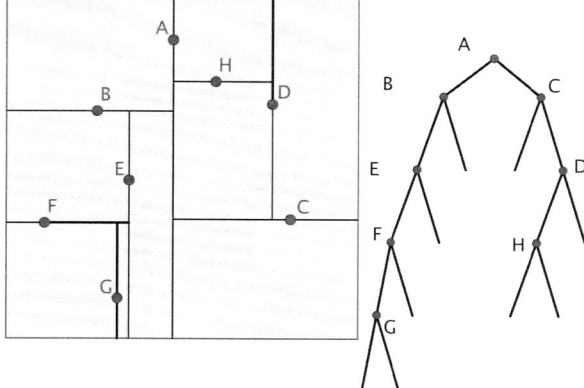

12.2.1 The kD Tree

The name of the kD tree comes from the fact that it is a multidimensional binary tree. It is a generalization of the bisection algorithm to n-dimensional space. The kD tree was discovered by Bentley in 1975 (see [4]). The kD tree is a binary tree, where each internal node contains one point and corresponds to a parallelepiped that has been defined by that point (see Fig. 12.2). A parallelepiped is a convex set bounded by convex polygons whose supporting hyperplanes are two-by-two parallel. The root node corresponds to the whole parallelepiped-like region of interest. The parallelepiped is divided in two parts by hyperplanes orthogonal to each one of the coordinate axes alternatively on the different levels of the kD tree. A rotation of the axes can be performed beforehand in order to match the Principal Components Analysis axes.

New points are inserted by descending the tree until a leaf is reached. The parallelepiped in which the new point is located is split into two parallelepipeds along the axis corresponding to the level of the kD tree. As an example, we suppose the last added point was E, and we are adding F. Since F is in the right child parallelepiped of E and it is separated from the other child of E by a line parallel to the x axis, it must become the right child of E and the left and right children parallelepipeds of F must be separated by a line parallel to the y axis of coordinates.

The main problem of the kD tree is that its shape depends on the order in which points were added. In the worst case, n points require n levels, and a search will take linear time.

In order to avoid this problem, a balanced version of the kD tree was introduced. The adaptive kD tree splits each set of points in subsets of roughly equal cardinalities forming two parallelepipeds (see Fig. 12.3). This ensures that the resulting adaptive kD tree is a balanced tree. Note that the bisection into two sets of roughly equal cardinalities requires median finding, but not necessarily sorting. Therefore, since median finding can be done in linear time, the complexity of the kD tree is $O(n \log n)$. Any search in an adaptive kD tree will require logarithmic time in the worst case. However, the balancing imposed by the adaptive kD tree makes it harder

Fig. 12.3 An adaptive 2D tree

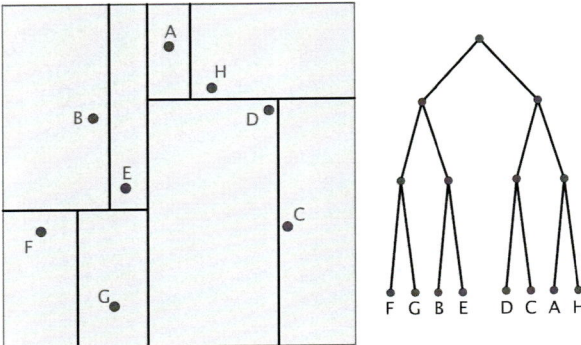

Fig. 12.4 A 3D tree (from [5])

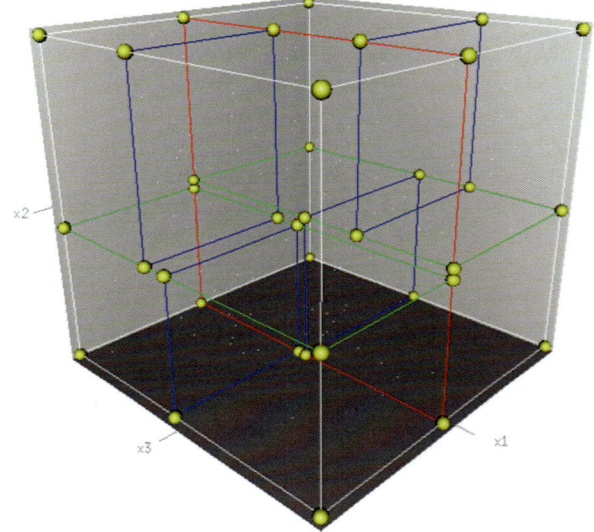

to insert or delete points, because standard balancing strategies based on rotations cannot be used because the rotations would destroy the rule that at a given depth of the tree, all the bisections are done with respect to a plane orthogonal to the same axis of coordinates. However, this rule has been relaxed in the most recent extension of kD trees known as the generalized kD tree. Therefore an adaptive generalized kD tree allows easier insertions and deletions and satisfies the balancing property.

A 3D tree is shown on Fig. 12.4.

Fig. 12.5 A BSP tree generated by hyperplanes

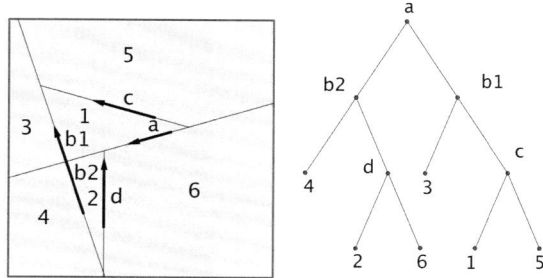

Fig. 12.6 A BSP tree generated by line segments composing the boundary of a polygon

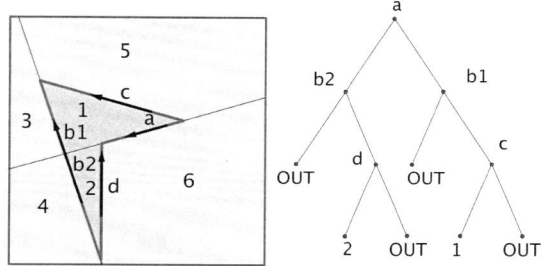

12.2.2 The Binary Space Partitioning Tree

The BSP-Tree was originally developed for 3D computer graphics by Fuchs, Kedem and Naylor in [6]. The BSP tree represents the organization of space by a set of half-spaces (see Fig. 12.5). It is very well suited for storing a collection of unrelated hyperplanes, but is can also be used to store line segments or polygons (see Fig. 12.6). The input hyperplanes or line segments will be called input objects hereafter. The BSP tree uses hyperplanes supporting the input objects to divide the space recursively into half-spaces. Nodes are added when a space is subdivided by the addition of an input hyperplane or line segment. A positive orientation is chosen together with the supporting hyperplane in order to define a half-space. The resulting half-space will be used in order to split all the existing half-spaces. Therefore, each addition of an half-space splits all the intersected half-spaces into two new half-spaces, and the object corresponding.

Note also that the kD tree is a special case of a BSP tree (planes perpendicular to one axis of coordinates, the axis alternates at each level change).

12.2.3 Quadtrees

Quadtree is the generic name for all kinds of tree built by the recursive division of the two-dimensional space into four quadrants. The foundational work was done by Finkel and Bentley in [7].

Fig. 12.7 A point quadtree

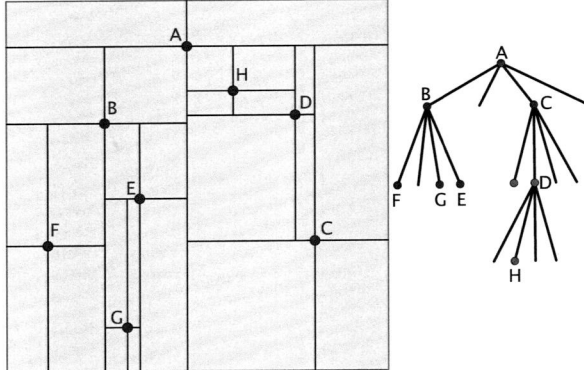

Fig. 12.8 A point-region quadtree

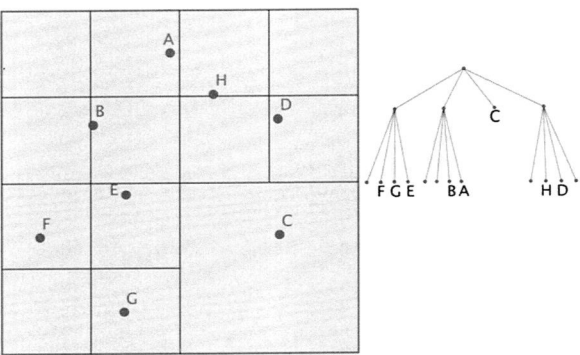

Quadtrees are always quaternary trees where the quadrants are named SW (for South–West), NW (for North–West), NE (for North–East) and SE (for South–East) and stored in that order, and all nodes at the same depth have the same area. At each step a quadrant is subdivided into four new quadrants. However, the steps need not to be in the order of the input: i.e., some input requires the addition of more than one level to the quadtree. Since quadtrees may not be balanced, any search may take linear time in the worst case.

The most common quadtree types are:
- the point quadtree (see Fig. 12.7);
- the point-region quadtree (see Fig. 12.8); and
- the region quadtree (see Fig. 12.9).

The difference between the different flavors of quadtrees lies in the way the space is subdivided into quadrants with respect to the input.

In the point quadtree, the quadrants are defined by the input points: the horizontal and vertical lines pass through the input points. In the point-region quadtree, the quadrants do not pass through input points, but exactly one of the four quadrants contains each input point.

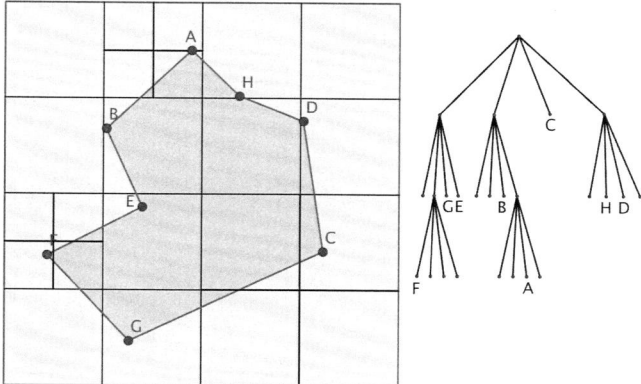

Fig. 12.9 A region quadtree

In the region quadtree, there are two kinds of leaves: the white leaves that are not covered by bounding line segments of the input regions; and the black leaves that are covered by bounding line segments of input regions. The region quadtree final decomposition induces a set of cells (quadrants of different sizes) such that the black quadrants approximate the input regions.

Quadtrees have found favor in many commercial Spatial Database Management Systems (SDBMS) because of their applicability to many data types, their ease of implementation and their good practical performance.

12.2.4 Octrees

Octrees are the analog of quadtrees in the three-dimensional space. They were first invented by Meagher in [8]. Octrees are 8-ary trees, where each node that is not a leaf has eight children (called octants). The same categories of quadtrees are found in octrees. Again, the search takes linear time in the worst case. An octree is depicted in Fig. 12.10.

12.3   Object-Driven Spatial Access Methods

The main object-driven spatial access method is the R tree, proposed by Guttman in [10]. Rather than divide space, R Trees (Rectangle Trees) group objects into a hierarchical organization using minimum bounding rectangles (*MBR*) or their n-dimensional analog (minimum bounding hyper-rectangles), see Fig. 12.11. Groups of points lines or polygons are indexed based on their MBR. Objects are added to the MBR that requires the least expansion of the R tree to accommodate the object, provided this will not exceed the memory page storage of the MBR. Internal nodes are guaranteed to contain at least a fixed percentage of their maximal capacity. If an MBR must expand beyond a preset parameter it is split into two

Fig. 12.10 An octree
(from [9])

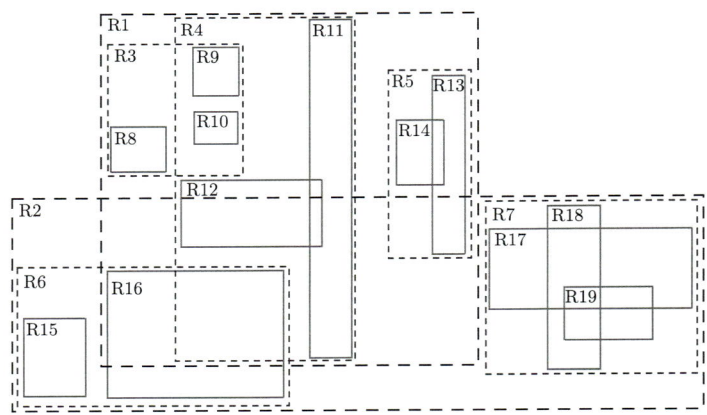

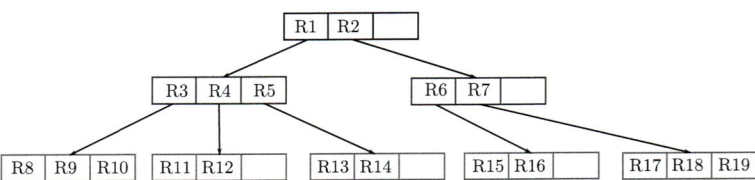

Fig. 12.11 A R tree (from [11])

MBRs. R trees have a static branching factor, which is the maximal number of children for any node of the R tree. R trees are very flexible and give good performance.

Objects are added the following way into a R tree. The MBR of the object or set of objects is added in the subtree of the node (hyper-rectangle) that will expand the least in area. If several hyper-rectangles expand least, the one that has the smallest area will be used for the insertion of the new hyper-rectangle. When the number of children equals the branching factor, the children of that node need to be redistributed between their current parent and the new internal node that will be created in order to store the remaining children.

R trees have some redundancy in the way they store objects. Indeed, the MBRs interiors may intersect, i.e. the MBRs *overlap*. Therefore, the same object can intersect more than one MBR. The area of a MBR is called *coverage*. Overlap and coverage are two major draw-backs of R trees, which limit their performance because of the redundancy they induce. R trees have been generalized in order to try to avoid the overlap between rectangles and the coverage by rectangles. R+ trees [12] remove the overlap by breaking down MBRs that overlap and storing the objects in those MBRs in two different places. They can be seen as a compromise between the kD tree and the R tree. R* trees [13] try to avoid both of them while storing both points and two-dimensional data. R trees have been extended in many different ways, including in the n-dimensional space. R trees have a good performance in practice. However, they were not asymptotically optimal. The Priority R tree or PR tree is the first optimal R tree (invented by Arge et al. [14]) in the sense that it can answer range queries in an asymptotically optimal number of input/output: $O((N/B)^{1-1/d} + T/B)$, where N is the number of d-dimensional (hyper-) rectangles stored in the R tree, B is the disk block size, and T is the output size.

Figure 12.12 shows a 3D R tree.

12.4 Conclusions

The purpose of spatial data indexing is to organize objects within a given space so as to aid efficient spatial retrieval, location and selection. There exists a wide variety of spatial indexing techniques. R tree indexing is considered to be the most efficient but this is at the expense of complexity. Most of current commercial Spatial DataBase Management Systems include spatial data indexing, and most advanced spatial database management systems offer the R tree spatial access method.

- IBM DB2 (Spatial Extenders) uses grids.
- Informix Universal Server (Spatial Datablade) uses R trees.
- Oracle spatial uses R trees (in 2D, 3D, and 4D) and quadtrees.

Finally the ANN library (A Library for Approximate Nearest Neighbor Searching) [15] implements an approximative nearest neighbor search in a n-dimensional space using the kD tree.

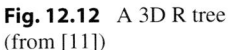

Fig. 12.12 A 3D R tree
(from [11])

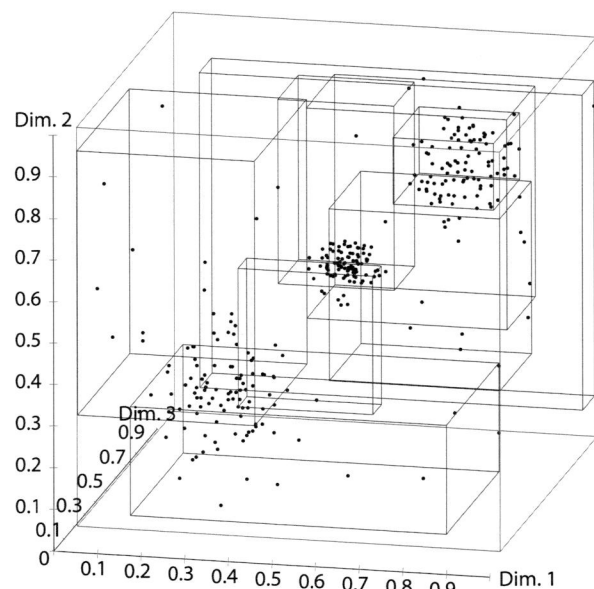

12.5   Exercises

Exercise 12.1 Write a kD tree as a BSP tree.

Exercise 12.2 Write an adaptive kD tree as a BSP tree.

Exercise 12.3 Write the different quadtrees as BSP trees.

Exercise 12.4 Implement the BSP tree using GEL.

Exercise 12.5 Implement the R tree using GEL.

References

1. Elmasri, R., Navathe, S.B.: Fundamentals of Database Systems. Benjamin-Cummings, Redwood City (1989)
2. Shekhar, S., Chawla, S.: Spatial Databases: A Tour. Prentice Hall, New York (2002). http://www.amazon.ca/exec/obidos/redirect?tag=citeulike09-20&path=ASIN/0130174807
3. Shekhar, S., Chawla, S., Ravada, S., Fetterer, A., Liu, X., Lu, C.-t.: Spatial databases-accomplishments and research needs. IEEE Trans. Knowl. Data Eng. **11**(1), 45–55 (1999). doi:10.1109/69.755614
4. Bentley, J.L.: Multidimensional binary search trees used for associative searching. Commun. ACM **18**(9), 509–517 (1975). doi:10.1145/361002.361007
5. Wikipedia: kD-tree. http://en.wikipedia.org/wiki/Kd-tree

6. Fuchs, H., Kedem, Z.M., Naylor, B.F.: On visible surface generation by a priori tree structures. In: SIGGRAPH '80: Proceedings of the 7th Annual Conference on Computer Graphics and Interactive Techniques, pp. 124–133. ACM, New York (1980). doi:10.1145/800250.807481

7. Finkel, R.A., Bentley, J.L.: Quad trees: a data structure for retrieval on composite keys. Acta Inform. **4**, 1–9 (1974)

8. Meagher, D.: Octree encoding: a new technique for the representation, manipulation and display of arbitrary three dimensional objects by computer. Technical report IPL,TR-80-111, Rensselaer Polytechnic Institute, Troy, NY, USA (1980)

9. Wikipedia: Octree. http://en.wikipedia.org/wiki/Octree

10. Guttman, A.: R-trees: a dynamic index structure for spatial searching. In: SIGMOD '84: Proceedings of the 1984 ACM SIGMOD International Conference on Management of Data, pp. 47–57. ACM, New York (1984). doi:10.1145/602259.602266

11. Wikipedia: Rtree. http://en.wikipedia.org/wiki/R-tree

12. Sellis, T.K., Roussopoulos, N., Faloutsos, C.: The R+-tree: a dynamic index for multi-dimensional objects. In: VLDB '87: Proceedings of the 13th International Conference on Very Large Data Bases, pp. 507–518. Morgan Kaufmann, San Francisco (1987)

13. Beckmann, N., Kriegel, H.-P., Schneider, R., Seeger, B.: The R*-tree: an efficient and robust access method for points and rectangles. In: SIGMOD '90: Proceedings of the 1990 ACM SIGMOD International Conference on Management of Data, pp. 322–331. ACM, New York (1990). doi:10.1145/93597.98741

14. Arge, L., de Berg, M., Haverkort, H.J., Yi, K.: The priority r-tree: a practically efficient and worst-case optimal r-tree. In: SIGMOD '04: Proceedings of the 2004 ACM SIGMOD International Conference on Management of Data, pp. 347–358. ACM, New York (2004). doi:10.1145/1007568.1007608

15. Mount, D.M., Arya, S.: ANN: a library for approximate nearest neighbor searching. http://www.cs.umd.edu/~mount/ANN/ (2010)

Convex Hulls

In this chapter, we present the notion of convex hull of a set, which is the smallest convex set enclosing that set, and is therefore, very closely connected to the notions of convexity and convex combination presented in Chap. 2. Convex hulls offer a construction algorithm for Delaunay triangulations, which will be presented in next chapter.

Convexity is a mathematical (geometric and analytical) notion that has many applications in computational geometry, but also in optimization, and thus potentially in many real-world applications. These applications cover a wide range of disciplines: control systems, signal processing, communications and networks, electronic circuit design, data analysis and modeling, statistics (optimal design), and finance. In Computational Geometry, convexity is closely related to the Delaunay triangulation and the Voronoi diagram, which are two very widely used mathematical constructs in Computational Geometry, and the subject of next chapter. In fact, both the Delaunay triangulation and the Voronoi diagram can be constructed from some particular convex hulls (see Chap. 14). Convexity is the basic framework for explaining convex hulls and their relationships with Delaunay graphs and Voronoi diagrams.

This chapter is organized as follows: Sect. 13.1 recalls the concept of convexity while Sect. 13.2 recalls the definition of convex hulls. Section 13.3 presents the algorithms for 2D convex hulls, while Sect. 13.4 presents the most efficient algorithms for convex hulls in 3D. Section 13.5 concludes this chapter.

13.1 Convexity

We have introduced the notion of affine spaces, barycenters and convexity in Chap. 2. We recall these notions here. It is not possible to define convexity without an affine space constructed from a vector space. Indeed, the notion of convexity is closely related to the notions of an infinite straight line and of barycenter, which are closely related to the notion of affine space. Moreover, the definition of convexity from the definition of an affine space allows one to understand why convexity is invariant by any affine transformation (any composition of translations and/or central similarities).

J.A. Bærentzen et al., *Guide to Computational Geometry Processing*, DOI 10.1007/978-1-4471-4075-7_13, © Springer-Verlag London 2012

Fig. 13.1 The convex hull of
a set of points in the plane as
intersection of the half-spaces
bounded by supporting
hyperplanes

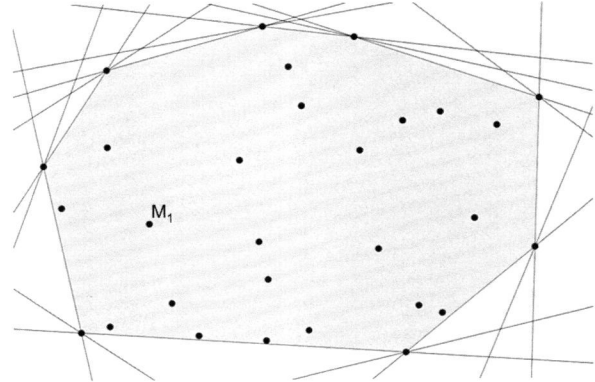

Definition 13.1 ([1, 2, Definition 11.1.1]) A subset S of an affine space X is called
convex if, for any $x, y \in S$, we have $[x, y] \subset S$, where $[x, y] = \{\lambda x + (1 - \lambda) y | \lambda \in
[0, 1]\}$.

Examples:
- intervals in $\mathbb{R}$,
- interval vectors in $\mathbb{R}^n$,
- half-planes, half-spaces,
- convex polyhedra (intersection of a finite set of closed half-spaces),
- polytopes[1] (compact convex polyhedra with non empty interior) like cubes, tetrahedra, pyramids, balls,
- contour sets of convex objects.

13.2   Convex Hull

Let us first define supporting hyperplanes, which can be used in some definitions of
the convex hull.

Definition 13.2 ([1, 2, Definition 11.5.1]) Let A be an arbitrary subset of an affine
space X. A supporting hyperplane for A is any hyperplane H containing a point
$x \in A$ and such that all points of A belong to one of the two half-spaces bounded
by H. We say that H is a supporting hyperplane for A at x (see Fig. 13.1).

Alternative definitions: The convex hull can be defined in the following (all prov-
ably equivalent) ways:

Definition 13.3 ([1, adapted from Sect. 11.1.8.1]) The convex hull of a subset A of
a affine space X is the intersection of all convex sets that contain A.

---

[1]In some countries, polytopes are defined as bounded polyhedra, without requiring them to be
convex.

Fig. 13.2 The convex hull of
a set of points in the plane

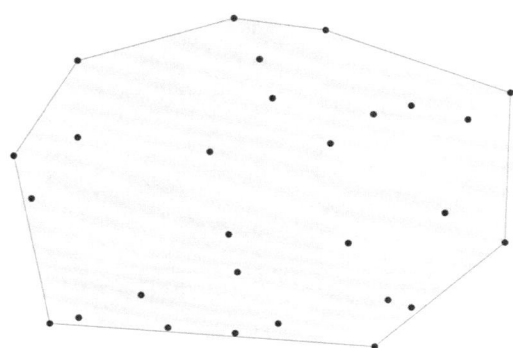

Proposition 13.1 ([1, 2, adapted from Sect. 11.1.8.1]) *The convex hull of a subset A of an affine space X is the smallest convex set that contains S.*

Proposition 13.2 ([1, 2, adapted from Proposition 11.5.5]) *The convex hull of a closed subset A of an affine space X is the intersection of all the closed half-spaces containing A.*

Proposition 13.3 ([1, 2, Proposition 11.1.8.4]) *The convex hull of a subset A of an affine space X is the set of barycenters of families of points of A with non-negative masses:* $\{\sum_{i \in I} \lambda_i x_i | x_i \in A, \lambda_i \geq 0, \sum_i \lambda_i = 1, I\,arbitrary\}$ *with* $\lambda_I = 0$ *except for finitely many indices.*

Examples:
- the convex hull of a finite set of points of $\mathbb{R}^n$ is called a polytope; in 2D it is called a polygon (see Fig. 13.2);
- the convex hull of $n + 1$ points of $\mathbb{R}^n$ that are not on a same hyperplane is called a simplex.

Notation: The convex hull of a subset A of an affine space X is denoted as $CH(A)$ hereafter in this chapter.

The computation of the convex hull of a set $S = S_1 \cup S_2$ can be subdivided into the computation of the convex hull of two (or more) smaller convex sets: the convex hulls of S_1 and S_2.

Theorem 13.1 *For any subset* $S = S_1 \cup S_2$ *of a real affine space* X, $CH(S) = CH(CH(S_1) \cup CH(S_2))$ *(see Fig. 13.3).*

Proof We will prove the equality of the two sets above by proving that each one of them is a subset of the other. Let us first prove that $CH(S_1 \cup S_2) \subseteq CH(CH(S_1) \cup CH(S_2))$. First, by Definition 13.3, it is clear that $A \subseteq B \implies CH(A) \subseteq CH(B)$. Clearly, $S_1 \cup S_2 \subseteq CH(S_1) \cup CH(S_2)$. Thus, $CH(S_1 \cup S_2) \subseteq CH(CH(S_1) \cup CH(S_2))$. Then, let us prove now that $CH(CH(S_1) \cup CH(S_2)) \supseteq CH(S_1 \cup S_2)$. Clearly $S_i \subseteq$

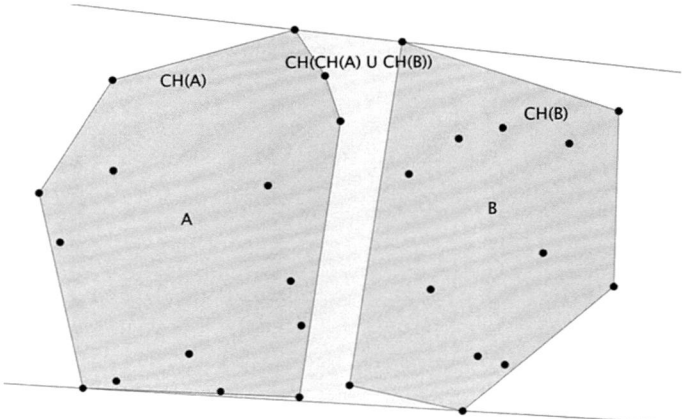

Fig. 13.3 The convex hull can be computed by decomposing the set into smaller sets recursively (until the set consists of 3 points), and then recombining the convex hulls by adding supporting hyperplanes for the union of convex hulls

S for $i = 1, 2$. Thus, $CH(S_i) \subseteq CH(S)$ for $i = 1, 2$. Hence, $CH(S_1) \cup CH(S_2) \subseteq CH(S)$ and as $CH(S)$ is convex, the same arguments as above proves $CH(CH(S_1) \cup CH(S_2)) \subseteq CH(S)$.   □

13.3 Convex Hull Algorithms in 2D

We consider the computation of the convex hull $CH(\mathcal{P})$ of a set of points $\mathcal{P} = p_1, \ldots, p_n$ in the affine Euclidean plane $\mathbb{R}^2$. We will present here some useful strategies for designing geometric algorithms that are widely used in Computational Geometry in order to design any geometric algorithm. These include the following:

- Incremental algorithms: construct the final result by inserting in turn each object of the input and computing the intermediary result after insertion of that object. The incremental algorithm for convex hulls is presented in Sect. 13.3.2. Even though it is not optimal in 2D, it can be faster in practice than optimal algorithms, and it is optimal in 3D.
- Divide and conquer: recursive approach: divide recursively the input until the problem can be solved trivially for that input size, and then merge the subproblem solutions. The divide and conquer algorithm for convex hull is presented in Sect. 13.3.3. It is optimal in any dimension.
- Scan: scan the input along one direction or around a point. A scan algorithm for convex hulls in 2D is presented in Sect. 13.3.1.
- Sweep line or plane-sweep: decompose the input into vertical strips, such that the information necessary for solving the problem is located on the vertical lines delimiting those vertical strips. There exists a sweep line algorithm for convex hulls in 2D.

   Sorting reduces to the convex hull in linear time. Indeed if the convex hull of a set of points on a parabola of equation $y = x^2$ is known, then by projection on the x-axis of coordinates (by keeping only the x-coordinates), the sorting of the corresponding projections in $\mathbb{R}$ is known. Thus, if the lower bound of the convex hull problem was lower than the lower bound of sorting, then by computing the convex hull of the projections of the points on the parabola and then projecting back the points on the straight line, one could get the sorting in better time than $\Omega(n \log n)$, a contradiction with the fact the lower bound of the sorting problem is $\Omega(n \log n)$. Thus, the lower bound of the convex hull problem is at least the same as the lower bound of the sorting problem, i.e. $\Omega(n \log n)$. We will see in this chapter that this lower bound is actually attained by several convex hull algorithms.

   Finally, the convex hull reduces to sorting in linear time. Indeed, if a list of points is sorted according to the former, their polar angle with respect to a point inside the polygon formed by the list of points, and if there are ties, according to their square distance to that point, then the convex hull of the set of points can be computed in linear time by removing from the sorted list all vertices whose internal angle is larger than π. This is the principle of Graham's scan algorithm, which we introduce now. The output of all the 2D convex hull algorithms is always an ordered list of vertices forming the boundary of the convex hull sorted in the same order as before.

13.3.1 Graham's Scan Algorithm

This algorithm is not dynamic nor semi-dynamic because it requires knowing all the points in the input before the beginning. We will describe the Graham scan, cf., Fig. 13.4, which is the most efficient scan algorithm. The algorithm starts with a start point $s \in \mathcal{P}$ which is by construction on the boundary of the convex hull (which is a polygon): e.g. the rightmost lowest x-coordinate point. It also uses a point o that is chosen to be in the interior of the convex hull. Such a point could be the centroid (barycenter with all masses being equal) of any three points in $\mathcal{P}$. Let L be a periodic infinite list (or a circular doubly linked in computer science terms) of the vertices p_i sorted in increasing lexicographic order of polar angle and squared polar distance around o with pointers to the preceding vertex (PRED) and to the following vertex (NEXT).[2]

   It performs a scan of the points and computes at each scanned point v, the relative position of $NEXT(NEXT(v))$ with respect to the vector $\mathbf{vNEXT(v)}$. If $NEXT(NEXT(v))$ is on the right of or on $\mathbf{vNEXT(v)}$, $NEXT(v)$ is eliminated from the convex hull and the scan starts again from $PRED(v)$ (this is called backtracking hereafter). The relative position is computed using the determinant $\det(\mathbf{vNEXT(v)}, \mathbf{vNEXT(NEXT(v))})$. A flag f is used in order to avoid scanning again the start vertex except in case of backtracking.

   Its pseudo-code is presented in Algorithm 13.1.

---

[2]The lexicographic ordering is the ordering of a language dictionary. In this case the vertices are first ordered according to their polar angle around o, and if there are ties, the vertices having the same polar angle around o are ordered according to their square distance to o.

Algorithm 13.1 Graham's scan 2D convex hull

Require: $\mathcal{P} = \{p_1, \ldots, p_n\}$
Ensure: $CH(\mathcal{P})$ is the polygon defined by the output L, where the vertices are
    stored in order of increasing polar angle in L.
  1: Compute the start vertex v as the rightmost lowest y coordinate by sorting the
    y coordinates of the points in $\mathcal{P} = \{p_1, \ldots, p_n\}$ (and the x coordinates of the
    points having the lowest y coordinate).
  2: Let L be a periodic infinite list (or a circular doubly linked in computer science
    terms) of the vertices p_i sorted in increasing lexicographic order of polar an-
    gle and squared polar distance around o with pointers to the preceding vertex
    (PRED) and to the following vertex (NEXT).
  3: Let $w := PRED(v)$
  4: $f :=$ **false**
  5: **while** $NEXT(v) \neq s$ or $f =$ **false do**
  6:     **if** $NEXT(v) = w$ **then**
  7:         $f :=$ **true**
  8:     **end if**
  9:     **if** $\det(\mathbf{vNEXT(v)}, \mathbf{vNEXT(NEXT(v))}) \leq 0\{$right turn$\}$ **then**
10:         Remove $NEXT(v)$ from L
11:         Let $v := PRED(v)$
12:     **else** $\{$left turn$\}$
13:         Let $v := NEXT(v)$.
14:     **end if**
15: **end while**
16: **return** L

    Proof of correctness: If $NEXT(NEXT(v))$ is on or on the right of $\mathbf{vNEXT(v)}$, then
$NEXT(v)$ is a reflex vertex (the internal angle at $NEXT(v)$ is reflex, i.e. $\geq \pi$), and
$NEXT(v)$ is internal to the triangle $OvNEXT(NEXT(v))$, thus the polygon defined
by the vertices and their ordering in L is not convex. Thus, it is not the convex hull
of $\mathcal{P}$. Thus, none of the internal angles $vNEXT(v)\widehat{NEXT}(NEXT(v))$ are reflex and
all the edges of the form $vNEXT(v)$ have all points in $\mathcal{P}$ on one side of them when
Graham's scan terminates. Thus, all hyperplanes through edges $vNEXT(v)$ touch
exactly two vertices of $\mathcal{P}$ and all points in $\mathcal{P}$ are on the same side with respect to
these hyperplanes. Thus all hyperplanes through edges $vNEXT(v)$ are supporting
hyperplanes. Thus, the polygon defined by L is the boundary of $CH(\mathcal{P})$.
    Analysis: The total cost of the while loop is linear since the operations are per-
formed for every vertex: line 13 (if the angle is not reflex) or lines 10 and 11 (if
the angle is reflex). The backtracking of line 11 must stop since the start vertex is
assumed to be on the convex hull, and once a vertex of the boundary of the convex
hull at which the internal angle is not equal to π has been visited, the backtracking
cannot go back before that vertex. The total number of visited vertices is thus linear.
The cost of the remainder of the algorithm is dominated by the cost of searching,

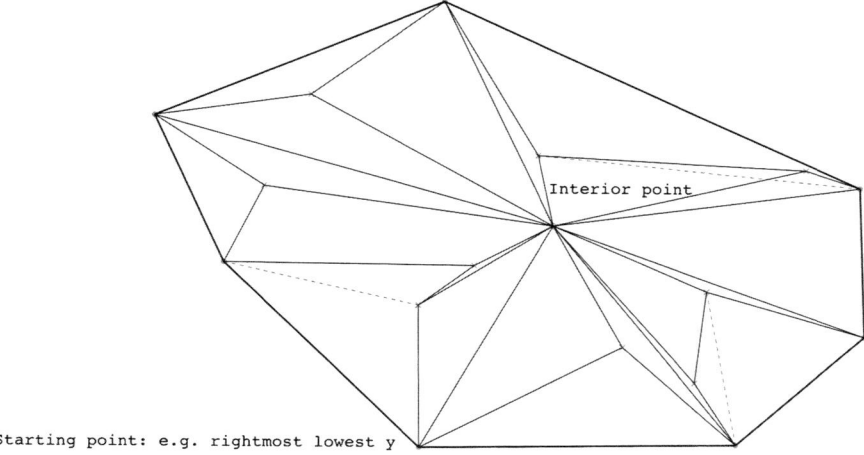

Starting point: e.g. rightmost lowest y

Fig. 13.4 The scan of the points from a point in the interior of the convex hull (Graham's scan): the order of the scan is given by the radial segments through the point in the interior of the convex hull; the edges between input points resulting from forward scanning are the plain non-radial edges; the edges between input points resulting from the backtracking are *dashed* for those that do not belong to the boundary of the convex *hull* and *thick* plain lines correspond to the edges of the boundary of the convex hull

which is $\Theta(n \log n)$. Thus the complexity of the algorithm is $O(n \log n)$. It matches the lower bound $\Omega(n \log n)$ for the convex hull problem, so it is optimal.

13.3.2 Incremental (Semi-dynamic) Algorithm

Such an algorithm is semi-dynamic because it allows one to add points and maintain the convex hull as points are added. The algorithm consists in starting from three points. The convex hull of a set of three points is trivial: it is the triangle that joins the three points (even in the case of a degenerate situation like the three points being collinear). Then, every point of the input set is added and the convex hull is maintained as follows.

Its pseudo-code is presented in Algorithm 13.2.

Proof of correctness: Since the polar angle of a point of the boundary of the convex hull with respect to $p(i + 1)$ (pole) and a given line (polar axis) is a continuous function, the image of the convex hull is an interval. Thus, the points at which the minimum and maximum are attained are on the boundary of the convex hull $CH(i + 1)$. Assume that there are two distinct points A and B on which the minimum is attained. Since A and B are aligned with $p(i)$, either all the points in between them are aligned with $p(i)$, or there is a reflex angle at one of such a points, but this is impossible since the convex hull is convex! If all the points defining the convex hull are such that no three of them are aligned, then, by the continuity of the polar angle function described above, we can assume that there is one and exactly

Algorithm 13.2 Incremental 2D convex hull (see Fig. 13.5)

Require: $\mathcal{P} = p(1), \ldots, p(n)$

Ensure: $CH(n) = CH(\mathcal{P})$, where the vertices are stored in the lexicographic order of increasing polar angle first, and square distance second.

1: $CH(3) = \{p(1), p(2), p(3)\}$
2: **for** $i = 4$ to n **do**
3:      Let CH_i be the convex hull of points $\mathcal{P}(i) = p(1), \ldots, p(i)$
4:      Let $p(i+1)$ be the point being inserted and
5:      The current convex hull $CH(i)$ is traversed either clockwise or counter-clockwise to determine whether $p(i+1)$ belongs to $CH(i)$ or not
6:      **if** $p(i+1) \in CH(i)$ **then**
7:          $CH(i+1) = CH(i)$
8:      **else** $\{ p(i+1) \notin CH(i) \}$
9:          Determine the 2 vertices $v1(i)$ and $v2(i)$ defining the new supporting hyperplanes together with $p(i+1)$, which are before and after $p(i+1)$ in the lexicographic order
10:         All edges and vertices between $v1(i)$ and $v2(i)$ are removed from the convex hull
11:         The straight line segments $v1(i)p(i+1)$ and $v2(i)p(i+1)$ are added to $CH(i)$
12:     **end if**
13: **end for**
14: **return** $CH(n)$

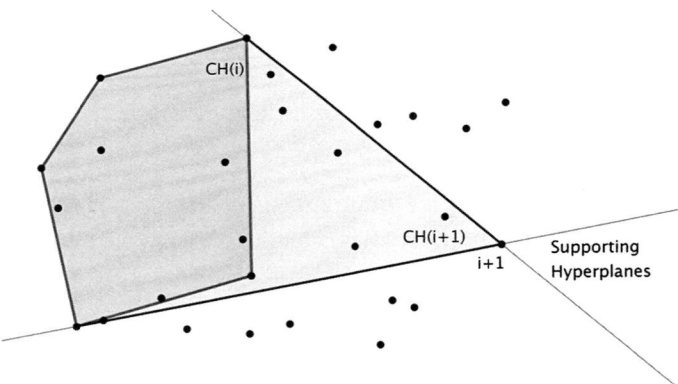

Fig. 13.5 The incremental construction of the convex hull

one vertex $v1(i)$ at which the minimum is attained. In the same way, we can prove that there is one and exactly one vertex $v2(i)$ at which the maximum is attained. Finally, all edges between $v1(i)$ and $v2(i)$ (in the order around $p(i+1)$) separate $CH(i)$ and $p(i+1)$. Thus, they are not on the boundary of the convex hull $CH(i+1)$

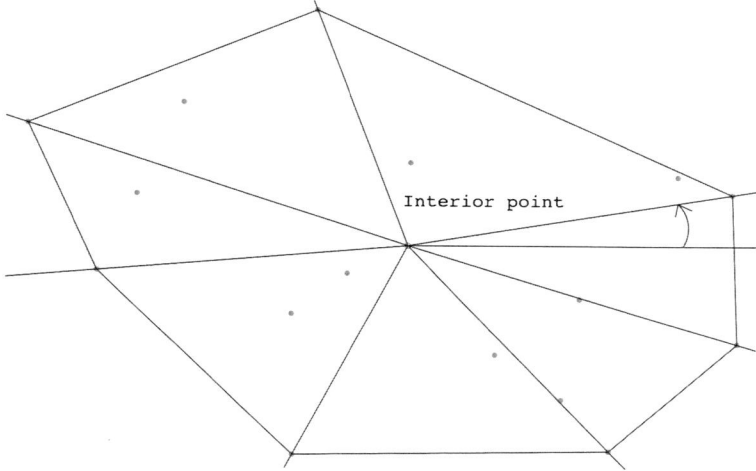

Fig. 13.6 The polar angles of the points on the boundary of $CH(\mathcal{P})$ with respect to a point in the interior of the convex hull

and, therefore, they must be replaced by the edges $v1(i)p(i+1)$ and $v2(i)p(i+1)$, which are on the boundary of $CH(i+1)$.

Analysis: Since at insertion of point p_i, the algorithm needs to visit at most n vertices, the algorithm is quadratic (i.e. $O(n^2)$). Since it is asymptotically higher than the lower bound for the convex hull problem, it is not optimal. However, it allows one to add points, which is not the case for the other optimal convex hull algorithms presented in this chapter.

13.3.3 Divide and Conquer Algorithm

Such an algorithm is not dynamic nor semi-dynamic because it requires knowing all the points in the input before beginning. The algorithm consists in recursively splitting the input into subsets of sizes differing by at most 1 (for example around the median in the x-coordinate) until the size of the input is 3 or less. The convex hull of a set of up to three points is trivial: it is the triangle that joins the three points (even in the case of a degenerate situation like the three points being collinear).

The pseudo-code is presented in Algorithm 13.3.

Proof of correctness: If $p1$ belongs to the interior of $CH2$, then the polar angle of the points defining the boundary of $CH2$ with respect to $p1$ (pole) and a given line (polar axis) is a monotonic function (see Fig. 13.6). Otherwise, according to the same reasoning as the one in the proof of correctness of the incremental 2D convex hull algorithm, there must exist two vertices $v1$ and $v2$ on the boundary of $CH2$ on which the polar angle of the points of the boundary of $CH2$ with respect to $p1$ (pole) and a given line (polar axis) attain, respectively, their minimum and their maximum. Also, according to a similar reasoning as in the proof already mentioned above, all

Algorithm 13.3 Divide and conquer 2D convex hull

Require: $\mathcal{P} = p_1, \ldots, p_n$
Ensure: CH is the convex hull of points $\mathcal{P} = p_1, \ldots, p_n$, where the vertices are
    stored in the lexicographic order of increasing polar angle first, and square dis-
    tance second.
  1: Sort the $\mathcal{P} = p_1, \ldots, p_n$
  2: Function *ConvexHull($\mathcal{P}$)*
  3: **if** *cardinality($\mathcal{P}$)* ≤ 3 **then**
  4:         **return** $\mathcal{P}$
  5: **else** {*cardinality($\mathcal{P}$)* > 3}
  6:         Split the set $\mathcal{P}$ into two sets $\mathcal{P}_{\text{left}}$ and $\mathcal{P}_{\text{right}}$
  7:         $CH1 = ConvexHull(\mathcal{P}_{\text{left}})$ and $CH2 = ConvexHull(\mathcal{P}_{\text{right}})$
  8:         Let $p1$ be any point in the interior of $CH1$
  9:         Determine whether $p1$ is in the interior of $CH2$
 10:         **if** $p1$ is not in the interior of $CH1$ **then**
 11:                 Determine the vertices $v1$ and $v2$ of $CH2$ on which the maximal and
                 minimal polar angles with respect to $p1$ are attained
 12:                 $v1$ and $v2$ separate two monotonic chains, one that is closer to $p1$
                 than the other one and the other one
 13:                 Discard the monotonic chain that is closer to $p1$ (these points separate
                 $CH1$ and $CH2$) from $CH2$
 14:                 Order all the points of $CH1$ and $CH2$ in the increasing order of the
                 polar angle with respect to $p1$
 15:         **end if**
 16:         Store the output of Graham's scan of $CH1 \cup CH2$ in CH
 17:         **return** CH
 18: **end if**

the edges between $v1$ and $v2$ separate $p1$ and $CH2$. Theorem 13.1 allows one to
conclude.

Analysis: Since the algorithm requires sorting first, it requires at least $O(n \log n)$.
The cost for computing the convex hull of the union of two convex hulls is linear in
the size of the input, because the ordering of the points of $CH1$ and of $CH2$ is linear
since both the points of $CH1$ and of $CH2$ are already sorted. Thus, the running time
follows the following law: $T(n) \leq 2T n/2 + O(n)$. Finally, by the master theorem
[3, Theorem 4.1 and proof in Sect. 4.4], $T(n) = O(n \log n)$.

13.4   3D Algorithms

We consider the computation of the convex hull $CH(\mathcal{P})$ of a set of points $\mathcal{P} =
p_1, \ldots, p_n$ in the affine Euclidean three-dimensional space $\mathbb{R}^3$. An example of a
convex hull of a set of points in 3D is shown in Fig. 13.7.

Fig. 13.7 The convex hull of
a set of points in 3D

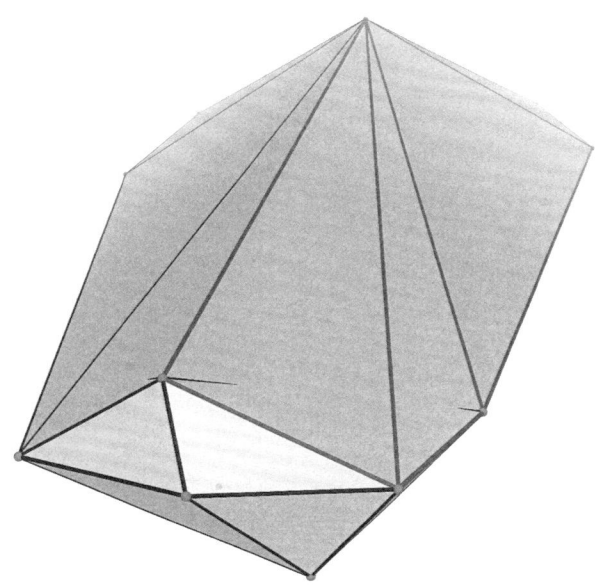

13.4.1 Incremental Algorithm

An incremental (dynamic) algorithm for computing the convex hull of points in
3D exists. It is a straightforward extension of the incremental 2D convex hull algorithm. The algorithm checks whether the point $p(i + 1)$ to be inserted is in the
current convex hull $CH(i)$ or not. If $p(i + 1) \in CH(i)$, there is nothing to do:
$CH(i + 1) = CH(i)$. If not, all the planar facets that separate $CH(i)$ from $p(i + 1)$
need to be identified. The set of all these planar facets is bounded by a cycle
$E = \{e(j), j = 1, \ldots, m\}$ of edges in $CH(i)$. The convex hull $CH(i + 1)$ is obtained
by removing all planar facets that separate $CH(i)$ from $p(i + 1)$; and adding all
the edges $e(j)p(i + 1)$, $j = 1, \ldots, m$. The complexity of the algorithm is quadratic,
since at each new insertion, at most a linear number of faces must be tested. Indeed, by Euler's formula ($v - e + f = 2$), the number of facets f and the number
of edges e are linear in the number of vertices v. It is therefore optimal, unlike in
the 2D case. It uses the same gift wrapping technique as the following divide and
conquer algorithm.

13.4.2 Divide and Conquer Algorithm

Another optimal 3D convex hull algorithm is the divide and conquer algorithm. The
general behavior of the algorithm is the same as in 2D. Only the merging of the
convex hulls of smaller sets is different due to the three-dimensionality. The merging of two convex hulls in 3D requires the construction of a triangulation that is
wrapping the two convex hulls with simplicial facets (triangles) whose edges and
vertices belong to different convex hulls, such that the plane that spans that triangle

Fig. 13.8 The gift wrapping
algorithm on the set of points
shown in Fig. 13.7

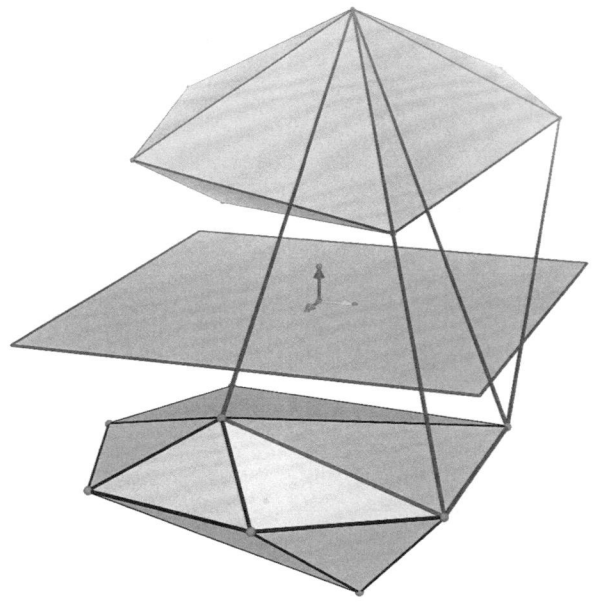

is a supporting hyperplane for $\mathcal{P}$. For this purpose, we begin by identifying an edge
of the convex hull of the union of the two convex hulls. This can be done by projecting the two convex hulls onto one of the planes of the equation $x = 0$ or $y = 0$ or
$z = 0$. Then, we choose a starting vertex as in the 2D case. The starting edge is the
edge connecting that starting vertex with another projection such that the hyperplane
passing through them is a supporting hyperplane for the set of projections of $\mathcal{P}$ onto
the chosen plane of projection. The corresponding edge in the three-dimensional
space is the starting edge. It links two vertices that do not belong to the same previous convex hull. The first supporting hyperplane will contain that starting edge and
its projection onto the chosen plane (by construction, it is a supporting hyperplane
for the set of original points).

The merging of the two convex hulls uses a "gift wrapping" algorithm (see
Fig. 13.8). Starting from the starting edge and the first supporting hyperplane, the algorithm rotates the starting hyperplane around the starting edge, until the first vertex
of the point set is hit. Thus, the first triangle of the gift wrapping has been defined.
Two new edges have also been defined: the edges that connect the extremities of the
starting edge with the newly hit vertex. Each time a new triangle is found, all the
old facets that are hidden by that new triangle are discarded from the convex hull.
Then, the search proceeds as follows: the next vertex of the convex hull is found by
rotating the hyperplane containing the last triangle found around the chosen newly
found edge. Since there are two new edges, there might be a branching with each
new branch of the search that corresponds to one of the two newly found edges
of the convex hull. The search stops each time it visits a triangle that has been al-

ready visited. Then, the search continues with the other neighbor of the last visited triangle.

The merging can be done in time proportional to the size of the input. Thus, again by the master theorem, the complexity of the divide and conquer algorithm is $O(n \log n)$.

13.5   Conclusions

This chapter has presented the notion of convex hulls through several algorithms that implement the general strategies for designing geometric algorithms. In the next chapter, we will see two very important spatial data structures: the Voronoi diagram and the Delaunay triangulation, which are strongly related to convex hulls.

13.6   Exercises

Exercise 13.1 Prove that the convex hull of a circle is itself together with its interior (prove that the disk bounded by the circle is the smallest convex set containing the circle). What are the supporting hyperplanes?

Exercise 13.2 Prove that the convex hull of an ellipse is itself together with its interior (prove that the closed region bounded by the ellipse is the smallest convex set containing the ellipse). What are the supporting hyperplanes?

Exercise 13.3 What is the convex hull of the graph of the function $y = 1/x$? What is the convex hull of the closed region bounded by the coordinate axes and the graph of the function $y = 1/x$?

Exercise 13.4 Implement the 2D convex hull algorithms described in this chapter using GEL or the quad-edge data structure code provided in [4] and plot the running time for different input sizes. Observe the running time behavior of these algorithms as a function of the input size.

Exercise 13.5 Implement the 3D gift-wrapping algorithm described in this chapter using GEL or the quad-edge data structure code provided in [4]. Use it to compute the Delaunay triangulation of a set of points (see Chap. 14).

References

1. Berger, M.: Géométrie, vol. 3. CEDIC, Paris (1977). Convexes et polytopes, polyèdres réguliers, aires et volumes. [Convexes and polytopes, regular polyhedra, areas and volumes]
2. Berger, M.: Geometry. I. Universitext, p. 428. Springer, Berlin (1987). Translated from the French by M. Cole and S. Levy

3. Cormen, T.H., Leiserson, C.E., Rivest, R.L., Stein, C.: Introduction to Algorithms, 2nd edn. MIT Press, Cambridge (2001). http://www.amazon.ca/exec/obidos/redirect?tag=citeulike09-20&path=ASIN/0262531968
4. Lischinski, D.: Incremental Delaunay triangulation. In: Heckbert, P. (ed.) Graphics Gems IV, pp. 47–59. Academic Press, Boston (1994)

Triangle Mesh Generation: Delaunay Triangulation

14

The generation of triangle meshes from points is a central issue when dealing with geometric data, e.g., where surface data are often created by triangulating a point set. Examples include the generation of a surface from points produced by an acquisition device such as a laser scanner, or the triangulation of a domain such that simulations can be performed on it via the finite element method.

This chapter deals with triangulation, primarily by the introduction of the 2D *Delaunay triangulation*, which in some sense is the archetypical triangulation. Hereby key concepts and general principles of triangulation are introduced, such that the reader hopefully appreciates the nature of meshing and is able to understand the general literature within the field with more ease. Good general references on the subject include [1–4].

This chapter is organized by first introducing key concepts of triangulation and Delaunay triangulation specifically, cf. Sects. 14.1 and 14.2, whereupon algorithmic details are covered in Sect. 14.3. Following these, general issues of triangulation are briefly presented in Sects. 14.4 and 14.5. The chapter ends with a brief overview of Voronoi diagrams, in Sect. 14.6, which are closely linked to the Delaunay triangulation.

In relation to implementation, it should be noted that the Quad-Edge data structure of Sect. 5.6.1 is in many aspects ideal for computing the Delaunay triangulation and the Voronoi diagram, because it can traverse both the Delaunay triangulation and the Voronoi diagram at the same time. Furthermore, it is the data structure used in [5], as discussed in the section on the divide and conquer algorithm.

14.1 Notation and Basic 2D Concepts

In 2D, a triangulation classically starts with a set of points, $\mathbf{p}_i \in \mathcal{P}$, which shall be connected with edges, $e_i \in \mathcal{E}$, such that these edges form triangles, $t_i \in \mathcal{T}$. These edges and triangles can formally be considered as sets of two or three points, respectively. The goal of the triangulation process is, intuitively, to form a unified geometric entity of the given points, $\mathcal{P}$. As the name states, this geometric entity takes the form of a triangular mesh. More formally:

J.A. Bærentzen et al., *Guide to Computational Geometry Processing*,
DOI 10.1007/978-1-4471-4075-7_14, © Springer-Verlag London 2012

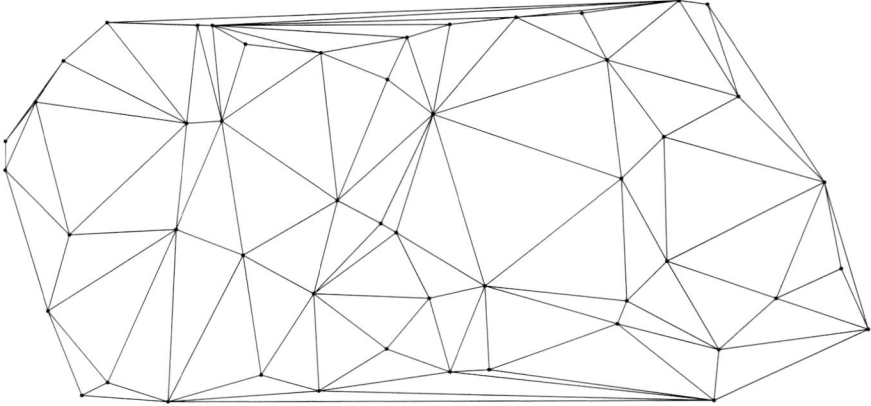

Fig. 14.1 The Delaunay triangulation of 50 random points

Definition 14.1 (Triangulation) A triangulation of a point set $\mathcal{P}$ is a set of triangles $\mathcal{T}$, such that we have the following:

- All points, $\mathbf{p}_i \in \mathcal{P}$, are vertices in at least one triangle.
- The interiors of any two triangles do not intersect.
- All the points $\mathbf{p}_i$ only intersect the triangles $t_i \in \mathcal{T}$ at the triangles vertices.
- The union of all the vertices of the triangles is $\mathcal{P}$.

The points, $\mathbf{p}_i$, are often referred to as vertices, $v_i \in \mathcal{V}$, as well, since they are 'corners' in polygons. In this chapter the two terms are used interchangeably, unless otherwise stated.

14.2   Delaunay Triangulation

The Delaunay Triangulation is a special triangulation of a point set $\mathcal{P}$, which has some additional nice properties—see Fig. 14.1. In fact the Delaunay triangulation is often thought of as the archetypical 'nice' triangulation. The concepts of Delaunay triangulation in 2D can be extended to 3D where the concept becomes the Delaunay tetrahedralization.

In this chapter, the classical concept from computational geometry, Delaunay triangulation, will first be defined, along side proofs of many of its properties. A vehicle for doing this is an algorithm for performing Delaunay triangulation, the flip algorithm. The presentation of this algorithm is naturally done via a string of constructive proofs, which give insight into the Delaunay triangulation. The derivation here closely follows that given in [3], but also [1, 2] provide good introductions to the subject. A motivation for including proofs of some of the properties is that the line of thought behind them is similar to those in the algorithms for Delaunay triangulation.

Fig. 14.2 Here edge e_i is
Delaunay

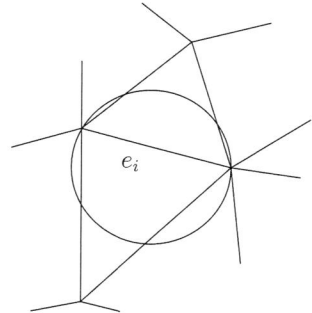

Definition 14.2 (Empty Circle) A circle is empty if and only if its interior contains no points of $\mathcal{P}$.

Definition 14.3 (Circumcircle of an Edge) A circumcircle of an edge from $\mathbf{p}_i$ to $\mathbf{p}_j$ is a circle going through $\mathbf{p}_i$ and $\mathbf{p}_j$.

Note that for a given edge there are infinitely many circumcircles.

Definition 14.4 (Delaunay Edge) An edge is Delaunay if and only it has **an** empty circumcircle.

Note, it is almost always possible to find a non-empty circumcircle of an edge, but this is of no consequence for this definition. See Fig. 14.2.

Definition 14.5 (Delaunay Triangulation) A triangulation, **D**, is a Delaunay triangulation, if an edge is Delaunay if and only if it is in **D**.

That is the Delaunay triangulation of a set of points is all the Delaunay edges related to those points.

This definition is only perfectly meaningful, in relation to triangulation, if no four points in $\mathcal{P}$ lie on a common circle, and no three points in $\mathcal{P}$ are aligned. This is referred to as the points being in *general position*. The non general case is considered in Sect. 14.2.2, where it is argued that in the non general case a choice has to be made as to which ambiguous Delaunay edges are not included in the triangulation.

Definition 14.6 (Circumcircle of a Triangle) The circumcircle of a triangle is the unique circle going through its three vertices.

See Fig. 14.3.

Definition 14.7 (Delaunay Triangle) A triangle is Delaunay if and only its circumcircle is empty.

Fig. 14.3 Here triangle t_i is Delaunay

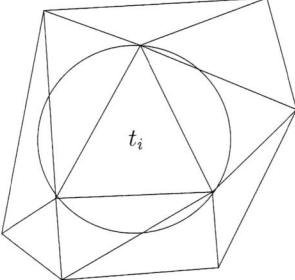

Fig. 14.4 It is seen that a circumcircle of e_i cannot be constructed such that it contains neither $\mathbf{p}_i$ or $\mathbf{p}_j$

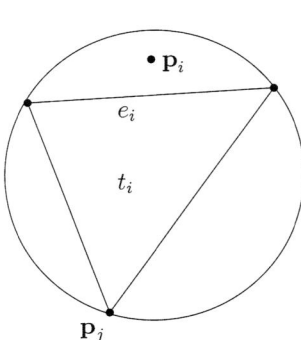

The concepts of Delaunay triangles and Delaunay edges are connected, since

Lemma 14.1 *For a given triangulation $\mathcal{T}$ of the point set $\mathcal{P}$, all the triangles of $\mathcal{T}$ are Delaunay if, and only if, all the edges are Delaunay.*

Proof Note that the circumcircle of a given triangle is also a circumcircle for the three edges of the triangle. Since all edges belong to a triangle, it is seen that if all triangles $t_i \in \mathcal{T}$ are Delaunay, then so are all the edges.

Conversely, assume that there is a non Delaunay triangle, t_i, in a given triangulation, where all the edges are Delaunay—in particular the edges of that triangle. This implies that there is at least one point, $\mathbf{p}_i$, located in the circumcircle of t_i. Since $\mathcal{T}$ is a triangulation the point $\mathbf{p}_i$ is located outside of t_i (see Fig. 14.4).

Denote the edge separating the inside of t_i and $\mathbf{p}_i$ by e_i—within the space of the circumcircle of t_i. Denote the vertex of t_i not connected to e_i by $\mathbf{p}_j$. Next, note that any circumcircle of e_i is defined by the endpoints of e_i and the center of the circle. The center of a given circumcircle is located on the line of points of equal distance to the endpoints of e_i. Consider the circumcircle of e_i that is the circumcircle of t_i, which has $\mathbf{p}_i$ in its interior. If a circumcircle of e_i is to be constructed, such that it does not contain $\mathbf{p}_i$ the center of the circumcircle has to move further away from e_i in the direction away from $\mathbf{p}_i$. This will, however, cause $\mathbf{p}_j$ to lie in the circumcircle. Hence e_i is *not* Delaunay.

So if a given triangle is not Delaunay one of its edges is not either, which is a contradiction to the assumption, thus concluding the proof.   □

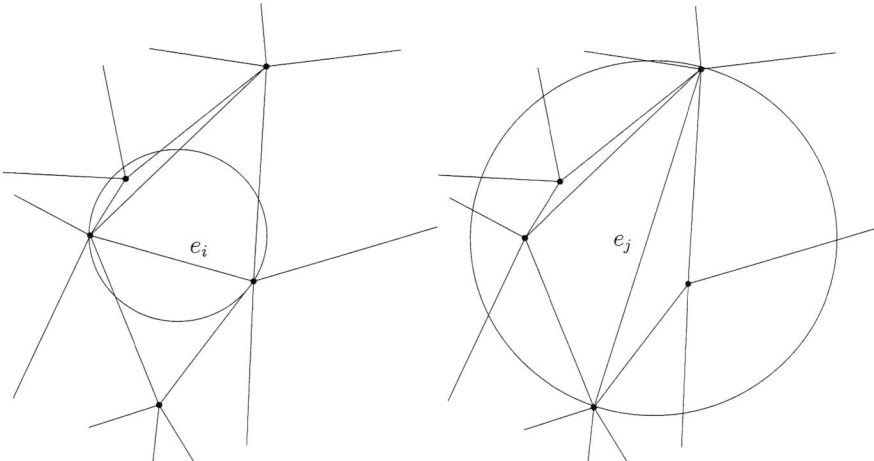

Fig. 14.5 On the *left* the edge e_i is Local Delaunay. If it is flipped, yielding the situation on the *right* where it is denoted e_j, it is *not* local Delaunay

A less restrictive definition, which is useful when constructing algorithms is the following definition.

Definition 14.8 (Local Delaunay Edge) In a triangulation $\mathcal{T}$, an edge, e_i is locally Delaunay if and only if it is either a boundary edge or if it is Delaunay with respect to the vertices of the two triangles that contain the edge e_i.

See Fig. 14.5. This definition is, however, quite strong and leads to the following.

Lemma 14.2 (Delaunay Lemma) *If all edges, $e_i \in \mathcal{E}$, of a triangulation, $\mathcal{T}$, are local Delaunay then all edges are also Delaunay.*

Proof Assume that all edges of $\mathcal{T}$ are local Delaunay, but that there exists an edge, e_1, which is *not* (globally) Delaunay. The proof of Lemma 14.1 implies that one of the triangles, t_0, which e_1 is incident on is not Delaunay and as such has a point, $\mathbf{p}'$, in its circumcircle. For ease of explanation, and without loss of generality, let us assume that the orientation is such that e_1 is part of a horizontal line, and that $\mathbf{p}'$ is located above this line and t_0 below it.

Construct a line, l, from the center of e_1 to $\mathbf{p}'$, depicted as the dashed line in Fig. 14.6.[1] Enumerate the sequence of edges that l intersects in order of occurrence, i.e., $[e_2, e_3, \ldots, e_n]$, and denote by t_i the triangle above e_i and by $\mathbf{p}_i$ the vertex of t_i not part of e_i. See Fig. 14.6. Note that $\mathbf{p}' = \mathbf{p}_n$.

---

[1]If l goes through another point, replace $\mathbf{p}'$ with that.

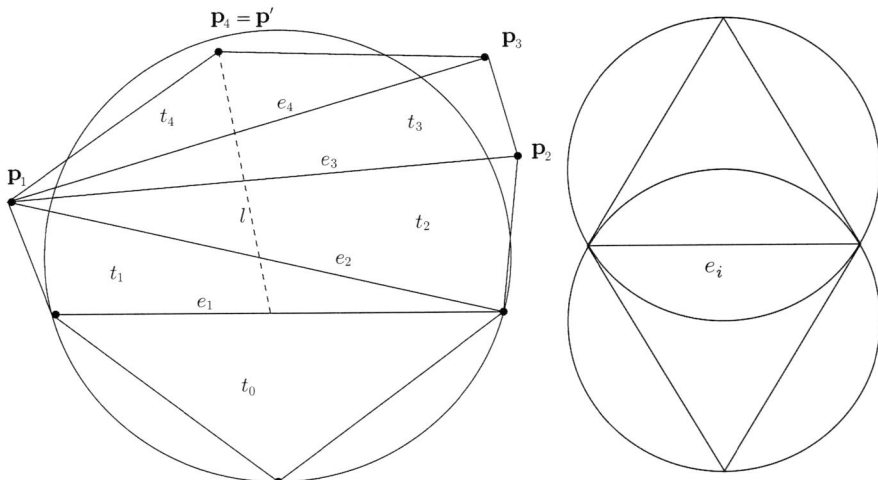

Fig. 14.6 *Left*: The notation and basic construction of Lemma 14.2. *Right*: The part of the circumcircle of the *lower triangle* above the common edge is completely in the circumcircle of the *upper triangle*, by the local Delaunay property

Consider the circumcircles of the two triangles, t_{i-1} and t_i, that contain the edge e_i. As seen in Fig. 14.6, if e_i is locally Delaunay the part of the lower circumcircle above e_i is completely in the interior of the upper circumcircle.

By construction $\mathbf{p}'$ is located in the circumcircle of t_0 above e_1, and by the above argument in the circumcircle of t_1. Since $\mathbf{p}'$ is located above all the edges $[e_1, e_2, \ldots, e_n]$, it is seen by induction that $\mathbf{p}'$ is located in the circumcircles of $[t_0, \ldots, t_n]$. The fact that $\mathbf{p}'$ is located in the circumcircles of t_{n-1}, and t_n is a contradiction to e_n being locally Delaunay, thus concluding the proof.   □

Hence, if all edges in a triangulation are local Delaunay, the triangulation is a Delaunay triangulation.

14.2.1 Refining a Triangulation by Flips

In this section, a theoretical outline of the flip algorithm for Delaunay triangulation is presented, where the general concept is that triangulations can be refined to become a Delaunay triangulation by successive flipping of edges. Hereby further properties of the Delaunay triangulation are also drawn forth. A more detailed description of this algorithm is given in Sect. 14.3.2.

The flip algorithm is an algorithm which, given a triangulation, optimizes it in a greedy fashion by flipping edges—hence the name. In the case of Delaunay triangulation, it searches for edges which are not local Delaunay and flips them, such that they become local Delaunay. The algorithm then terminates when there are no more edges to flip.

Fig. 14.7 Here e_i is not flippable since the associated quadrilateral is concave

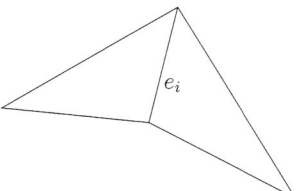

Fig. 14.8 The quadrilateral of edge e_i with endpoints a and b

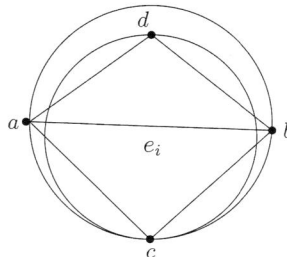

Definition 14.9 (Edge Flip) An edge is flipped by erasing it and inserting an edge in the other diagonal of the formed quadrilateral. An edge is only *flippable* if the quadrilateral associated with the edge is convex, see Figs. 14.5 and 14.7.

Flipping of edges does 'improve' the triangulation, in that

Lemma 14.3 *Let e_i be an edge of a triangulation $\mathcal{T}$, then either e_i is local Delaunay or e_i is flippable and the edge created is locally Delaunay.*

Proof Let the quadrilateral associated with e_i have the four vertices a, b, c and d where the ends of e_i are a and b—see Fig. 14.8. Consider now the circumcircle of triangle $\triangle abc$. If d is not in this circumcircle e_i is local Delaunay, thus satisfying the Lemma.

If d is inside the circumcircle of $\triangle abc$ it can be shown that by construct the quadrilateral is convex. To see this, note that, since d is bounded by the circumcircle of $\triangle abc$ and is on the opposite side of e_i w.r.t. c, no angle can be greater than $180°$. Thus in the case where d is inside the circumcircle of $\triangle abc$, e_i is flippable.

Also, a circle passing through c and d, which is also tangent to the circumcircle of $\triangle abc$ at c, contains neither a nor b. Hence the edge created by a flip of e_i is local Delaunay, see Fig. 14.8. □

It follows that the flip algorithm is correct.

Lemma 14.4 *Given a triangulation $\mathcal{T}$ of n points, $\mathbf{p}_i \in \mathcal{P}$, then the flip algorithm terminates after $O(n^2)$ flips, and the result is a triangulation whose edges are all Delaunay.*

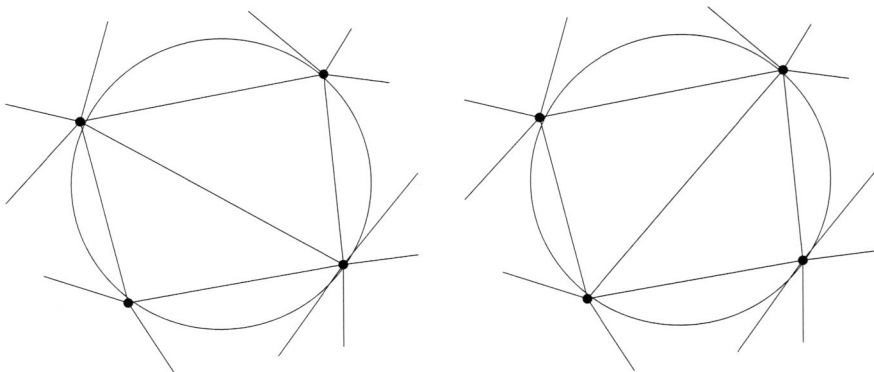

Fig. 14.9 Four points on a common circle. It is seen that two triangulations are possible

For a proof, see [3]. As an immediate consequence of this the following can be shown.

Theorem 14.1 *Let $\mathcal{P}$ be a set of three or more points in the plane, where no three points lie on a common line and no four points lie on a common circle. Then the Delaunay triangulation, $\mathcal{D}$, of $\mathcal{P}$ is a triangulation, and the flip algorithm produces it. This is the case provided that an initial triangulation of $\mathcal{P}$ exists.*

Proof Because no three points are lie on a common line and no four points lie on a common circle (i.e. points are in general position), there exists a Delaunay triangulation of the points. By Lemma 14.4, the application of the flip algorithm to any triangulation produces a triangulation whose edges are all Delaunay.

It is also seen that no other edge is Delaunay. Consider an edge $e_i \notin \mathcal{D}$ with endpoints $\mathbf{p}_i$ and $\mathbf{p}_j$ both in $\mathcal{P}$. Since $\mathcal{D}$ is a triangulation, e_i must cross some edge $e_j \in \mathcal{D}$. The edge e_j has an empty circumcircle, since it is in $\mathcal{D}$, where either $\mathbf{p}_i$ or $\mathbf{p}_j$ lies strictly outside, since no four points are on a common circle. It follows that e_i does not have an empty circumcircle and as such is not Delaunay. □

14.2.2 Points *not* in General Position

In the above, it has been assumed that no four points are located on a common circle and no three points are aligned. If this is not the case, Delaunay edges intersecting themselves at points that do not belong to the point set or degenerate (flat) triangles can—potentially—be included in the Delaunay graph, and the concept of Delaunay triangulation becomes ambiguous. This is illustrated in Fig. 14.9, where there are two possible triangulations of a point set all only including Delaunay edges. However, in the non general case, we cannot include *all* Delaunay edges and still have a triangulation, e.g., if we included both possible edges in Fig. 14.9 they would cross and the result would *not* be a triangulation. Hence a compromise has to be made.

- Either all Delaunay edges will be included, and the result will be a Delaunay graph (but not a Delaunay triangulation),
- or a consistent choice of Delaunay edges is included, and the result is indeed a triangulation consisting of Delaunay edges, albeit not all such Delaunay edges are included in the triangulation.

In practice, all Delaunay edges are typically not included, and a triangulation consisting of only—but not all—Delaunay edges is formed, i.e., the latter choice.

14.2.3 Properties of a Delaunay Triangulation

The Delaunay triangulation of a point set has some nice properties, apart from the circumcircles of all edges and triangles being empty. It is these properties that make the Delaunay triangulation the typical triangulation of choice. Some of the more central properties are briefly mentioned here.

Min Max Angle   A common problem in triangulations is that some triangles become too skinny, in that one of their angles becomes too small. This, among other, relates to finite element simulations, where very skinny triangles can cause numerical instability. To avoid this—as best as possible—the Delaunay triangulation is a good choice, since the following can be shown.

Lemma 14.5 (Min Max Angle) *Among all triangulations of a point set, $\mathcal{P}$, the Delaunay triangulation maximizes the minimal angle.*

See e.g., [2] for a proof.

Lifted Circle   Considering the circumcircle of three points $\mathbf{p}_1$, $\mathbf{p}_2$ and $\mathbf{p}_3$. Then a fourth point $\mathbf{p}_4$, location w.r.t. this circumcircle has an alternative interpretation if the points are projected onto the surface $f : (x, y) \to (x, y, x^2 + y^2)$ (which is a paraboloid of revolution, see Fig. 14.10).

Lemma 14.6 *Given the 3D points $\bar{\mathbf{p}}_1 = (\mathbf{p}_{1x}, \mathbf{p}_{1y}, \mathbf{p}_{1x}^2 + \mathbf{p}_{1y}^2)$, $\bar{\mathbf{p}}_2 = (\mathbf{p}_{2x}, \mathbf{p}_{2y}, \mathbf{p}_{2x}^2 + \mathbf{p}_{2y}^2)$, $\bar{\mathbf{p}}_3 = (\mathbf{p}_{3x}, \mathbf{p}_{3y}, \mathbf{p}_{3x}^2 + \mathbf{p}_{3y}^2)$ and $\bar{\mathbf{p}}_4 = (\mathbf{p}_{4x}, \mathbf{p}_{4y}, \mathbf{p}_{4x}^2 + \mathbf{p}_{4y}^2)$. Then the point $\mathbf{p}_4$ lies in the circumcircle of $\mathbf{p}_1$, $\mathbf{p}_2$, and $\mathbf{p}_3$, if, and only if, $\bar{\mathbf{p}}_4$ lies below the plane spanned by $\bar{\mathbf{p}}_1$, $\bar{\mathbf{p}}_2$ and $\bar{\mathbf{p}}_3$. The vertical direction is the z-coordinate.*

Proof Consider the plane spanned by $\bar{\mathbf{p}}_1$, $\bar{\mathbf{p}}_2$ and $\bar{\mathbf{p}}_3$, $z = k_1 x + k_2 y + k_3$ and the paraboloid of revolution, $z = x^2 + y^2$, as functions of x and y. It is seen that for the intersection of these two surfaces the x and y values are given by

$$k_1 x + k_2 y + k_3 = x^2 + y^2$$
$$\Leftrightarrow \frac{k_1^2 + k_2^2}{4} + k_3 = x^2 - 2\frac{k_1}{2}x + \left(\frac{k_1}{2}\right)^2 + y^2 - 2\frac{k_2}{2}y + \left(\frac{k_2}{2}\right)^2$$
$$= \left(x - \frac{k_1}{2}\right)^2 + \left(y - \frac{k_2}{2}\right)^2,$$

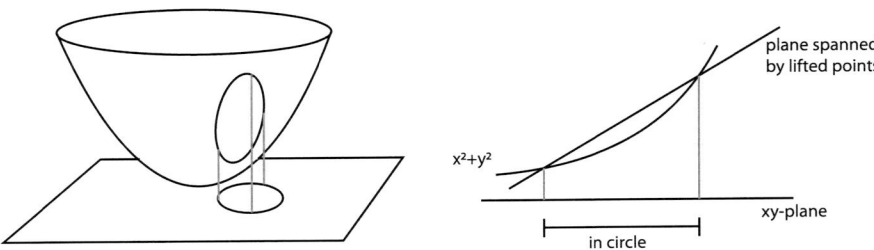

Fig. 14.10 Illustration of the lifted circle principle. On the *left* it is illustrated that three points on the parabola $x^2 + y^2$ define a plane, and the intersection of that plane with the parabola define a 2D ellipsoid. If this ellipsoid is projected into 2D it will be the circumcircle of the three points projected into 2D. Thus any point in the circumcircle of the three points will project to a point below the before mentioned plane. On the *right* this is illustrated by a 2D vertical slice through the parabola, where any point in the "in circle" area will project to a point on the parabola below the plane

which is a circle. It is this circle which is the projection of the intersection of the two surfaces onto the xy-plane.

Denote the circumcircle of $\mathbf{p}_1$, $\mathbf{p}_2$ and $\mathbf{p}_3$ by C. Since C is unique and $\bar{\mathbf{p}}_1$, $\bar{\mathbf{p}}_2$ and $\bar{\mathbf{p}}_3$ naturally lie on the geometric loci defined by p and f, it is seen that the C projects to the intersection of the geometric loci defined by f and p. Since f defines a paraboloid and p a plane, it is obvious that the points inside C project to the only points on the paraboloid defined by f which are below the hyperplane given by p. This can also be seen by stating the fact the point (x, y) is interior to the circle C centered on (x_0, y_0) of radius r_0 if its distance to the circle center is less than the radius of the circle: $(x - x_0)^2 + (y - y_0)^2 < r_0^2$. Separating the second degree terms (in the variables x and y) from the others, we get $x^2 + y^2 < 2xx_0 + 2yy_0 + (x_0^2 + y_0^2 + r^2)$. The left hand side of the previous inequality corresponds to the z coordinate of the lifting of the point (x, y) on the paraboloid; while the right hand side corresponds to the z coordinate of the lifting of the point (x, y) on the hyperplane defined by p. Thus we conclude the proof. $\square$

This is an issue of theoretical importance, which gives insight into the properties of the Delaunay triangulation, and is the basis of proofs concerning this triangulation, e.g., the `InCircle` predicate defined later. In particular it states that if we project the 2D points, $\mathbf{p}_i \in \mathcal{P}$, onto the parabola $z = x^2 + y^2$, resulting in the 3D points $\bar{\mathbf{p}}_i \in \bar{\mathcal{P}}$, then the Delaunay triangulation corresponds to the convex hull of these 3D points $\bar{\mathbf{p}}_i$, cf. Chap. 13.

14.3 Delaunay Triangulation Algorithms

As indicated above, Delaunay triangulation is very widely used and has been around for some time, hence there is a multitude of algorithms for achieving it with varying running and implementation complexity. A survey of algorithms is beyond the scope here. For an overview the reader is referred to [1, 3, 4].

The most efficient Delaunay triangulation algorithms run in $\mathcal{O}(n \log n)$ time, where n is the number of vertices. Arguably, the fastest of these algorithms is the divide and conquer algorithm of Guibas and Stolfi described in [5], which is relatively easy to implement. Another popular class of algorithms is the sweepline type algorithms, which have similar properties.

The divide and conquer algorithm does not generalize to dimensions higher than two, which is e.g., the case of the flip algorithm mentioned previously in this chapter. The flip algorithm is of the incremental insertion type is very easy to implement and will run in $\mathcal{O}(n^2)$ time; cf. e.g., [3, Sect. 2.1.5] [2, 5].

This section starts out by presenting the algorithmic details of some of the most common geometric primitives used for triangulation, such as the in circle test. Following this, a detailed description of the flip algorithm is given in Sect. 14.3.2, whereafter the divide and conquer algorithm is outlined.

14.3.1 Geometric Primitives

In most algorithms for triangulation there are two basic computational primitives that prevail. These are tests of if a point is located to the left of a line spanned by two other points LeftOf, and the test of if a point is located in the circumcircle of three other points, InCircle. Hence these will be described here and will be dealt with again in Sect. 14.4.

Definition 14.10 (LeftOf) The predicate LeftOf$(\mathbf{p}_1, \mathbf{p}_2, \mathbf{p}_3)$ determines if the point $\mathbf{p}_3$ lies to the left of the line spanned by $\mathbf{p}_1$ and $\mathbf{p}_2$.

A more symmetrical definition is that the points $\mathbf{p}_1$, $\mathbf{p}_2$, and $\mathbf{p}_3$ lie in a counterclockwise order. A concrete way to calculate this predicate is by determining the sign of a determinant.

Lemma 14.7 *The predicate* LeftOf$(\mathbf{p}_1, \mathbf{p}_2, \mathbf{p}_3)$ *can be calculated by the sign of the following determinant*:

$$\begin{vmatrix} 1 & \mathbf{p}_{1x} & \mathbf{p}_{1y} \\ 1 & \mathbf{p}_{2x} & \mathbf{p}_{2y} \\ 1 & \mathbf{p}_{3x} & \mathbf{p}_{3y} \end{vmatrix} > 0. \tag{14.1}$$

Proof First note that the determinant becomes zero if, and only if, the points are collinear implying that the 3×3 matrix loses rank. Consider then the case of $\mathbf{p}_1 = (0, 0)$, $\mathbf{p}_2 = (1, 0)$ and $\mathbf{p}_3 = (1, 1)$, which clearly are in a counterclockwise order. In this case the determinant is given by

$$\begin{vmatrix} 1 & 0 & 0 \\ 1 & 1 & 0 \\ 1 & 1 & 1 \end{vmatrix} = 1 > 0.$$

Note that for the case of $\mathbf{p}_1 = (0, 0)$, $\mathbf{p}_2 = (1, 1)$, and $\mathbf{p}_3 = (1, 0)$, which clearly are in a clockwise order

$$\begin{vmatrix} 1 & 0 & 0 \\ 1 & 1 & 1 \\ 1 & 1 & 0 \end{vmatrix} = -1 < 0.$$

Interpreting (14.1) as a function on $\mathbb{R}^6$ it is seen that it is smooth, and that it is only zero when the underlying three points, $\mathbf{p}_1$, $\mathbf{p}_2$ and $\mathbf{p}_3$, are collinear. Also, any smooth deformation of a given set of points, $\mathbf{p}_1$, $\mathbf{p}_2$ and $\mathbf{p}_3$, can only go from being counter-clockwise to being clockwise by being collinear. So with the two above examples it is seen that (14.1) is positive if the points are counterclockwise and negative if they are clockwise.   □

Definition 14.11 (InCircle) The predicate `InCircle`$(\mathbf{p}_1, \mathbf{p}_2, \mathbf{p}_3, \mathbf{p}_4)$ determines if point $\mathbf{p}_4$ lies inside of the circumcircle of points $\mathbf{p}_1$, $\mathbf{p}_2$ and $\mathbf{p}_3$, where it is assumed that the points $\mathbf{p}_1$, $\mathbf{p}_2$ and $\mathbf{p}_3$ are in counterclockwise order.

The reason that $\mathbf{p}_1$, $\mathbf{p}_2$ and $\mathbf{p}_3$ are assumed to be in counterclockwise order is that `InCircle` is also calculated as the sign of a determinant. This sign naturally depends on the ordering of points $\mathbf{p}_1$, $\mathbf{p}_2$ and $\mathbf{p}_3$, which would change of the point were ordered in a clockwise fashion. This implies that if the points $\mathbf{p}_1$, $\mathbf{p}_2$ and $\mathbf{p}_3$ are ordered in a clockwise fashion, the result of `InCircle` should be negated. This can naturally be incorporated in the implementation of the `InCircle` predicate if so desired. A concrete way to calculate `InCircle` is as follows.

Lemma 14.8 *The predicate* `InCircle`$(\mathbf{p}_1, \mathbf{p}_2, \mathbf{p}_3, \mathbf{p}_4)$ *can be calculated by the sign of the following determinant:*

$$\begin{vmatrix} \mathbf{p}_{1x} & \mathbf{p}_{1y} & \mathbf{p}_{1x}^2 + \mathbf{p}_{1y}^2 & 1 \\ \mathbf{p}_{2x} & \mathbf{p}_{2y} & \mathbf{p}_{2x}^2 + \mathbf{p}_{2y}^2 & 1 \\ \mathbf{p}_{3x} & \mathbf{p}_{3y} & \mathbf{p}_{3x}^2 + \mathbf{p}_{3y}^2 & 1 \\ \mathbf{p}_{4x} & \mathbf{p}_{4y} & \mathbf{p}_{4x}^2 + \mathbf{p}_{4y}^2 & 1 \end{vmatrix} < 0. \tag{14.2}$$

Proof Considering Lemma 14.6 and its proof, it is seen that (14.2) is the signed volume of the tetrahedron spanned by $\bar{\mathbf{p}}_1$, $\bar{\mathbf{p}}_2$, $\bar{\mathbf{p}}_3$ and $\bar{\mathbf{p}}_4$. It is noted the $\bar{\mathbf{p}}_4$ is below the plane spanned by $\bar{\mathbf{p}}_1$, $\bar{\mathbf{p}}_2$ and $\bar{\mathbf{p}}_3$ if $\mathbf{p}_4$ is inside the circumcircle of $\mathbf{p}_1$, $\mathbf{p}_2$ and $\mathbf{p}_3$ and above otherwise. Hence with the traditional sign conventions the volume is positive when $\mathbf{p}_4$ is inside and negative when it is outside—given the counterclockwise orientation of $\mathbf{p}_1$, $\mathbf{p}_2$ and $\mathbf{p}_3$. It is also seen that the determinant in (14.2) only becomes zero when $\bar{\mathbf{p}}_1$, $\bar{\mathbf{p}}_2$, $\bar{\mathbf{p}}_3$, and $\mathbf{p}_4$ become coplanar implying that $\mathbf{p}_1$, $\mathbf{p}_2$, $\mathbf{p}_3$, and $\mathbf{p}_4$ are on a common circle.   □

As an example, consider the points $\mathbf{p}_1 = (0,0)$, $\mathbf{p}_2 = (1,0)$, $\mathbf{p}_3 = (1,1)$, and $\mathbf{p}_4 = (\frac{2}{3}, \frac{1}{3})$, where $\mathbf{p}_4$ is clearly inside the circumcircle of $\mathbf{p}_1$, $\mathbf{p}_2$, and $\mathbf{p}_3$ which are ordered in a counterclockwise fashion. Inserting into (14.2) gives

$$
\begin{vmatrix} 0 & 0 & 0 & 1 \\ 1 & 0 & 1 & 1 \\ 1 & 1 & 2 & 1 \\ \frac{2}{3} & \frac{1}{3} & \frac{5}{9} & 1 \end{vmatrix} = \begin{vmatrix} 0 & 0 & 0 \\ 1 & 0 & 1 \\ 1 & 1 & 2 \end{vmatrix} - \begin{vmatrix} 0 & 0 & 0 \\ 1 & 1 & 2 \\ \frac{2}{3} & \frac{1}{3} & \frac{5}{9} \end{vmatrix} + \begin{vmatrix} 0 & 0 & 0 \\ 1 & 0 & 1 \\ \frac{2}{3} & \frac{1}{3} & \frac{5}{9} \end{vmatrix} - \begin{vmatrix} 1 & 0 & 1 \\ 1 & 1 & 2 \\ \frac{2}{3} & \frac{1}{3} & \frac{5}{9} \end{vmatrix}
$$

$$
= 0 - 0 + 0 - \left(-\frac{4}{9}\right) = \frac{4}{9} > 0,
$$

where the matrix is expanded by minors w.r.t. the last column. Note that in 3D similar predicates exist.

14.3.2 The Flip Algorithm

As outlined in Sect. 14.2.1 an arbitrary triangulation can be refined into a Delaunay triangulation via a succession of flips. This property can be used to construct an algorithm for computing a Delaunay triangulation of a given set of points, $\mathcal{P}$, here assumed to be in 2D. This construction process primarily consists of addressing a number of practical issues brushed over in Sect. 14.2.1.

The first practical issue is that Sect. 14.2.1 deals with the refinement of triangulations, and here we do not have a triangulation but just a set of points. The way this is dealt with is by using an incremental insertion approach. That is, we first construct a triangle, which is so large that it is certain to include all points, $\{\mathbf{p}_1, \ldots, \mathbf{p}_n\} \in \mathcal{P}$. Hereby we add three extra points $\mathbf{p}_{n+1}, \mathbf{p}_{n+2}, \mathbf{p}_{n+3}$, being the corners of this new 'super' triangle. It is seen that the triangulation of $\mathbf{p}_{n+1}, \mathbf{p}_{n+2}, \mathbf{p}_{n+3}$ only is a Delaunay triangulation, since there are only three points. Following this, each of the points are added to the triangulation one by one. As seen in Fig. 14.11, a point, $\mathbf{p}_i$ is added by

1. finding the triangle, t_j, the point $\mathbf{p}_i$ is located within;
2. adding edges from $\mathbf{p}_i$ to the three corners of t_j, whereby t_j is subdivided;
3. refining the resulting triangulation into a Delaunay triangulation via edge flips;

whereafter $\mathbf{p}_{i+1}$ is added in a similar manner. Since the triangulation is obviously a Delaunay triangulation after Step 3, after the last point, $\mathbf{p}_n$, has been added, the resulting triangulation is a Delaunay triangulation of the original points $\mathcal{P}$ plus $\mathbf{p}_{n+1}, \mathbf{p}_{n+2}, \mathbf{p}_{n+3}$. The last part of the flip algorithm thus consists of removing points $\mathbf{p}_{n+1}, \mathbf{p}_{n+2}, \mathbf{p}_{n+3}$, and all edges connected to them.

Other practical issues concerning the flip algorithm should also be mentioned. Firstly concerning Step 1, it is noted that the added points are never on an edge, due to the assumption of points being in a general position. If this is not the case decisions have to be made on how to deal with this issue, cf. Sect. 14.2.2.

There is also the issue of how to efficiently find the triangle t_j containing $\mathbf{p}_i$. Due to the close links to the Voronoi diagram, cf. Sect. 14.6, it is seen that a corner of t_j must be the inserted point already inserted closest to $\mathbf{p}_i$. Efficient techniques for

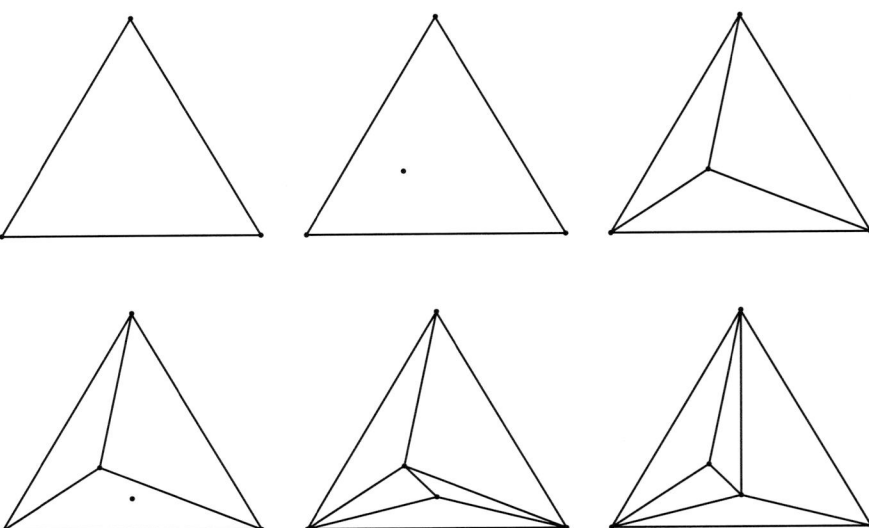

Fig. 14.11 An illustration of the first few steps of the flip algorithm. *Top left*: A large triangle encompassing all the points $\mathcal{P}$ is formed. *Top middle*: The first point, $\mathbf{p}_1$ is inserted. *Top right*: Edges from $\mathbf{p}_1$ to the corners of its encompassing triangle are added. Since this is already a Delaunay triangulation no refinement is illustrated. *Bottom left*: The second point $\mathbf{p}_2$ is inserted. *Bottom middle*: Edges from $\mathbf{p}_2$ to the corners of its encompassing triangle are added. *Bottom right*: The triangulation is refined, here by a single flip

doing this are found in Chap. 12. A more specialized solution is to form a search tree based on the incremental triangulation history, cf. [2]. For a basic implementation, all triangles can of course be tested in a sequential manner.

In relation to the refinement of Step 3, it is noted that if all edges are locally Delaunay, then the whole triangulation is Delaunay and vice versa. Also before the insertion in Step 1 all edges are locally Delaunay. Thus after the insertion the only possible non locally Delaunay edges are the ones from t_j and the three edges added in Step 2. Thus, initially only these six edges have to be considered for flipping.[2] These edges can be inserted into a heap—containing edges for consideration—where subsequent edges of triangles effected by a flip should be inserted into. When this heap is empty, all edges are locally Delaunay.

Lastly, it should be mentioned that the amortized running time of the algorithm will be improved if the points are inserted in a random order [6].

---

[2]Actually, of these six edges the three new ones will be locally Delaunay. To see this, note that before the insertion of the point the triangle t_j is Delaunay, and thus has an empty circumcircle. It is seen that a circumcircle of one of the new edges can be constructed completely within this old circumcircle (e.g., by shrinking the old circumcircle), whereby the new edges are locally Delaunay at the outset.

Table 14.1 Pseudo code for the central recursion in the divide and conquer algorithm for Delaunay triangulation

$\bar{\mathcal{T}} =$**Delaunay_Split_and_Merge**($\bar{\mathcal{P}}$)

  If $|\bar{\mathcal{P}}| > 3$

    1. Split $\bar{\mathcal{P}}$ into two equal parts L and R, such that all points in L have a lower y value than the points in R.

    2. $\mathcal{T}_l =$ **Delaunay_Split_and_Merge**(L).

    3. $\mathcal{T}_r =$ **Delaunay_Split_and_Merge**(R).

    4. $\bar{\mathcal{T}} =$ **Merge**($\mathcal{T}_l, \mathcal{T}_r$).

  Else If $|\bar{\mathcal{P}}| = 3$ form triangle.

  Else form a line

  Return triangulation, $\bar{\mathcal{T}}$.

14.3.3 The Divide and Conquer Algorithm

As mentioned above, the arguably most efficient (i.e., best) algorithm for computing the Delaunay triangulation in 2D is the divide and conquer algorithm by Guibas and Stolfi [5], which is relatively simple to implement, albeit not as simple as the flip algorithm. Another difference from the flip algorithm is that the divide and conquer algorithm does not generalize to higher dimensions than two. A discussion of the divide and conquer strategy in general is found in Chap. 13.

The basis of this divide and conquer algorithm is that it is relatively simple and efficient to merge two *non*-overlapping Delaunay triangulations into one Delaunay triangulation. Also, if the points are sorted according to y value it is straight forward to divide the point sets where the points in one set all have higher y value than the other, and thus must have non-overlapping triangulations. Lastly, if a set of points has two or three points it is easy to form a Delaunay triangulation by forming a line or a triangle (although it could be argued if a line is a triangulation, but this is of no consequence here).

The algorithm starts out by sorting the 2D points $\mathcal{P}$ relative to their y value, whereby the points can easily be divided according to increasing y value, and the recursion of Table 14.1 is performed on the points $\mathcal{P}$. For details on implementation, especially on the merge step, the interested reader is referred to [5].

14.4 Stability Issues

An issue in many algorithms from computational geometry, and with triangulation in particular, is that the algorithms are not stable w.r.t. numerical noise. The reason is that the algorithms work by taking some discrete action based on the value of some predicate. By discrete is meant that there is no smooth transition between the actions. The problem arises when the predicate value approaches the 'border' between implying two actions, and the calculated value is uncertain due to round off

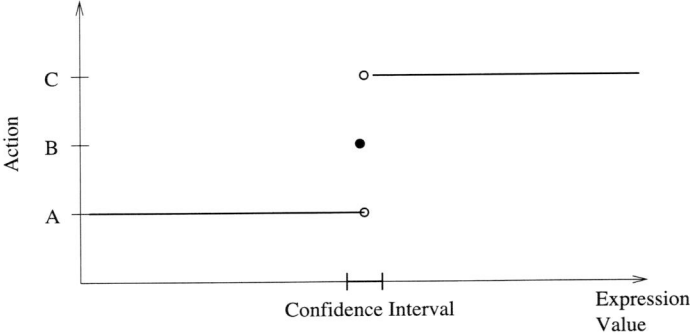

Fig. 14.12 With the given certainty of the predicate it is not possible to determine which action to take

Fig. 14.13 Four points close
to being on a common circle

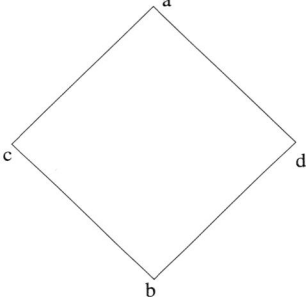

error, as illustrated in Fig. 14.12. In this regard it should be noted that the rounding of the error can vary for the semantically same predicate expressed in two different syntaxes.

As an example consider four points, a, b, c, d, close to being on a common circle in the context of a Delaunay triangulation algorithm—see Fig. 14.13. Here the predicate `InCircle` is likely to be used for determining if the diagonal should go from a to b or from c to d. However, it is likely that it will say both or neither, in the face of numerical noise.

It should be noted that this issue is not just of theoretical interest. In fact most algorithms will fail too frequently for comfort if this stability issue is not addressed. Hence *for real triangulation applications these issues need to be considered* and in many applications should be dealt with.

There are a couple of different ways of addressing this problem, a few of which will be discussed here. One way is to use exact arithmetic in the algorithm instead of the traditional floats or doubles. This, however, has the drawback of slowing the computations down considerably—approximately 70 times in some applications [7].

Another approach is to monitor if a predicate value comes close to the 'border' (see Fig. 14.12). By close is meant close to numerical precision. In these few cases the predicate is calculated with increased precision, in order that the issue ceases to exist. There are a couple strategies for doing this, two of which are [7, 8].[3]

Another issue is that for some predicate values there is no appropriate action. E.g. if a triangulation is to be made of three points located on a common line, in which case (14.1) will be zero. A solution is to permute the points slightly—making them non collinear—whereby an appropriate action can be taken, cf. e.g., [2].

14.5   Other Subjects in Triangulation

As mentioned, mesh generation can easily fill a whole course, and as such there are many subjects related to this matter not covered here. In order to make the student aware of their existence two of the most central will be covered briefly here.

14.5.1 Mesh Refinement

One major subject in mesh generation is mesh refinement. The idea is that if it is not possible to make a nice enough triangulation of a point cloud, e.g., via a Delaunay triangulation, then final triangle mesh can be improved by inserting strategic points via some rule or another, cf. e.g., [3].

An example where this could be desirable is the triangulation of a point cloud, with the intent of simulation. Here some triangles sometimes become two slim for numerical comfort, and extra points can be inserted to break up such triangles.

14.5.2 Constrained Delaunay Triangulation

A constrained Delaunay triangulation is a triangulation of a set of points, with the constraint that the triangulation should contain certain edges—which are called constrained edges, and which need not be Delaunay—and that all Delaunay edges that do not cross any of those constrained edges (except at their extremities) belong to the triangulation, cf. e.g., [1–3]. This necessitates a slight modification of the Delaunay properties of triangles and edges termed constrained Delaunay. Indeed, the constrained edges are flagged so that they may not be changed by subsequent triangle flips.

An example of this is the triangulation of the interior of a polygon. Here the edges of that polygon should naturally be in the final triangulation, regardless of them being Delaunay or not. Note that the triangulation of a polygon is a special case of triangulation, and as such, special algorithms have been constructed for this purpose, cf. e.g., [1, 9], including a $\mathcal{O}(n)$ running time algorithm; cf. e.g., [10].

---

[3]Both papers contain further references to the subject.

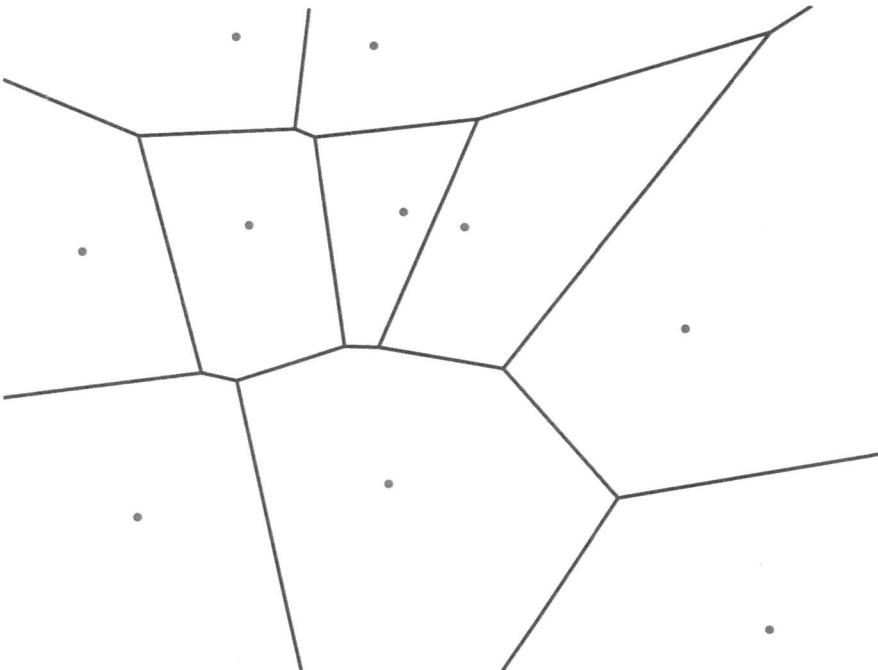

Fig. 14.14 The Voronoi diagram of ten random points, $\mathcal{P}$. Note that the regions, V_i, denote exactly the parts of $\mathbb{R}^2$ which are closest to its respective point $\mathbf{p}_i \in \mathcal{P}$

Recently, constrained Delaunay triangulations have been extended in order to handle moving points that may move along constrained trajectories. The result is a kinetic constrained Delaunay triangulation [11], where the moving point may be part of a constrained Delaunay edge. Its main application is in the field of Geospatial Information Systems; however, it can also be applied to navigation and robotics.

14.6   Voronoi Diagram

A problem closely related to Delaunay triangulation is that of determining which one of a number of given locations, $\mathcal{P}$, an arbitrary position is closest to, e.g., which hospital is closest to you. This extends to the problem of computing a diagram or map of indicating for all points, e.g., in 2D, which point in $\mathcal{P}$ is the closest, cf. Fig. 14.14. This diagram is called the Voronoi Diagram, after Georgy Voronoi, and is seen to consist of regions, V_i, associated with each point, $\mathbf{p}_i$, formally defined as follows.

Definition 14.12 (Voronoi Region in 2D) Given a set of points $\{\mathbf{p}_1, \ldots, \mathbf{p}_n\} \in \mathcal{P}$ the Voronoi regions are defined by

$$V_i = \left\{ x \in \mathbb{R}^2 \mid \|x - \mathbf{p}_i\| \le \|x - \mathbf{p}_j\|, \ \forall \mathbf{p}_j \in \mathcal{P} \right\}, \tag{14.3}$$

i.e., the region of points in $\mathbb{R}^2$ closest to $\mathbf{p}_i$.

In computing the Voronoi diagram, the edges are naturally of special interest, in that they define the Voronoi regions, V_i. These edges or boundaries occur when two Voronoi regions, V_i and V_j, are adjacent, and points on these have equal distance to $\mathbf{p}_i$ and $\mathbf{p}_j$, and no other points are closer. This is equivalent to being able to construct a circle

- centered at the edge between the Voronoi regions V_i and V_j;
- traversing $\mathbf{p}_i$ and $\mathbf{p}_j$;
- which is empty.

This is equivalent to having an empty circumcircle of an edge from $\mathbf{p}_i$ to $\mathbf{p}_j$, which is again equal to the definition of a Delaunay edge from $\mathbf{p}_i$ to $\mathbf{p}_j$. A definition of a Delaunay triangulation of a point set $\mathcal{P}$ is then that it is the dual of the Voronoi diagram of the same point set, $\mathcal{P}$. By this is meant that there is a Delaunay edge from $\mathbf{p}_i$ to $\mathbf{p}_j$ if, and only if, there is an edge between V_i and V_j. It can be shown that this is equivalent to the previous definition of Delaunay triangulations given in this chapter, cf. [2]. An example is given in Fig. 14.15.

In practice, the Voronoi diagram of a point set $\mathcal{P}$ is often computed by first computing the Delaunay triangulation of $\mathcal{P}$, because efficient algorithms exist for doing the latter, cf. e.g., [9]. In this regard it should be noted that the corners of the Voronoi regions correspond to points which have equal distances to three points in $\mathcal{P}$, and as such correspond to the centers of the circumcircles of the Delaunay triangles.[4] Thus a Voronoi diagram can be constructed from a Delaunay triangulation, by adding a vertex at the center of each circumcircle of the Delaunay triangles, and then connecting two such vertices if, and only if, the corresponding triangles share an edge.

14.7 Exercises

Exercise 14.1 The aim of this exercise is to implement an algorithm for Delaunay triangulation of points. Hereby it is hoped that a greater understanding of the theory behind Delaunay triangulation is obtained. It is thus important that a working algorithm is achieved; however, an industrial strength one is not needed.

What algorithm you choose to implement is up to you—as long as it is a variant of one that someone has argued to be correct. The simplest algorithm we can recommend is an incremental insertion algorithm, cf. e.g., Sect. 14.3.2. A more challenging algorithm—with a more applicable running time—is a divide and conquer type algorithm. A good description of such an algorithm is found in [5].

---

[4]Note that the center of a Delaunay triangle's circumcircle need not necessarily be located within that triangle.

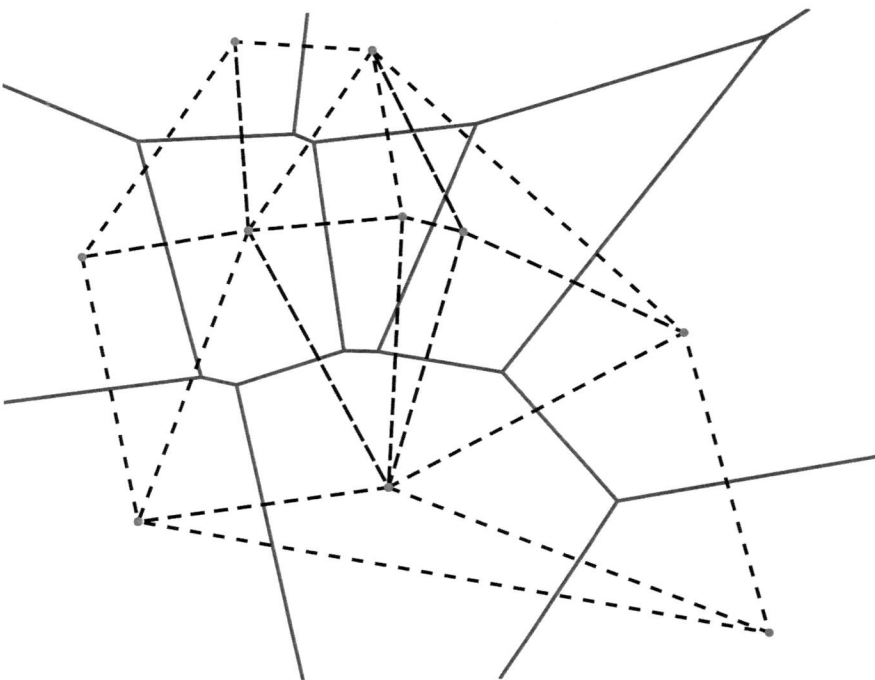

Fig. 4.15   The Voronoi diagram from Fig. 14.14, with the Delaunay triangulation added

We have included two data sets on which you should apply your algorithms. An easier one consists of 10 random points. A more realistic one is containing 3D terrain data. The 3D terrain data are to be projected into the xy-plane and triangulated there. This induces a triangulation in 3D which is the result you should achieve.

Deliverables: Illustration of the triangulated data sets, and interpretation of the results.

Resources: The Gel framework, the files containing the two point sets, `data.txt` and `kotel.txt` and a small demonstration to get you started, `delaunay.cpp`.

References

1. Bern, M., Eppstein, D.: Mesh generation and optimal triangulation. In: Computing in Euclidean Geometry. Lecture Notes Ser. Comput., vol. 1, pp. 23–90. World Scientific, River Edge (1992)
2. Edelsbrunner, H.: Geometry and Topology for Mesh Generation. Cambridge Monographs on Applied and Computational Mathematics, vol. 7, p. 177. Cambridge University Press, Cambridge (2006). 978-0-521-68207-7; Reprint of the 2001 original
3. Shewchuck, J.R.: Lecture notes on Delaunay mesh generation. Technical report, UC Berkeley (1999). http://www.cs.berkeley.edu/~jrs/meshpapers/delnotes.ps.gz
4. Su, P., Drysdale, R.S.: A comparison of sequential Delaunay triangulation algorithms. Comput. Geom. **7**, 361–385 (1997)

5. Guibas, L., Stolfi, J.: Primitives for the manipulation of general subdivisions and the computation of Voronoi diagrams. ACM Trans. Graph. **4**, 74–123 (1985)
6. Guibas, L.J., Knuth, D.E., Sharir, M.: Randomized incremental construction of Delaunay and Voronoi diagrams. Algorithmica (New York) **7**(4), 381–413 (1992)
7. Pion, S., Devillers, O.: Efficient exact geometric predicates for Delaunay triangulations. In: 5th Workshop on Algorithm Engineering and Experiments (ALENEX 03) (2003)
8. Shewchuk, J.R.: Adaptive precision floating-point arithmetic and fast robust geometric predicates. Discrete Comput. Geom. **18**, 305–363 (1997)
9. de Berg, M., van Kreveld, M., Overmars, M., Schwarzkopf, O.: Computational Geometry. Algorithms and Applications, p. 365. Springer, Berlin (1997)
10. Chazelle, B.: Triangulating a simple polygon in linear time. Discrete Comput. Geom. **6**, 485–524 (1991)
11. Gold, C.M., Mioc, D., Anton, F., Sharma, O., Dakowicz, M.: A methodology for automated cartographic data input, drawing and editing using kinetic Delaunay/Voronoi diagrams. In: Studies in Computational Intelligence, vol. 158 (2009). doi:10.1007/978-3-540-85126-4

3D Surface Registration via Iterative Closest Point (ICP)

15

Many geometric 3D models of real objects originate from a 3D scanning process, such as laser scanning. An often occurring issue in this scanning process is that it is only practically possible to capture partial scans of the object in question, cf. Fig. 15.1. So in order to get a complete 3D model these partial 3D scans or surfaces have to be combined into one. A major task in doing this is getting the partial surfaces into a common reference frame or coordinate system, in that it is seldom known with sufficient accuracy how the object moved relative to the scanner between scans.

The typical process for getting these partial surfaces into the same reference frame is by first registering them to each other, such that corresponding points are identified. Following this, a transformation is found which minimizes the distance between the correspondences. Following this, the surfaces, e.g., triangular meshes, are merged into one combining data structure.

The main subject of this chapter is that of registration, and the iterative closest point method (ICP) in particular. The merging of partial scans is, however, briefly covered in Sect. 15.5 with links to the methods presented through out this book. It should also be noted that registration of 3D surfaces has a wide range of applications beyond the merging of partial scans, e.g., *object recognition* and searching of 3D object databases, cf. e.g., [1].

15.1 Surface Registration Outline

Registration of data sets, be they signals, images, surfaces, etc., is one of the main problems in computer science and signal analysis. For 3D surfaces representing (partially) overlapping parts of the same (or a similar) object, the aim is to find the locations, e.g., points, on the surfaces corresponding to the same entity on the underlying object, cf. e.g., Fig. 15.2.

One way of addressing the surface registration problem is to first find salient features and then try to match them to each other by pattern recognition, [2]. This is an approach much used in image analysis, cf. e.g., [3, 4]. Instead of relying on being

J.A. Bærentzen et al., *Guide to Computational Geometry Processing*,
DOI 10.1007/978-1-4471-4075-7_15, © Springer-Verlag London 2012

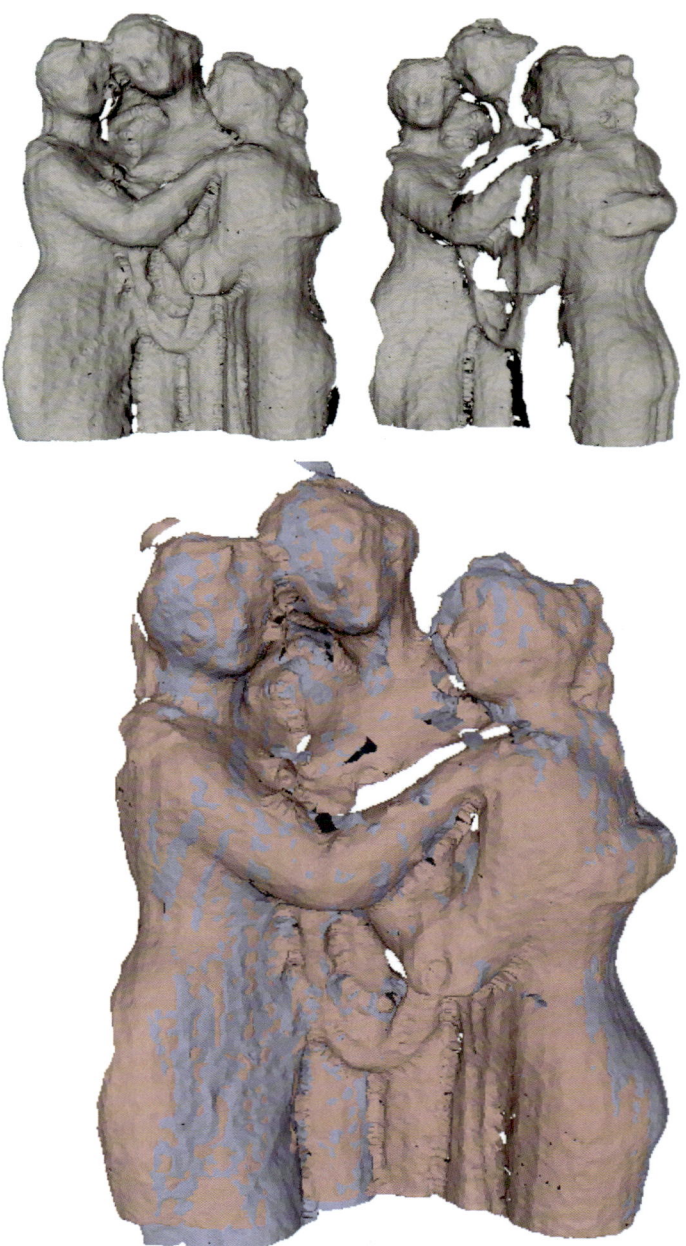

Fig. 15.1 An illustration of partial scan registration. *Top*: Two partial scans of a statue taken from different angles by a structured light scanner. *Bottom*: The two surfaces or meshes registered and aligned to each other using ICP

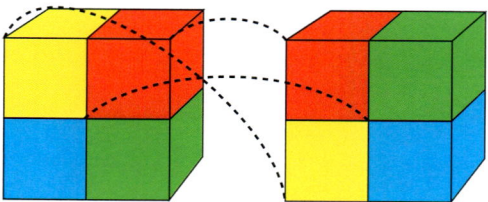

Fig. 15.2 A general illustration of 3D surface registration, where correspondences (*dashed lines*) are made between the part of the surface corresponding to the same underlying part of the object (here illustrated by color)

able to solve the registration by devising a pattern recognition approach, a popular approach in geometric processing is the iterative closest point method (ICP), which is covered here. The ICP method is used to match two surfaces and determines corresponding points as the closest one between surfaces. Following this, the two surfaces are aligned so as to minimize the distance between the corresponding points, and the process is repeated until convergence, e.g., the closest point correspondences does not change.

The ICP is thus a quit simple, albeit rather successful, heuristic for solving the problem. The ICP algorithm also uses a greedy optimization strategy. As such its success relies on adequate initialization, i.e., the two surfaces not being too far from each other. If this is not the case the feature-based methods mentioned above could be considered, or else several random initializations could be attempted. In the case of aligning partial scans from a scanner, a rough estimate is often available. Such a rough estimate is e.g. achieved from laser scanners, which move relatively little between scans to have a large overlap, by initially assuming that the object had *not* moved relative to the scanner.

15.2 The ICP Algorithm

To summarize the above, the ICP algorithm is used to align one mesh $\mathcal{M}_1$ to another $\mathcal{M}_2$. The assumption is that the two meshes are a realization/discretization of the same underlying surface, possibly with noise added, and that the mesh $\mathcal{M}_1$ has undergone a rigid transformation (rotation and translation), cf. Fig. 15.3. The problem is to estimate this rigid transformation, which will map $\mathcal{M}_1$ onto $\mathcal{M}_2$, in an 'optimal' way.

The ICP algorithm does this in the following way, cf. [5]:

1. For all vertices $\mathbf{p}_{1i} \in \mathcal{M}_1$, find the closest vertex $\mathbf{p}_{2i} \in \mathcal{M}_2$, cf. Fig. 15.4.
2. Find the rigid transformation (rotation $\mathbf{R}$ and translation $\mathbf{t}$) that minimizes the distance between the transformed $\mathbf{p}_{1i}$ and the $\mathbf{p}_{2i}$, i.e.,

$$\min_{\mathbf{R},\mathbf{t}} \sum_i \left\| (\mathbf{R}\mathbf{p}_{1i} + \mathbf{t}) - \mathbf{p}_{2i} \right\|^2. \tag{15.1}$$

3. Apply the estimated rigid transformation to $\mathcal{M}_1$.

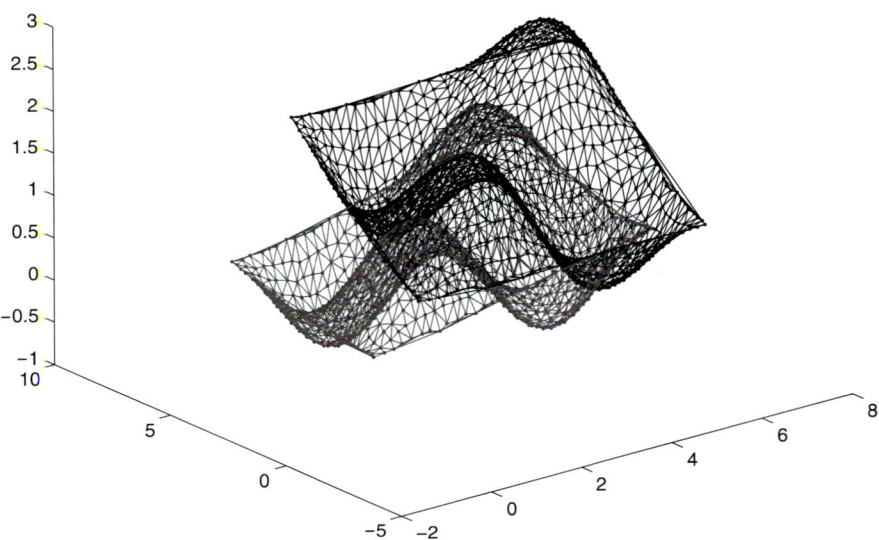

Fig. 15.3 Two similar meshes, one rotated and translated w.r.t. the other

Fig. 15.4 Two stylized meshes, where the distances between point pairs are denoted by the *dotted lines*

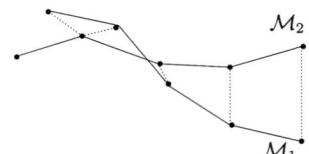

4. If convergence has not been reached go to 1. A typical measure of convergence is that the closest point pairings in step 1 are unchanged.

A simple illustration of the workings of the ICP algorithm is given in Fig. 15.5. An interpretation of this algorithm is; that we iteratively assume that the closest point is the correct correspondence for $\mathbf{p}_{1i}$, estimate and apply the appropriate transformation. It should also be noted that the fact that the least squares norm, $\| \cdot \|_2^2$, is used can be interpreted as the noise corrupting the meshes is Gaußian.

ICP is in essence a greedy algorithm, and will not always converge to the global optimum, but it has obtained popularity, because it gives good results, with decent data and initial guesses.

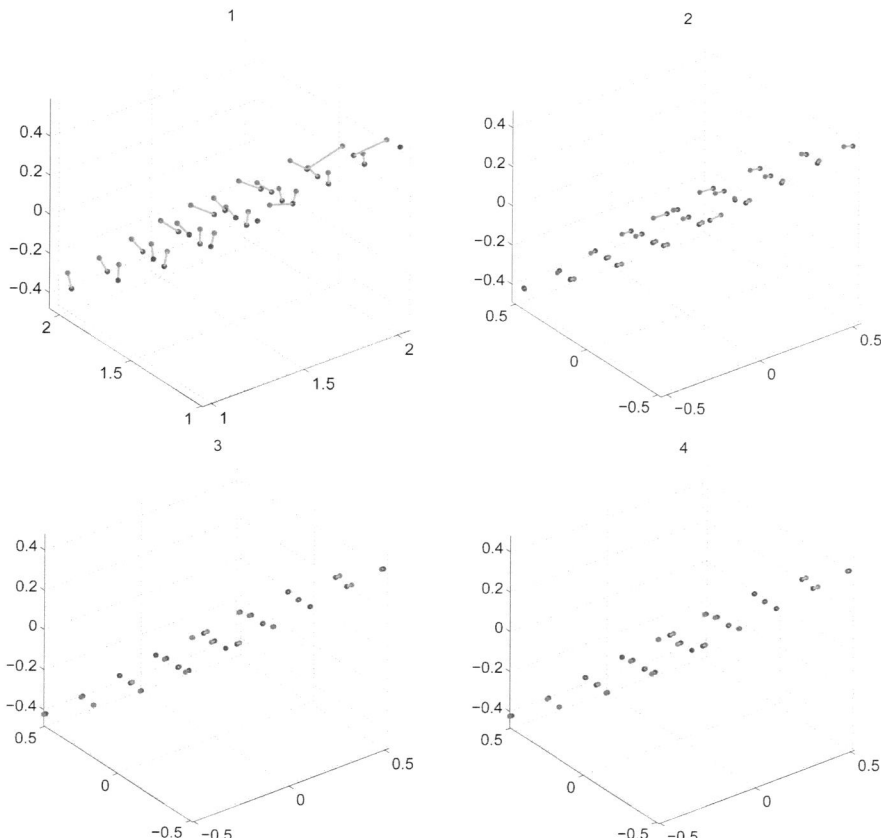

Fig. 15.5 Four iterations of the ICP algorithm where the *red point* set is aligned to the *blue*. The closest point relations are denoted by the *green lines*

15.2.1 Implementation Issues

There are several issues to be dealt with when implementing the ICP, and naturally there are different preferences in this area. Firstly, there is the issue of finding the closest point/vertex v_{2i} to v_{1i}. The 'obvious' thing to do would be for all $\mathbf{p}_{1i}$ to traverse all the vertices in $\mathcal{M}_2$. However, this is a $\mathcal{O}(n^2)$ operation, and becomes a disadvantage in that this algorithm is often run on meshes containing thousands of vertices. Instead a kD tree, or similar, should be used, see e.g., [6]. Such spatial data structures are the subject of Chap. 12.

Another thing is that the matchings or closest points are *not* required to be unique. That is, more than one vertex in $\mathcal{M}_1$ can have a given vertex in $\mathcal{M}_2$ as a closest point. This can in part be motivated by not wanting to endure the cost of enforcing the uniqueness constraint (e.g., using methods like [7]), but also by the fact that the two meshes might not be equally sampled. E.g., a small triangle might be inserted in

$\mathcal{M}_1$ instead of a single vertex in $\mathcal{M}_2$. This non-uniqueness also implies that different closest point pairs will be achieved depending on whether $\mathcal{M}_1$ is aligned to $\mathcal{M}_2$ or vice versa. Hence the general algorithm is non-symmetric depending on which mesh is aligned to which. However, if the algorithm is successful the difference should be negligible.

A last main part of implementing an ICP algorithm is how to find the rigid transformation ($\mathbf{R}$ and $\mathbf{t}$) that minimizes the distance between the closets point pairs, cf. (15.1). The straightforward solution is to apply a nonlinear optimization algorithm to (15.1), e.g., [8]. There are, however, more direct methods for matching point sets, cf. e.g., [9], which will be presented next.

15.2.2 Aligning Two 3D Point Sets

Here the method of [9] will be outlined, although there are other direct methods achieving the same goal. Denote the closest point pairs corresponding to $\mathbf{p}_{1i}, \mathbf{p}_{2i}$ as $\mathbf{p}_i, \mathbf{p}'_i$—the distinction is made, because all vertices might not be used, and some may occur more than once. As such we rewrite (15.1) as

$$\min_{\mathbf{R},\mathbf{t}} \sum_i \left\| (\mathbf{R}\mathbf{p}_i + \mathbf{t}) - \mathbf{p}'_i \right\|^2. \tag{15.2}$$

The first result needed is that the optimal $\mathbf{t}$ is that which connects the center of mass of the two point sets, i.e., (see [9] for a proof)

$$\mathbf{t} = \frac{1}{n}\sum_{i=1}^{n}\mathbf{p}'_i - \frac{1}{n}\sum_{i=1}^{n}\mathbf{p}_i. \tag{15.3}$$

For the further analysis, aimed at estimating the 3 by 3 rotation matrix $\mathbf{R}$, we will assume that the two point sets have been translated such that their center of mass is at the origin. That is, define

$$\mathbf{q}_i = \mathbf{p}_i - \frac{1}{n}\sum_{j=1}^{n}\mathbf{p}_j,$$

$$\mathbf{q}'_i = \mathbf{p}'_i - \frac{1}{n}\sum_{j=1}^{n}\mathbf{p}'_j.$$

It can then be shown that $\mathbf{R}$ can be found via the SVD[1] of the following matrix:

$$\mathbf{H} = \sum_{i=1}^{n}\mathbf{q}_i\mathbf{q}'^{T}_i.$$

---

[1] Singular Value Decomposition, cf. Chap. 2.

Denote the SVD by $\mathbf{H} = \mathbf{U}\Sigma\mathbf{V}^T$, then

$$\mathbf{R} = \mathbf{U}\mathbf{V}^T.$$

It should, however, be verified[2] that the determinant of $\mathbf{R}$ is 1 and not -1. In the latter case the algorithm has failed, which should only happen seldomly and in the face of extreme noise.

There is an important detail concerning the calculation of $\mathbf{H}$, seen by expanding the expression:

$$
\begin{aligned}
\mathbf{H} &= \sum_{i=1}^{n} \mathbf{q}_i \mathbf{q_i}'^{T} \\
&= \sum_{i=1}^{n} \left(\mathbf{p}_i - \frac{1}{n}\sum_{j=1}^{n}\mathbf{p}_j \right)\left(\mathbf{p}_i' - \frac{1}{n}\sum_{k=1}^{n}\mathbf{p}_k' \right)^{T} \\
&= \sum_{i=1}^{n} \mathbf{p}_i \mathbf{p}_i'^{T} - \left(\sum_{i=1}^{n}\mathbf{p}_i \right)\left(\frac{1}{n}\sum_{k=1}^{n}\mathbf{p}_k' \right)^{T} \\
&\quad - \left(\frac{1}{n}\sum_{j=1}^{n}\mathbf{p}_j \right)\left(\sum_{i=1}^{n}\mathbf{p}_i' \right)^{T} + n\left(\frac{1}{n}\sum_{j=1}^{n}\mathbf{p}_j \right)\left(\frac{1}{n}\sum_{k=1}^{n}\mathbf{p}_k' \right)^{T} \\
&= \sum_{i=1}^{n} \mathbf{p}_i \mathbf{p}_i'^{T} - \frac{1}{n}\left(\sum_{j=1}^{n}\mathbf{p}_j \right)\left(\sum_{k=1}^{n}\mathbf{p}_k' \right)^{T}.
\end{aligned}
$$

This implies that $\mathbf{H}$ (and thus $\mathbf{R}$) and $\mathbf{t}$ can be calculated by one pass of the closest point sets, instead of the two indicated by the original formula. In other words, by updating

$$ n, \quad \sum_{i} \mathbf{p}_i \mathbf{p}_i'^{T}, \quad y\sum_{j}\mathbf{p}_j, \quad \sum_{k}\mathbf{p}_k' $$

as step 1 is performed in the ICP algorithm outlined above, all the information needed in order to calculate $\mathbf{t}$, $\mathbf{H}$ and thus $\mathbf{R}$ is collected. Hence there is no need for explicitly saving the closest point pairs. An outline of how to estimate the rigid transformation, also known as an *Euclidean similarity transformation* is given in Table 15.1.

---

[2] If the determinant is -1 $\mathbf{R}$ is a reflection and not a rotation.

Table 15.1   Pseudo code for aligning two point sets via a rigid transformation

Given a set of 3D point correspondences, $\mathbf{p}_i, \mathbf{p}'_i, i \in [1, \ldots, n]$, find the rigid transformation, $\mathbf{R}, \mathbf{t}$ that minimizes

$$\min_{\mathbf{R},\mathbf{t}} \sum_i \|(\mathbf{R}\mathbf{p}_i + \mathbf{t}) - \mathbf{p}'_i\|^2.$$

Compute

$$\sum_i \mathbf{p}_i \mathbf{p}'^T_i, \ \sum_j \mathbf{p}_j, \ \sum_k \mathbf{p}'_k.$$

This can be done in one pass through the data. Then set

$$\mathbf{t} = \frac{1}{n}\sum_{i=1}^{n}\mathbf{p}'_i - \frac{1}{n}\sum_{i=1}^{n}\mathbf{p}_i$$

$$\mathbf{H} = \sum_{i=1}^{n}\mathbf{p}_i\mathbf{p}'^T_i - \frac{1}{n}\left(\sum_{j=1}^{n}\mathbf{p}_j\right)\left(\sum_{k=1}^{n}\mathbf{p}'_k\right)^T$$

$$\mathbf{U}\mathbf{\Sigma}\mathbf{V}^T = \mathbf{H} \qquad \text{The SVD of } \mathbf{H}$$

$$\mathbf{R} = \mathbf{U}\mathbf{V}^T$$

Verify that the determinant of $\mathbf{R}$ is plus one.

15.2.3 Degenerate Problems

It should be noted that if the two meshes are planar, then a unique solution is not achievable. This is clear since two planes can be perfectly aligned in many different ways. Thus such cases, and cases that are close to these, will cause the algorithm problems. As an obvious extension of this is if the meshes are located on a line or close to it, the problems get even worse. If large parts of the meshes are planar then weighting of the data points might be in order cf. [10].

15.3   ICP with Partly Overlapping Surfaces

An underlying assumption of the ICP algorithm is that upon convergence, for all points the closest point is the correct correspondence. It follows from this assumption that the two surfaces or meshes to be registered capture the same underlying 3D geometry, in that for more or less all points the corresponding point on the other surface should exist. With partially overlapping surfaces, e.g., aligning different partial scans, this assumption is violated to a point where the basic ICP algorithm breaks down, cf. Fig. 15.6.

This issue of aligning partially overlapping meshes, can be addressed by incorporating a heuristic into the ICP algorithm whereby points, which do not have corresponding points on the other surface, are not included in the alignment. One such heuristic, which has proven highly successful, is proposed in [11]. This heuristic

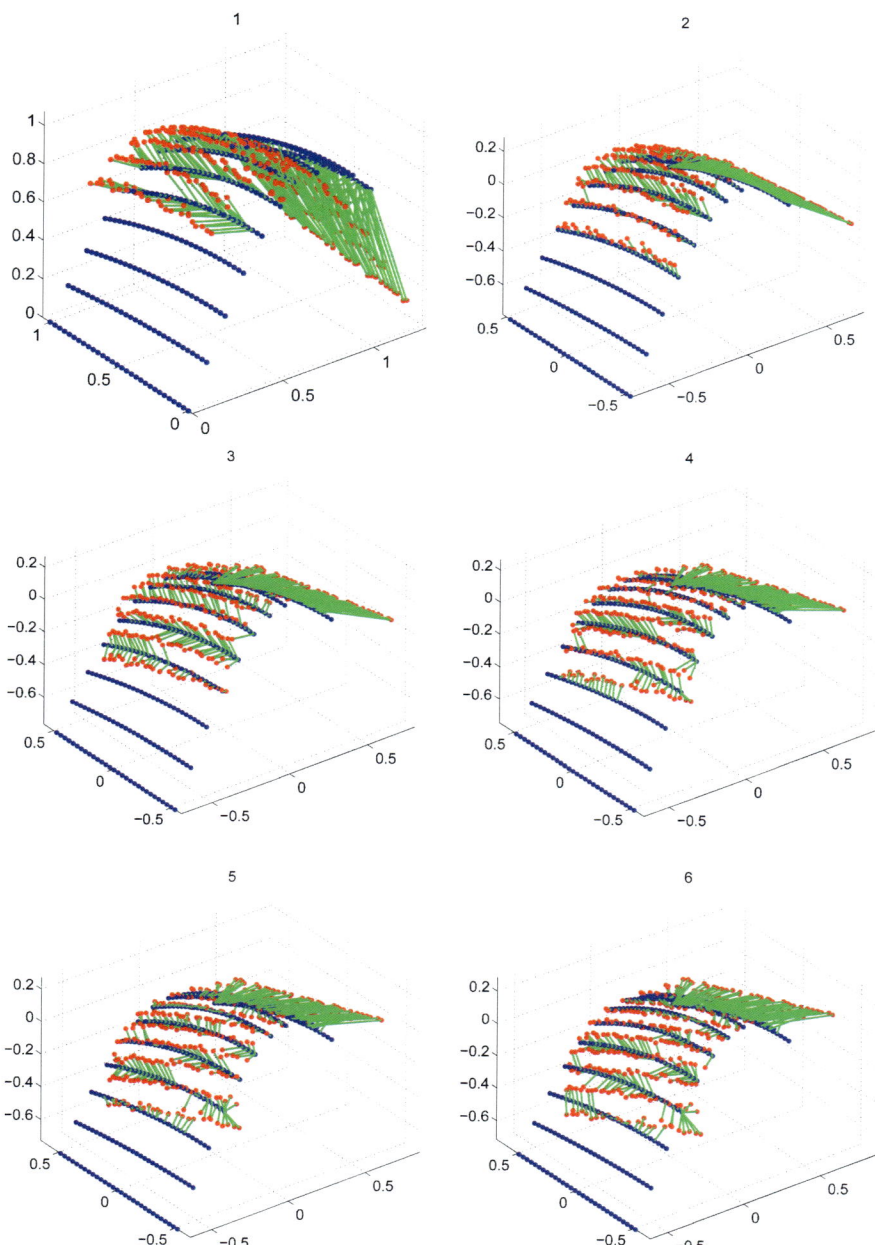

Fig. 15.6 *Six* iterations of the *standard* ICP algorithm of the *red point* set to the *blue*. Since the point sets, or surfaces, are only partially overlapping, it is seen that the large amount of *red points* overwhelm the algorithm giving erroneous results

labels points for which the corresponding point on the other surface is an edge point as not having a corresponding point. As such these points should not be included in the alignment, cf. Fig. 15.7. The underlying assumption about this heuristic of [11], is that if the surfaces are located on top of each other, points in non-overlapping regions will be closest to edges on the other surface. This assumption often holds, cf. Figs. 15.5 and 15.1.

Thus adapting the standard ICP algorithm to working with partially overlapping meshes in this manner, requires that edge points can be computed. If the surface is represented as a triangular mesh this can e.g., be done by identifying points that only share *one* triangle with another point. Two points only sharing one triangle implies that one of the edges of that triangle does not have an opposing triangle, and as such is an edge of the surface.

15.4   Further Extensions of the ICP Algorithm

Although the ICP is a very successful algorithm, many extensions have been proposed to it in order to enhance its performance and to enable it to address more situations, e.g., for partially overlapping meshes as described above. A few more of these extensions will briefly be presented here, in order for the reader to be aware of their existence. However, for a more in depth discussion refer to [10].

15.4.1 Closest Point *not* a Vertex

The difference between the ICP method proposed in [5] and [12], is that in [12] the closest point to $\mathbf{p}_{1i}$ on $\mathcal{M}_2$ need not necessarily be a vertex. That is the point closest to $\mathbf{p}_{\_i}$ might be on a face or an edge of the mesh.

The advantage is obviously that if the two meshes are sampled differently and the faces are large, a much better estimate of the closest point will be obtained. The disadvantage is that it is more cumbersome to do, and is less robust towards hard problems, i.e., close to degenerate, cf. [10].

15.4.2 Robustness

There has also been work done on making the ICP more robust towards erroneous data, cf. [13–15], and choosing weighting schemes to better depict the reliability of the different parts of the mesh, cf. [10, 14, 16].

Also additional information associated with the mesh can be used. E.g., if the meshes has been generated from stereo images, there will also be a texture associated with it, cf. e.g., [17]. This texture can naturally be used for finding good candidates for the point on $\mathcal{M}_2$ corresponding to a $\mathbf{p}_{1i} \in \mathcal{M}_1$, i.e., a point should both be close *and* have a similar texture to be paired with $\mathbf{p}_{1i}$.

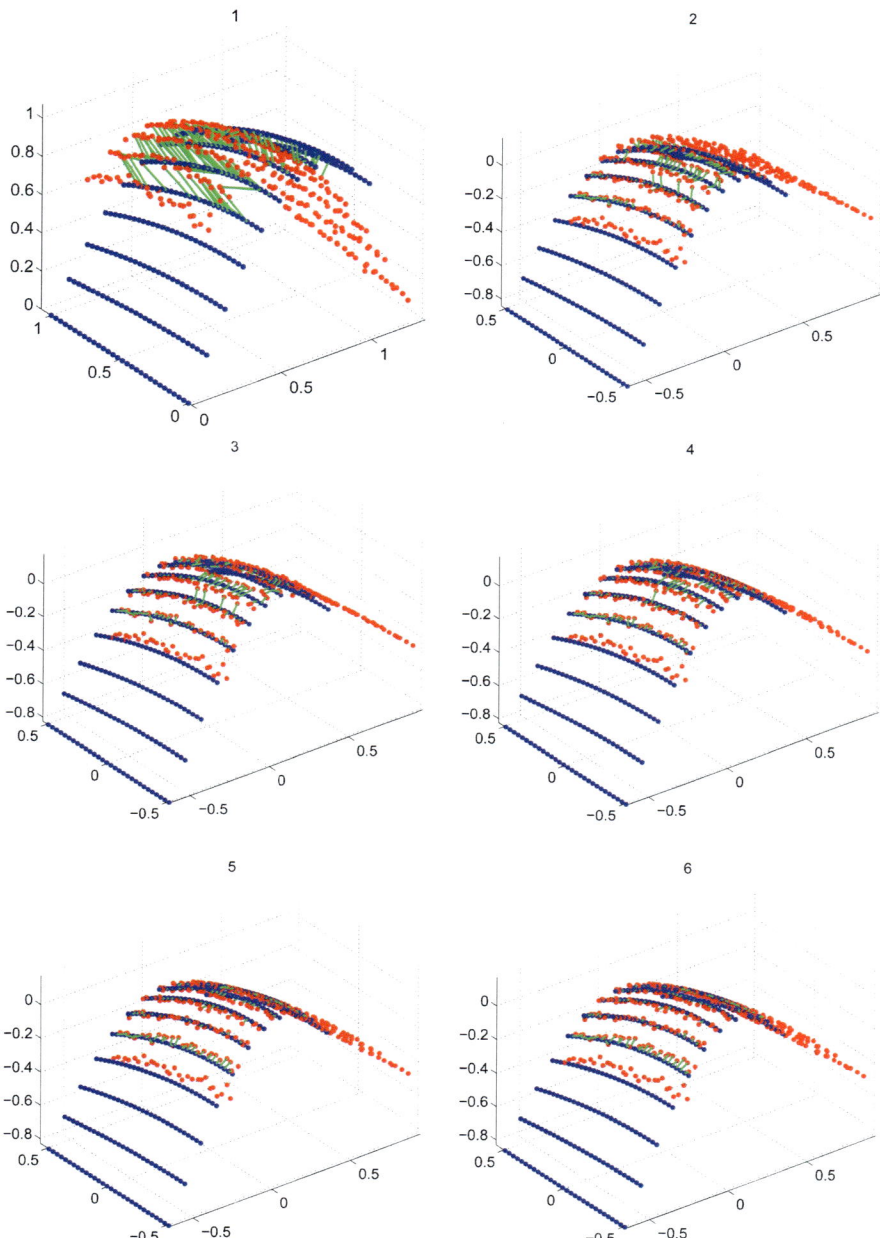

Fig. 15.7 *Six* iterations of the ICP algorithm *adapted to handle partial overlapping surfaces*, on the same data as Fig. 15.5. It is seen that no registrations, in the form of *green lines*, involve the edge of the *blue points*. Also the algorithm is seen to perform satisfactorily as opposed to the standard algorithm as illustrated in Fig. 15.5

15.5   Merging Aligned Surfaces

As mentioned in the beginning of this chapter, the typical task in which the ICP is a part, is that of merging two meshes. A way of doing this is as follows.

1. Align the two surfaces, e.g., via ICP, as described in this chapter.
2. Convert the two meshes into point clouds, e.g., by only using the vertices and not the faces.
3. Use the theory form Chap. 17 to reconstruct a surface from the points via a volumetric method.
4. If needed, use the methodology from Chap. 18 to convert the surface from the previous step into a mesh.

A few points should be made about this approach for surface merging. Firstly, it will also work on raw, un-meshed, point clouds. That is, there is no reason to form a point cloud into a mesh, before this method is run, in that ICP works on raw point clouds as well.

Secondly, the reason why two meshes are merged by via a conversion to an iso-surface representation, as opposed to directly merging the meshes, is that it generally works better. The issue with such a direct merging of meshes, is that non-manifold results are hard to avoid, in that there is no simple way to interpolate between meshes (with considerable curvature and possible holes). With iso-surfaces, i.e., volumetric methods, such interpolation between surfaces is simple and straightforward.

Lastly, if more than two scans are to be aligned—partly overlapping or not—methods exist which deal with making ICP work in such settings, cf. e.g., [18, 19]

15.6   Exercises

Exercise 15.1 The aim of this exercise is to implement a version of the ICP algorithm, used for the registration of meshes. Hereby it is hoped that a greater understanding of the relevant theory is obtained. It is thus important that a working algorithm is achieved, however, an industrial strength one is not needed.

We have included two data sets, on the books homepage, on which you should apply your algorithms, an easy and a realistic. These two data sets are named mesh1.x3d and mesh2.x3d. The first mesh is a simple one to develop your algorithm on, the second is more realistic.

Deliverables: Illustration of the aligned meshes, and interpretation of the results.

[GEL Users]   The Gel framework is a valuable aid. Note that the GEL framework contains an implementation of a kD-tree.

References

1. Johnson, A.E., Hebert, M.: Surface matching for object recognition in complex three-dimensional scenes. Image Vis. Comput. **16**(9–10), 635–651 (1998). doi:10.1016/S0262-8856(98)00074-2

2. Salti, S., Tombari, F., Stefano, L.D.: A performance evaluation of 3d keypoint detectors. In: International Conference on 3D Imaging, Modeling, Processing, Visualization and Transmission, pp. 236–243 (2011)
3. Aanæs, H., Dahl, A., Steenstrup Pedersen, K.: Interesting interest points. International Journal of Computer Vision, 1–18 (2011). doi:10.1007/s11263-011-0473-8
4. Tuytelaars, T., Mikolajczyk, K.: Local Invariant Feature Detectors: A Survey. Now Publishers, Hanover (2008)
5. Besl, P.J., McKay, H.D.: A method for registration of 3D shapes. IEEE Trans. Pattern Anal. Mach. Intell. **14**, 239–256 (1992)
6. Bentley, J.L.: Multidimensional binary search trees in database applications. IEEE Trans. Softw. Eng. **5**, 333–340 (1979)
7. Jonker, R., Volgenant, A.: A shortest augmenting path algorithm for dense and sparse linear assignment problems. Computing **38**, 325–340 (1987)
8. Marquardt, D.: An algorithm for least-squares estimation of nonlinear parameters. J. Soc. Ind. Appl. Math. **11**, 431–441 (1963)
9. Arun, K.S., Huang, T.S., Blostein, S.D.: Least-squares fitting of two 3D point sets. IEEE Trans. Pattern Anal. Mach. Intell. **9**, 698–700 (1987)
10. Rusinkiewicz, S., Levoy, M.: Efficient variants of the icp algorithm. In: Proceedings for the Third International Conference on 3D Digital Imaging and Modeling, pp. 145–152 (2001)
11. Turk, G., Levoy, M.: Zippered polygon meshes from range images. In: Computer Graphics Proceedings. Annual Conference Series, SIGGRAPH, pp. 311–318 (1994)
12. Chen, Y., Medioni, G.: Object modelling by registration of multiple range images. Image Vis. Comput. **10**, 145–155 (1992)
13. Dorai, C., Wang, G., Jain, A.K., Mercer, C.: Registration and integration of multiple object views for 3D model construction. IEEE Trans. Pattern Anal. Mach. Intell. **20**, 83–89 (1998)
14. Fitzgibbon, A.W.: Robust registration of 2D and 3D point sets. In: British Machine Vision Conference, pp. 662–670 (2001)
15. Zhang, Z.: Iterative point matching for registration of free-form curves and surfaces. Int. J. Comput. Vis. **13**, 119–152 (1994)
16. Dorai, C., Weng, J., Jain, A.K.: Optimal registration of object views using range data. IEEE Trans. Pattern Anal. Mach. Intell. **19**, 1131–1138 (1997)
17. Martins, F.C.M., Shiojiri, H., Moura, J.M.F.: 3D–3D registration of free formed objects using shape and texture. In: Proceedings of the SPIE—The International Society for Optical Engineering, pp. 263–274 (1974)
18. Bergevin, R., Soucy, M., Gagnon, H., Laurendeau, D.: Towards a general multi-view registration technique. IEEE Trans. Pattern Anal. Mach. Intell. **18**, 540–547 (1996)
19. Levoy, M., Rusinkiewcz, S., Ginzton, M., Ginsberg, J., Pulli, K., Koller, D., Anderson, S., Shade, J., Curless, B., Pereira, L., Davis, J., Fulk, D.: The digital Michelangelo project: 3D scanning of large statues. In: Conference Proceedings on Computer Graphics Proceedings. Annual Conference Series 2000, SIGGRAPH 2000. pp. 131–144 (2000)

Surface Reconstruction using Radial Basis Functions

16

In this chapter, we discuss methods for reconstructing surfaces from *scattered* points by interpolation using *radial basis functions* (RBF). By "scattered" we understand simply that the points do not lie on a regular grid. Interpolation is the process of finding the value at any point in space of a function sampled at the scattered points. For a general discussion of interpolation of scattered points, see [1].

It turns out that interpolation is a simple and effective method for reconstruction of surfaces from scattered points. In fact there are two different approaches both of which are discussed in this chapter. Taking the first approach, the surface becomes the graph of a function of two variables. If we take the second approach, the surface is the level set or isosurface of a function of three variables. The two approaches are illustrated (in 2D) in Fig. 16.1.

To be more precise, the first approach, which we discuss in Sect. 16.3, creates a surface, $S = \{\mathbf{x} = [x \ y \ z]^T \in \mathbb{R}^3 | z = s(x, y)\}$, which is given by a function s of x and y. To make S interpolate our set of scattered points, $\{\mathbf{x}_i\}$, we require of s that for any point i in our data set, $z_i = s(x_i, y_i)$. The 2D case is illustrated in Fig. 16.1 (left). In some cases we may want to relax the interpolation requirement, and then we obtain a surface which approximates rather than interpolates our points as indicated by the dotted line in the figure.

However, this approach does not always apply. In some cases, our points do not lie on the graph of a function of two variables as illustrated (again for the 2D case) on the right in Fig. 16.1. Interpolation using radial basis functions is still an effective tool, but the surface must now be defined implicitly, i.e., $S = \{\mathbf{x} \in \mathbb{R}^3 | s(\mathbf{x}) = \tau\}$, where s is a function of x, y, and z, and τ is known as the isovalue. We require of s that $s(\mathbf{x}_i) = \tau$. In other words, we find the implicit surface (cf. Sect. 3.10) that interpolates our points. The astute reader will have noticed an issue: the constant function that maps any point to τ fulfills the condition, but does not give us an isosurface. Consequently, we need to impose other conditions to use this method, which is discussed in Sect. 16.4.

In the following section (Sect. 16.1), we motivate the use of radial basis functions by discussing some limitations of the alternatives, and in Sect. 16.2, we discuss the RBF method in detail.

J.A. Bærentzen et al., *Guide to Computational Geometry Processing*,
DOI 10.1007/978-1-4471-4075-7_16, © Springer-Verlag London 2012

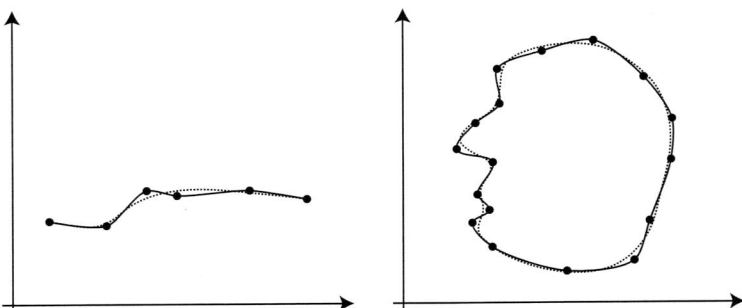

Fig. 16.1 A 2D illustration of the two approaches to creating a shape from scattered points. On the *left* the *solid curve* is a graph of a function, which interpolates the data points. The *dotted curve* is the graph of a function that approximates the points. On the *right*, the *solid curve* is an isocurve or level set of a function which interpolates the data points. Again, we can find a *curve* (*dotted*) which only approximates the *points*. The advantage to approximation is that we often get a smoother curve (or surface in 3D)

16.1   Interpolation of Scattered Data

Scattered data interpolation is a very general tool that is widely used for a range of applications. We have already discussed Delaunay triangulation and the Voronoi diagram in Chap. 14. Recall that the Voronoi cell of a given point $\mathbf{x}_i \in X$ where X is a is simply the region in space where every point is closer to $\mathbf{x}_i$ than to any other point in X. Thus, if we want a piecewise constant interpolation function s we can define it by simply finding the closest point and then using the data value for that point. This corresponds to finding out in what Voronoi cell the point is located. Implementation-wise, we would construct a spatial database, e.g., a kD tree, and use that to locate the nearest point.

However, if we wish a piecewise linear interpolation, we might want to use the dual of the Voronoi diagram instead, i.e., the Delaunay triangulation. We now define $s(\mathbf{x})$ by finding the triangle containing $\mathbf{x}$ and we linearly interpolate the data values at the corners to the point $\mathbf{x}$. Thus, we have a piecewise linear interpolation.

While these techniques are probably very suitable for a range of applications, we would often like an interpolation that is more smooth than piecewise constant or linear. Moreover, it is far from trivial to construct a Delaunay triangulation in higher dimensions. In other words, we would like a more general tool.

One simple technique is often referred to as *Shepard interpolation* [1]. The basic idea behind Shepard interpolation is to compute a spatially varying weighted average of the data points.

$$s(\mathbf{x}) = \frac{\sum_i w(\|\mathbf{x} - \mathbf{x}_i\|) f_i}{\sum_i w(\|\mathbf{x} - \mathbf{x}_i\|)}, \tag{16.1}$$

Fig. 16.2 A simple example
of Shepard interpolation

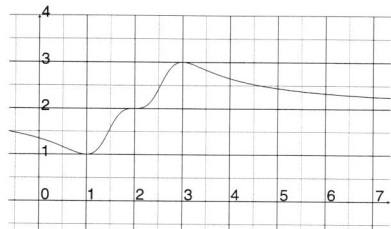

where, for instance, we let $w(r) = \frac{1}{(r+\epsilon)^2}$ using $\epsilon \ll 1$. This function will assign a
large weight to a point if we are very close, and a small weight otherwise. However,
it is clear that the method is only approximately interpolating. Precise interpolation
requires $w(0) = \infty$ but since that is not possible, we tend to settle for simply having
a large value at zero. However, this tends to lead to visible artifacts in the interpolation function. In Fig. 16.2, we see an example where $w(r) = \frac{1}{(r+10^{-5})^2}$. It is fairly
easy to guess where the data points are.

The problem with Shepard's method is that to get approximate interpolation, we
need a function w with very large derivatives which tends to show in the solution.
Thus, we tend to get better results with *radial basis functions*, the method to which
we devote the remainder of this chapter.[1]

16.2   Radial Basis Functions

In the following discussion, we often follow Powell's course notes on radial basis
functions [2], which is an introductory text on general interpolation using radial
basis functions, and [3], which is one of the early papers on the applications of
RBFs to surface reconstruction.

The idea behind interpolation using radial basis functions is to choose a basis
of functions with which we can represent a class of functions in the domain over
which we interpolate. Specifically, we choose a basis consisting of radially symmetric functions, and we associate one basis function with each data point. Our
interpolation function is represented in this basis, i.e., as a linear combination of the
radial functions associated with each data point.

Suppose, we are given a set of points $\{\mathbf{x}_i \in \mathbb{R}^d\}$ and associated data values f_i.
Furthermore, let $\psi = \exp(-\alpha r^2)$. In that case, we can write down a linear system

$$f_i = \sum_j \lambda_j \psi\left(\|\mathbf{x}_i - \mathbf{x}_j\|\right), \qquad (16.2)$$

whose solution is the set of coefficients λ_i. In matrix form

$$\mathbf{f} = \mathbf{\Psi}\boldsymbol{\lambda}, \qquad (16.3)$$

---

[1]Since w is also a function of point distance, one might say that the functions used in Shepard's
method are also "radial". Sometimes nomenclature can be a bit misleading.

where $\Psi_{ij} = \psi(\|\mathbf{x}_i - \mathbf{x}_j\|)$. Clearly, we cannot have two data points at the precise same location since that would lead to two identical rows in Ψ. Assuming we do not, then, having solved for λ, we can define a function

$$s(\mathbf{x}) = \sum_j \lambda_j \psi(\|\mathbf{x} - \mathbf{x}_j\|), \qquad (16.4)$$

which by construction interpolates the data values. However, it is not immediately clear whether (16.2) has a solution. It turns out, however, that because of our choice of ψ, the system is positive definite, i.e., that $\lambda^T \Psi \lambda > 0$ if $\lambda \neq 0$.

For most functions other than the Gaußian ψ, we need the coefficients to fulfill the condition that $\sum_j \lambda_j P(\mathbf{x}_j) = 0$ for *any* polynomial P whose order depends on the choice of ψ. If this condition is fulfilled for a vector $\lambda \neq 0$, we know that $\lambda^T \Psi \lambda > 0$.

Typically, P is either a constant or a linear polynomial. We can express the condition $\sum_j \lambda_j P(\mathbf{x}_j) = 0$ in matrix notation. For the sake of simplicity, assume that we just need a linear polynomial. Now, let $\mathbf{P}$ be a matrix where each row is $[1\ x_i\ y_i\ z_i]$ and where x_i, y_i, and z_i correspond are the coordinates for one of the points we wish to interpolate. $\mathbf{P}$ looks as follows:

$$\mathbf{P} = \begin{bmatrix} 1 & x_1 & y_1 & z_1 \\ 1 & x_2 & y_2 & z_2 \\ \cdots & & & \\ 1 & x_n & y_n & z_n \end{bmatrix}.$$

If λ is orthogonal to each column, i.e., $\mathbf{P}^T \lambda = 0$, λ is also orthogonal to any linear combination of columns and thus to any linear polynomial.

For reasons which will be explained shortly, we also add a polynomial term to the sum of radial basis functions. Thus, we need to solve

$$f_i = \sum_j \lambda_j \psi(\|\mathbf{x}_i - \mathbf{x}_j\|) + P(\mathbf{x}_i), \qquad (16.5)$$

where P is a polynomial with coefficients $\mathbf{c}$. In matrix form,

$$\begin{bmatrix} \Psi & \mathbf{P} \end{bmatrix} \begin{bmatrix} \lambda \\ \mathbf{c} \end{bmatrix} = \begin{bmatrix} \mathbf{f} \end{bmatrix}. \qquad (16.6)$$

Now, if we add the condition $\mathbf{P}^T \lambda = 0$ to the linear system, we get the system that we normally have to solve when dealing with RBF interpolation. The following theorem shows that this system is not singular, since that would lead to a contradiction.

Theorem 16.1 *Given a matrix of RBF coefficients, Ψ, and a matrix, $\mathbf{P}$, as defined above, where the order of the polynomial is chosen such that for a non-zero coefficient vector, λ, $\mathbf{P}^T \lambda > 0 \implies \lambda^T \Psi \lambda > 0$,*

$$\begin{bmatrix} \Psi & \mathbf{P} \\ \mathbf{P}^T & 0 \end{bmatrix} \begin{bmatrix} \lambda \\ \mathbf{c} \end{bmatrix} = \begin{bmatrix} \mathbf{f} \\ 0 \end{bmatrix} \qquad (16.7)$$

Table 16.1 This table contains various radial basis functions along with the required order of polynomial to use in (16.5)

Name	$\psi(r)$	Polynomial
Gaußian	$\psi(r) = \exp(-\alpha r^2)$	N/A
Linear	$\psi(r) = r$	Constant
Thin Plate Spline	$\psi(r) = r^2 \log(r)$	Linear
Cubic	$\psi(r) = r^3$	Linear

is non-singular for $\mathbf{f} \neq 0$.

Proof Assume that for a non-zero coefficient vector, $[\boldsymbol{\lambda}\ \mathbf{c}]^T$,

$$\begin{bmatrix} \boldsymbol{\Psi} & \mathbf{P} \end{bmatrix} \begin{bmatrix} \boldsymbol{\lambda} \\ \mathbf{c} \end{bmatrix} = \mathbf{0}$$

$$\boldsymbol{\Psi}\boldsymbol{\lambda} + \mathbf{P}\mathbf{c} = \mathbf{0}$$

$$\boldsymbol{\lambda}^T\boldsymbol{\Psi}\boldsymbol{\lambda} + (\boldsymbol{\lambda}^T\mathbf{P})\mathbf{c} = 0$$

$$\boldsymbol{\lambda}^T\boldsymbol{\Psi}\boldsymbol{\lambda} = 0$$

which is a contradiction since $\mathbf{P}^T\boldsymbol{\lambda} = 0$ implies that $\boldsymbol{\lambda}^T\boldsymbol{\Psi}\boldsymbol{\lambda} > \mathbf{0}$. To get the last row, we used $\boldsymbol{\lambda}^T\mathbf{P} = \mathbf{P}^T\boldsymbol{\lambda} = 0$. Thus, the system is not singular. □

Unsurprisingly, it is not trivial to choose a polynomial order such that $\mathbf{P}^T\boldsymbol{\lambda} > 0 \implies \boldsymbol{\lambda}^T\boldsymbol{\Psi}\boldsymbol{\lambda} > 0$. What order to use depends on the type of RBF. For a more detailed discussion, the reader is referred to [2]. Table 16.1 adapted from [2] summarizes the polynomial orders needed for popular selections of RBFs.

To sum up, for a given choice of radial basis functions, we can construct a system of linear equations of the form (16.7). Having solved that system, we obtain an interpolation function

$$s(\mathbf{x}) = \sum_j \lambda_j \psi\big(\|\mathbf{x} - \mathbf{x}_j\|\big) + P(\mathbf{x}). \tag{16.8}$$

16.2.1 Regularization

Typically, interpolating functions oscillate more than functions which merely approximate the data and if we assume that our data are not completely noise free, it is not necessarily a good idea to enforce interpolation in any case. Regularization simply relaxes the interpolation requirement.

If the system is ill-conditioned, regularization also improves the condition number of the system we need to solve. The principle is simply to add a constant to the diagonal of the linear system which we need to solve in order to obtain the RBF coefficients,

$$f_i = \sum_j \lambda_j \psi\big(\|\mathbf{p}_i - \mathbf{p}_j\|\big) + P(\mathbf{p}_i) + k\lambda_i, \tag{16.9}$$

where k is the regularization constant. In matrix form, we obtain

$$\begin{bmatrix} \boldsymbol{\Psi} + k\mathbf{I} & \mathbf{P} \\ \mathbf{P}^T & 0 \end{bmatrix} \begin{bmatrix} \lambda \\ \mathbf{c} \end{bmatrix} = \begin{bmatrix} \mathbf{f} \\ 0 \end{bmatrix}. \tag{16.10}$$

Note that the constant k should generally be small in comparison to the extent of the data. Note also that from a statistical point of view, this type of regularization can be seen as trading a small bias for a reduction in variance.

16.3   Surface Reconstruction

As described, the RBF method can be directly applied to reconstruct a surface if the surface can be seen as the 2D graph of a function. In this case, the value we interpolate is the height value, i.e., $f_i = z_i$. What then are the choices of ψ and what polynomials should be used? A few commonly used RBFs are shown in Table 16.1, adapted from [2], along with the required polynomial terms.

If we are given 2D points with associated height values, a particularly common choice is the so called *thin plate spline* RBF, $\psi(r) = r^2 \log(r)$. The thin plate spline solution has the nice property that it minimizes the linearized bending energy (cf. Sect 9.6)

$$E[s] = \frac{1}{2} \int s_{xx}^2 + 2s_{xy}^2 + s_{yy}^2 \, dx \, dy. \tag{16.11}$$

Thus, for 2D implementations, the thin plate spline basis function is an obvious choice. In Fig. 16.3 is an example showing a tessellated height map which is produced from a set of discrete height points by interpolation using thin plate splines.

16.4   Implicit Surface Reconstruction

However, this procedure only works for 2.5D data. What do we do if we need to reconstruct a surface from points which cannot be assumed to be sampled from a height map? The normal procedure is to use the points to reconstruct an implicit representation of the surface. This method was popularized by Turk and O'Brien [4].

If we know all the points lie on the surface we wish to reconstruct, an obvious strategy is to assign the data value 0 to all the points. However, there is an obvious problem, namely that $s(x, y, z) = 0$ would interpolate the points. To fix this problem, we need to have points which do not lie on the isosurface. Provided we have estimated surface normals together with the points, the most common procedure is to move a fixed amount in the positive (or negative) normal direction and create a new point at that position. An example of this is shown in Fig. 16.4.

Examples of RBF surfaces reconstructed in this fashion are shown in Fig. 16.5. Note that we have cheated a bit in this example since the points are simply the vertices of a mesh. Of course, that makes it a lot easier to obtain the normal. In the

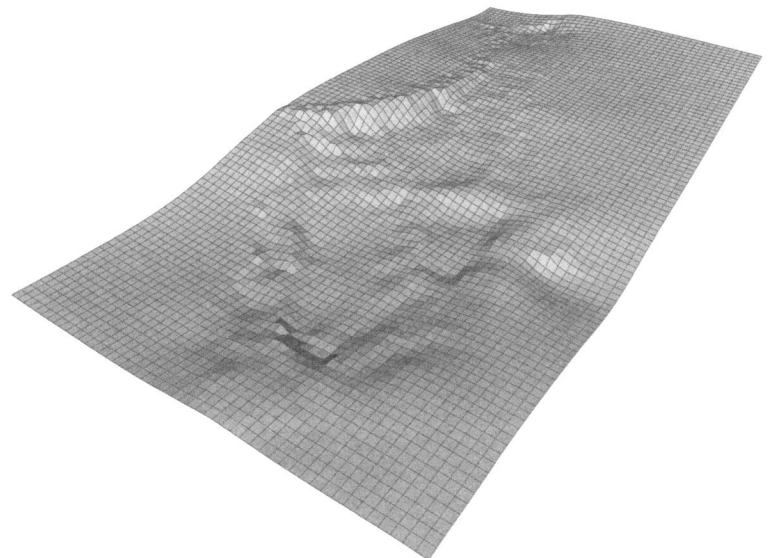

Fig. 16.3 A terrain reconstructed from scattered points using thin plate splines

Fig. 16.4 A set of points (*solid*) and auxiliary points (*hollow*) obtained by offsetting along the normal. Also shown is a surface which interpolates the point set

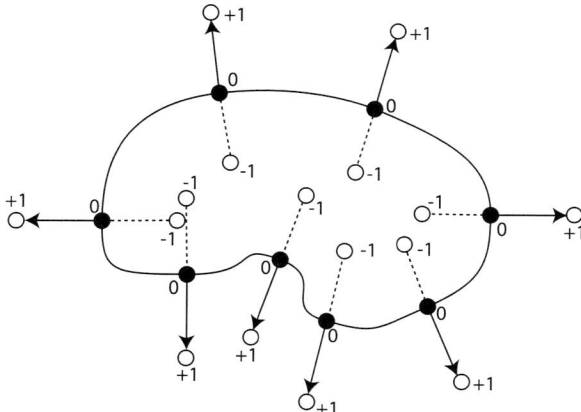

case of laser scanned data, normals are often estimated by locally fitting a plane to the point cloud. The normal at a given point is then the normal of the local plane estimate. To obtain the surface orientation, we can observe that the camera used by the laser scanner must have been able to see the point. Thus, we know a direction to something on the "outside", and this direction can be used to orient the normals. For a detailed discussion of how to estimate point normals, see Sect. 17.1.1.

An example including regularization is shown in Fig. 16.6. Note that the surface is more wavy in the reconstruction without regularization.

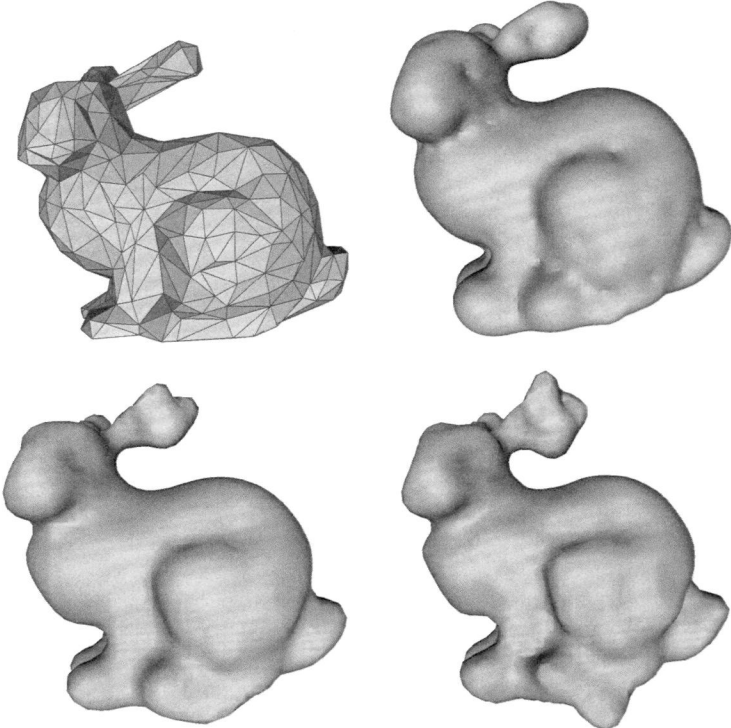

Fig. ˜6.5 Reconstruction from points in 3D using various radial basis functions. The original points are simply the mesh vertices (*top left*) plus a set of points offset in the normal direction. The radial basis functions are linear (*top right*), cubic (*bottom left*), and Gaußian (*bottom right*)

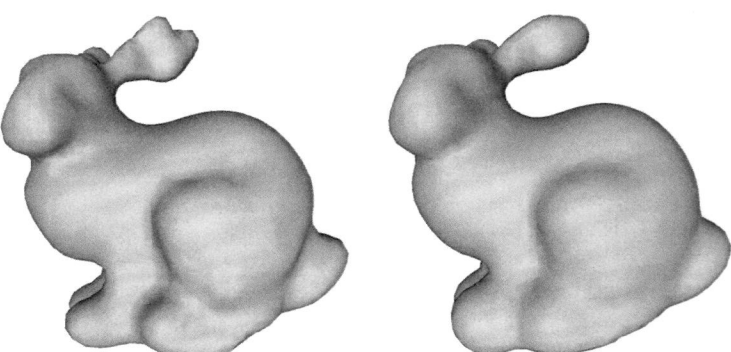

Fig. ˜6.6 The result using cubic radial basis functions without regularization (*left*) and with regularization (*right*)

16.5 Discussion

There are many methods for reconstruction of surfaces from point clouds. The methods discussed in this chapter are extremely simple and highly effective—but only for a modest number of points, since the matrices involved are dense. A remedy for this issue is to use compactly supported radial basis functions [5]. The volumetric methods which are discussed in the next chapter also solve the problem. This is because they largely decouple the reconstruction problem from the input points.

Capturing objects with sharp edges remains a problem, though. Shen et al. proposed a method for reconstruction of implicit surfaces from points and triangles where the reconstructed function will interpolate an entire triangle and uses the triangle's normal field as an interpolation condition [6]. This allows the authors to fairly accurately interpolate a triangle mesh without letting go of the advantages of implicit surfaces: in particular that the implicit surface closes holes in the surface given by the input points and triangles.

Unfortunately, the great simplicity of the method is easily lost when we aim to speed up the method or achieve a more exact approximation. Another issue is that of outliers. Regularization allows us to find a surface that approximates the input points rather than interpolating them. That is highly useful if the point cloud is noisy, but what if it contains outliers that are not even close to the true surface? A somewhat different method known as the Eigencrust method [7] was developed by Kolluri et al. In their paper, the authors created a Delaunay tetrahedralization of the input points and used a graph cut algorithm to select the subset of the tetrahedra considered to be interior. This method is able to deal with a significant fraction of outliers.

16.6 Exercises

Exercise 16.1 Obtain a set of 2D points with associated height values. One such data set is provided on the homepage accompanying this book. Use the thin plate splines basis function, $\phi(r) = r^2 \log(r)$, and find the coefficients of the resulting function, $s(\mathbf{x})$. To visualize the resulting function, generate a triangle mesh sampled on a regular 2D grid: the x, y values lie on the grid and $z = s(x, y)$.

[GEL Users] A demo program to get you started on this exercise and the next is provided in the GEL examples package.

Exercise 16.2 Next, use $\phi(r) = r^3$ to do interpolation between points in 3D with an associated scalar value. Load a small triangle mesh and for each vertex create a point with associated value 0 at the position of the vertex. Also, place some additional points inside and outside the mesh and give these points values different from zero. Points inside should all be greater than zero and points outside should all be less than zero (or the other way around). Use the RBF method to create a function s that interpolates these constraints and whose zero-level set is a surface passing through the original points. Visualize the results using an implicit-surface polygonizer (cf. Chap. 18).

The following are possible extensions to the basic exercise.

- Use mesh normals to automatically generate the inside and outside points in the last part of the exercise.
- Add some regularization to relax the precise interpolation requirement.
- Try different radial basis functions, but bear in mind that the linear system may not have full rank unless you use a different polynomial term.

References

1. Nielson, G.M.: Scattered data modeling. IEEE Comput. Graph. Appl. **13**(1), 60–70 (1993)
2. Powell, M.J.: Five lectures on radial basis functions. Technical report, Informatics and Mathematical Modelling, Technical University of Denmark, DTU, Richard Petersens Plads, Building 321, DK-2800 Kgs. Lyngby (2005). http://www2.imm.dtu.dk/pubdb/p.php?3600
3. Turk, G., Dinh, H.Q., O'Brien, J.F., Yngve, G.: Implicit surfaces that interpolate. In: SMI 2001 International Conference on Shape Modeling and Applications. IEEE Press, New York, pp. 62–71 (2001)
4. Turk, G., O'Brien, J.F.: Modelling with implicit surfaces that interpolate. ACM Trans. Graph. **21**(4), 855–873 (2002)
5. Ohtake, Y., Belyaev, A., Seidel, H.-P.: A multi-scale approach to 3d scattered data interpolation with compactly supported basis functions. In: Proceedings of Shape Modeling International 2003, Seoul, Korea (2003)
6. Shen, C., O'Brien, J.F., Shewchuk, J.R.: Interpolating and approximating implicit surfaces from polygon soup. In: Proceedings of ACM SIGGRAPH 2004, pp. 896–904. ACM Press, New York (2004). http://graphics.cs.berkeley.edu/papers/Shen-IAI-2004-08/
7. Kolluri, R., Shewchuk, J.R., O'Brien, J.F.: Spectral surface reconstruction from noisy point clouds. In: Symposium on Geometry Processing, July, pp. 11–21. ACM Press, New York (2004). http://graphics.cs.berkeley.edu/papers/Kolluri-SSR-2004-07/

Volumetric Methods for Surface Reconstruction and Manipulation

17

In the previous chapter, we discussed a method for surface reconstruction based on the implicit representation, i.e., a representation where the surface S is given as the set of points $\mathbf{x} \in \mathbb{R}^3$ such that $\Phi(\mathbf{x}) = \tau$, where τ is denoted the *isovalue*. As a convention, we usually choose $\tau = 0$. Another convention is the choice that $\Phi > 0$ outside the surface and $\Phi < 0$ inside.

In the previous chapter, Φ was represented by a set of radial basis functions. The central idea connecting the material in this chapter is that we can represent Φ as a dense collection of samples on a regular 3D grid as illustrated in Fig. 17.1 (right) for the 2D case. To change S we only need to change the samples defining Φ. If a sample (a pixel in 2D) in a uniformly positive region is made negative, a small lump of material is created. Thus, while changes to the surface are made indirectly, it is easy to make changes because each sample controls Φ, and hence the shape, only in a very small region of space.

In this context, the samples are usually denoted *voxels* in analogy to the pixels of a regular 2D grid, and the grid is often called a *volume* while this type of implicit representation is generally termed the *volume representation* or *volumetric representation*.

Initially, in Sect. 17.1 a volumetric method for reconstruction of surfaces from point clouds is presented. This method requires that we have a local linear approximation of the surface at each point via a point normal. In Sect. 17.1.1 we discuss the topic of computing point normals. Next, in Sect. 17.2 we discuss and compare to a similar method, which has gained broad acceptance, namely the *Poisson Reconstruction Method*.

In Sect. 17.3 we turn to the *Level Set Method*. The LSM may also be used for reconstruction, but its main use is manipulating existing surfaces. When it comes to surface manipulation, the main advantage of the implicit representation is that it is easier to deal with changes in topology. This may not be obvious, but hopefully the illustration in Fig. 17.2 makes it more clear. If we represent a surface explicitly, for instance as a triangle mesh, changes to topology involve cutting holes in the object and connecting the boundary loops of the holes with a tubular surface. If a surface is implicitly represented, it is given as the level set or isosurface of a function. When

J.A. Bærentzen et al., *Guide to Computational Geometry Processing*,
DOI 10.1007/978-1-4471-4075-7_17, © Springer-Verlag London 2012

Fig. 17.1 The shape is given implicitly by a function Φ which is represented by its samples on a voxel grid

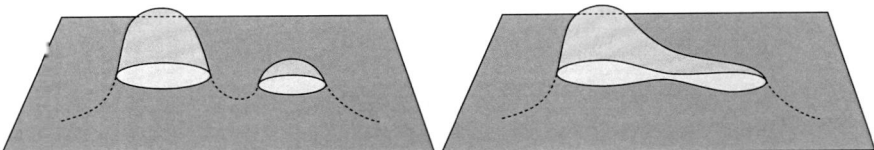

Fig. 17.2 The principle of the implicit surface (or curve) representation is that the surface (or curve) is a level set of a function one dimension higher. The figure illustrates the 2D case where we can think of the function as a terrain height map and of the level set as where the water surface intersects the terrain. By modifying the function (changing the terrain) we indirectly change the level set and thus the implicit curve. Even changes to topology are easy, as illustrated above where the *right* image shows the merge of the two curves resulting from a change to the function

this function is represented by samples on a grid, it is very easy to make these topological changes.

For the purposes of the LSM, the Φ function is usually an (approximate) signed distance field. What this means is that the absolute value of Φ at a point in space is the distance to the closest point on a surface. The sign of Φ then determines whether the point is inside or outside the surface. Finally, in Sect. 17.4 we discuss conversion of triangle meshes to signed distance fields.

17.1 Reconstructing Surfaces by Diffusion

Reconstruction of surfaces from point clouds using radial basis functions as discussed in Chap. 16 ceases to be a manageable approach when we have to deal with many thousand points. This is where volumetric methods come in. Rather than having a few basis functions for each point, we have a 3D grid of voxels. That may seem like a step in the wrong direction (since there are typically many more voxels), but we will no longer need to solve a *dense* linear system. Instead the volumetric methods lead to extremely sparse systems which we can solve with simple iterative schemes.

Fig. 17.3 Samples of the distance to surface. Light means *outside* and dark means *inside*. The monotone pale *blue areas* are of unknown status

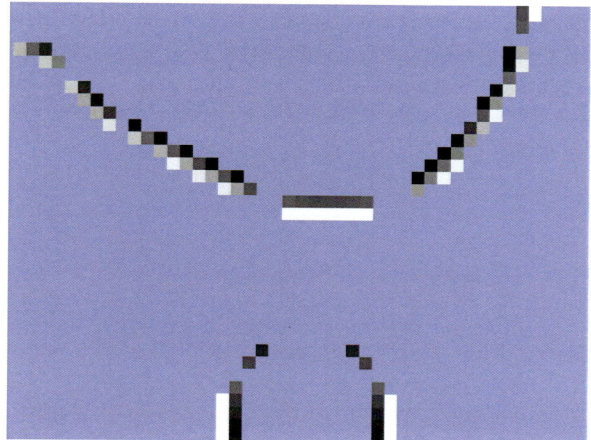

Very close to the surface, we assume that the surface is approximately planar. Of course, "close" is relative, but in practice a distance on the order of the distance between adjacent voxels is usually a good choice. We also assume that for each input point we have a normal vector tangent to the plane that approximates the surface at that point. Fortunately, obtaining a normal estimate is usually possible for optical scans, and, indeed, most surface reconstruction methods do require normals. Alternatively, a method for computing surface normals is discussed in Sect. 17.1.1.

With these definitions in place, the procedure is simple: For each voxel, we sample the distance to the plane given by the position of the point and the normal associated with each point that is close. If no points are close, the distance at the voxel is undefined. If several points are close, we compute the weighted average of the distances, typically using a Gaußian kernel.

Thus, we obtain a voxel grid where most voxels have unknown distances but a set of voxels close to the original points are seeded with distance values. See Fig. 17.3 for a 2D example. The next step is to extend the distance function from the sparse set of voxels near the input points to the rest of the voxels in the volume. A method for computing the distance values in the entire volume was suggested by Davis et al. [1]. The basic idea is to keep the values of the known voxels fixed as constraints while blurring the volume to obtain the values at all other voxels.

A simple example of a 2D implementation of a similar[1] algorithm is shown in Fig. 17.4. The black and white image has been sampled at locations with high gradient values. Some of the sampled pixels are shown in Fig. 17.3. It is clear from this figure that the contour of the object is not entirely defined by known pixels: we can go from inside to outside by a path consisting entirely of pixels where the value of Φ is unknown. It is reasonable to ask how we can be sure that the gaps are filled in the expected way? The answer is that blurring creates a smooth function that interpolates the known pixels and this smooth function will have a smooth zero

---

[1]We say 'similar' because the blurring is done a bit differently, but the principle is the same.

level set that probably matches our expectations. However, if the gaps are too large and the original surface very erratic, we can certainly arrive at a situation where the result is unexpected.

Of course, smooth is a rather vague term. To be more precise, the reconstruction is performed by solving

$$\Delta \Phi = 0 \qquad (17.1)$$

subject to the constraint that Φ is unchanged at the selected samples. If $\Delta \Phi = 0$ we have minimized the membrane energy,

$$\int \|\nabla \Phi\|^2 \, dx \, dy,$$

which is a measure of how smooth the function is. In practice this is simply a principled approach to blurring.

Inserting the discrete version of the Laplacian Δ, we obtain

$$\frac{1}{4}\big(\Phi[i-1,j] + \Phi[i+1,j] + \Phi[i,j-1] + \Phi[i,j+1] - 4\Phi[i,j]\big) = 0. \quad (17.2)$$

This leads to a very sparse linear system we can solve by a simple update procedure for each pixel (or voxel in 3D). However, some care must be taken to ensure stability, and one typically adds a damping constant k. The final update applied to each pixel looks as follows:

$$\Phi[i,j] \leftarrow \Phi[i,j] + \frac{k}{4}\Bigg(\sum_{(l,n)\in N_{i,j}^{2D}} \Phi[l,n] - \Phi[i,j] \Bigg), \qquad (17.3)$$

where

$$N_{i,j}^{2D} = \left\{ \begin{bmatrix} i-1 \\ j \end{bmatrix}, \begin{bmatrix} i+1 \\ j \end{bmatrix}, \begin{bmatrix} i \\ j-1 \end{bmatrix}, \begin{bmatrix} i \\ j+1 \end{bmatrix} \right\}.$$

This formula is iteratively applied to each pixel. However, for the pixels where we have a known value, we simply copy that value back. Putting all of this together, we arrive at the pseudocode shown in Algorithm 17.1. The results of an experiment are shown in the aforementioned Fig. 17.4.

In a volumetric setting, the formula is nearly identical except that we now have six neighboring voxels instead of four neighboring pixels, so

$$\Phi[i,j,k] \leftarrow \Phi[i,j,k] + kL[i,j,k], \qquad (17.4)$$

where

$$L[i,j,k] = \frac{1}{6} \sum_{(l,n,m)\in N_{i,j,k}^{3D}} \Phi[l,n,m] - \Phi[i,j,k], \qquad (17.5)$$

Algorithm 17.1 Pseudocode for 2D reconstruction

```
for (N iterations)
{
  for(each known pixel at i,j)
        Phi(i,j) = known_pixels(i,j);

  for(each pixel i,j)
  {
   L = 0
   for(l,n in N_2D(i,j))
     L += Phi(l,n)/4.0;
   L -= Phi(i,j);

   Phi_tmp(i,j) = Phi(i,j) + k * L;
  }
  Phi = Phi_tmp;
}
```

and

$$N_{i,j,k}^{3D} = \left\{ \begin{bmatrix} i-1 \\ j \\ k \end{bmatrix}, \begin{bmatrix} i+1 \\ j \\ k \end{bmatrix}, \begin{bmatrix} i \\ j-1 \\ k \end{bmatrix}, \begin{bmatrix} i \\ j+1 \\ k \end{bmatrix} \begin{bmatrix} i \\ j \\ k-1 \end{bmatrix}, \begin{bmatrix} i \\ j \\ k+1 \end{bmatrix} \right\}.$$

Thus, for $k = 1$ the algorithm simply amounts to iteratively replacing each voxel with its neighbors while keeping voxels of known value fixed. However, Algorithm 17.1 would converge very slowly and be quite sensitive to noise. These two issues will be addressed in the following.

Any acquisition method is subject to measurement noise. Volumetric methods for surface reconstruction from points are relatively resilient to noise. However, in some cases we get actual outliers, i.e., points which are far from the other points due to error—not because the point lies on a small feature captured by the scanner. The Laplacian smoothing algorithm just discussed is likely to create small spurious surface components if outliers are present as illustrated in Fig. 17.6. Thus, a practical algorithm should remove these outliers.

For the examples in this chapter, we find the neighbor set of points within some radius of a given point, **p**. If the number of neighbors is very small, we take it that **p** is isolated and reject it. Otherwise, we compute the average of the neighbors. If the distance from **p** to the average is greater than some constant times the distance to the farthest neighbor, we discard the point. Since normal estimation also requires us to find the set of points close to a given point, it is convenient to remove outliers as a part of the normal estimation, which is discussed below in Sect. 17.1.1. A different approach is adopted in the CGAL package: points are sorted in increasing order of average square distance to the n nearest neighbors [2]. With this ordering, the outliers are the points at the end of the list, and one may easily reject a user specified

Fig. 7.4 The original image (*top left*) was sampled as described in the text. Solving a discrete version of the Laplace equation with these values as constraints, we obtain the *top right* image after 100 iterations, the *bottom left* image after 50000 iterations. Thresholding the *bottom left* image produces a fairly good reconstruction shown *bottom right*

percentage of the points. It is observed that the constant n should be greater than the size of clusters of outliers. Clearly, both approaches require tweaking parameters since, unfortunately, there is no simple test for whether a point is an outlier. To some extent it is an arbitrary choice how aggressively to remove points.

An important performance concern is that the algorithm might take many iterations to converge. In the 2D example in Fig. 17.4, 50000 iterations were used. If the algorithm is run at a single resolution, it is likely that we might need the same order of iterations for 3D reconstruction. However, if we use a multi-resolution approach, we can do much better. For instance, if we run the algorithm on a very coarse voxel grid, say $8 \times 8 \times 8$ voxels, we can obtain a coarse solution in just a few iterations. Now, we can double the resolution of the coarse solution to $16 \times 16 \times 16$ voxels by linearly interpolating the voxels of the coarse grid and then use this interpolated solution to initialize the algorithm. Since we are already close to a solution, it typically takes only a few iterations to get a good result at which point we again double the resolution until the voxel grid is fine enough to capture the details of the model.

Algorithm 17.2 Pseudocode for 3D reconstruction

```
DIM = MIN_DIM
while(DIM<=MAX_DIM)
{
  for (N iterations)
  {
    for(each point p with normal n)
    {
      Plane P(p,n);
      for(each voxel i,j,k close to p)
        Phi(i,j,k) = P.distance(i,j,k)
    }
    for(each voxel i,j,k)
    {
      L = 0
      for(l,n,m in N_3D(i,j,k))
        L += Phi(l,n,m)/6.0;
      L -= Phi(i,j,k);

      Phi_tmp(i,j,k) = Phi(i,j,k) + k * L;
    }
    Phi = Phi_tmp;
  }
  DIM = DIM * 2;
  Phi = double_resolution(Phi);
}
Mesh = polygonize_zero_levelset(Phi);
```

The final step of the algorithm is the isosurface polygonization used to obtain a triangle mesh. Algorithms for isosurface polygonization are the topic of Chap. 18.

Putting the pieces together, we obtain the relatively simple Algorithm 17.2. Results from an implementation of this algorithm are shown in Fig. 17.5. Here 10 iterations are used with a damping constant $k = 1$ (i.e., voxels are replaced by a straight average of their neighbors) for each level or reconstruction except for the final level ($256 \times 256 \times 256$) where only five iterations are used and a smaller damping of $k = 0.8$ is used to keep things sharp.

17.1.1 Computing Point Normals

As mentioned, most reconstruction algorithms require that we have a so-called *normal* associated with each input point. Together with the point itself, the normal defines a local, planar approximation to the surface we are trying to reconstruct.

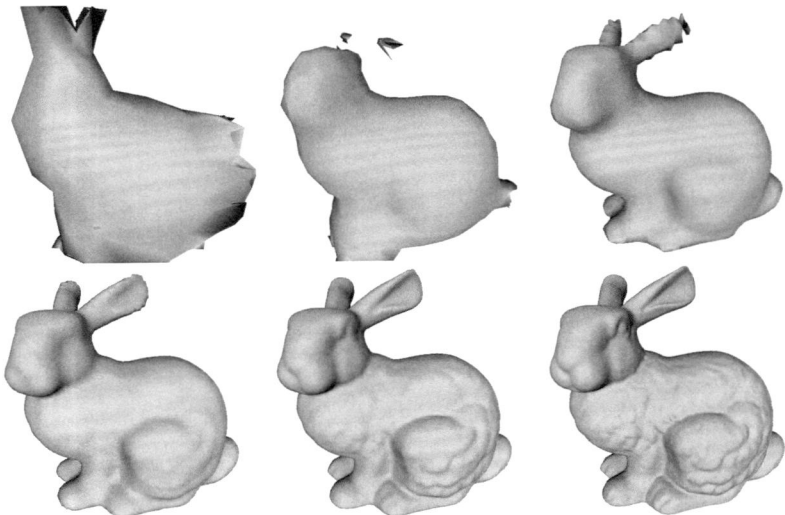

Fig. 17.5 *Top*: The Stanford bunny reconstructed from the original scans at volume resolutions of $8 \times 8 \times 8$, $16 \times 16 \times 16$, and $32 \times 32 \times 32$ voxels. *Bottom*: Reconstructions at resolutions of $64 \times 64 \times 64$, $128 \times 128 \times 128$, and $256 \times 256 \times 256$ voxels

One way of obtaining the normals is to project the points into 2D, compute the Delaunay triangulation (cf. Chap. 14), and then compute the vertex normals using the method described in Sect. 8.1. This is an excellent approach if there is a direction we can use for a 2D projection. Unfortunately, this is not always the case. Moreover, computing the Delaunay triangulation just to obtain normals may be deemed too computationally expensive, and, also, we may want to use a larger neighborhood of points than just those which are connected via the edges of a triangulation in order to obtain a smoother normal estimate.

For these reasons, we typically compute the normal of a given point by least squares fitting a plane to the points in its vicinity. Using a spatial data structure (generally a kD tree, cf. Chap. 12), we can find the set of points within a given radius or, for some number N, the N closest points. Let us denote a point in this set by $\mathbf{p}_i = [\, x_i \; y_i \; z_i \,]^T$.

The plane, P, which we seek, is defined by a point, $\mathbf{p}_o$, lying in the plane and a normal vector, $\mathbf{n}$, perpendicular to the plane. Given these two definitions, the distance to plane is

$$d_P(\mathbf{p}) = \mathbf{n}^T (\mathbf{p} - \mathbf{p}_o). \tag{17.6}$$

Thus $P = \{\mathbf{p} \in \mathbb{R}^3 \,|\, d_P(\mathbf{p}) = 0\}$. The choice of $\mathbf{p}_o$ is easy since we know that the average of the points, $\bar{\mathbf{p}} = [\, \bar{x} \; \bar{y} \; \bar{z} \,]^T = \frac{1}{N} \sum_{i=1}^{N} \mathbf{p}_i$ minimizes the sum of square distances to the plane for any normal direction.

Lemma 17.1 *Given a set of points, $\{\mathbf{p}_i\}$, and a plane, P, with any normal, $\mathbf{n}$, choosing $\mathbf{p}_o = \bar{\mathbf{p}}$ minimizes the sum of square distances, $\sum_{i=1}^{N} d_P^2(\mathbf{p}_i)$, to the plane.*

Proof To simplify things, we let $d = \mathbf{n}^T \mathbf{p}_o$. Since $\mathbf{n}$ and the points are given, the sum of square distances is a quadratic function of d alone:

$$f(d) = \sum_{i=1}^{N} d_P^2(\mathbf{p}_i) = \sum_{i=1}^{N} \left(\mathbf{n}^T (\mathbf{p}_i - \mathbf{p}_o)\right)^2 = \sum_{i=1}^{N} \left(\mathbf{n}^T \mathbf{p}_i - d\right)^2. \qquad (17.7)$$

Taking the derivative, $f'(d) = 2dN + 2\sum_{i=1}^{N} \mathbf{n}^T \mathbf{p}_i$, and requiring that $f'(d) = 0$, we arrive at $d = \mathbf{n}^T \frac{1}{N} \sum_{i=1}^{N} \mathbf{p}_i$. □

We can now turn to finding the normal $\mathbf{n}$ that minimizes the sum of square distances to the plane centered at $\mathbf{p}_o = \bar{\mathbf{p}}$.

Lemma 17.2 *Given a set of points, $\{\mathbf{p}_i\}$, and a plane, P, containing $\mathbf{p}_o = \bar{\mathbf{p}}$, the plane normal, $\mathbf{n}$, which minimizes $\sum_{i=1}^{N} d_P^2(\mathbf{p}_i)$ is the eigenvector corresponding to the smallest eigenvalue of the covariance matrix of $\{\mathbf{p}_i\}$.*

Proof

$$\mathbf{n} = \begin{bmatrix} n_x \\ n_y \\ n_z \end{bmatrix} = \underset{\mathbf{n}}{\arg\min} \sum_{i=1}^{N} \left((x_i - \bar{x})n_x + (y_i - \bar{y})n_y + (z_i - \bar{z})n_z\right)^2$$

$$= \underset{\mathbf{n}}{\arg\min} \sum_{i=1}^{N} \left(\mathbf{n}^T (\mathbf{p}_i - \bar{\mathbf{p}})\right)^2$$

$$= \underset{\mathbf{n}}{\arg\min} \, \mathbf{n}^T \left(\sum_{i=1}^{N} (\mathbf{p}_i - \bar{\mathbf{p}})(\mathbf{p}_i - \bar{\mathbf{p}})^T\right) \mathbf{n}$$

$$= \underset{\mathbf{n}}{\arg\min} \, \mathbf{n}^T \mathbf{M} \mathbf{n}$$

$$= \underset{\mathbf{n}}{\arg\min} \, f(\mathbf{n}), \qquad (17.8)$$

subject to the constraint that $\|\mathbf{n}\| = 1$. Observe that $\mathbf{M} = \sum_{i=1}^{N} (\mathbf{p}_i - \bar{\mathbf{p}})(\mathbf{p}_i - \bar{\mathbf{p}})^T$ is a symmetric positive definite (or semidefinite) matrix, which means that its eigenvectors are mutually orthogonal and its eigenvalues positive or zero. Therefore, we can write $\mathbf{n}$ as a linear combination of its eigenvalues $\mathbf{n} = \sum_{i=k}^{3} \alpha_k \mathbf{e}_k$ where $\mathbf{e}_k$ is the kth eigenvector and α_k the corresponding weight in the linear combination. The unit length constraint can be expressed as $\sum_{k=1}^{3} \alpha_k^2 = 1$. If we let ξ_k be the kth eigenvalue and plug this expansion into $f(\mathbf{n})$, we obtain

$$
\begin{aligned}
f(\mathbf{n}) &= f\left(\sum_{k=1}^{3} \alpha_k \mathbf{e}_k\right) \\
&= \left(\sum_{k=1}^{3} \alpha_k \mathbf{e}_k\right)^T \mathbf{M}\left(\sum_{k=1}^{3} \alpha_k \mathbf{e}_k\right) \\
&= \left(\sum_{k=1}^{3} \alpha_k \mathbf{e}_k\right)^T \left(\sum_{k=1}^{3} \alpha_k \xi_k \mathbf{e}_k\right) \\
&= \sum_{k=1}^{3} \alpha_k^2 \xi_k \\
&\geq \sum_{k=1}^{3} \alpha_k^2 \xi_3,
\end{aligned}
\tag{17.9}
$$

where ξ_3 is the smallest eigenvalue. Clearly, the last inequality is an equality iff $\alpha_3 = 1$ and $\alpha_{\{1,2\}} = 0$. Thus, f is minimal if $\mathbf{n}$ is the eigenvector corresponding to the smallest eigenvalue of $\mathbf{M}$. From its definition, it is clear that $\mathbf{M}$ is indeed the covariance matrix of the set of points $\{\mathbf{p}_i\}$.   □

What we have done is really to apply *principal component analysis* to the points, $\{\mathbf{p}_i\}$ The normal vector, $\mathbf{n}$, is the direction of the axis of least variation in $\{\mathbf{p}_i\}$. From a practical point of view, we normally compute the eigenvectors of the two largest eigenvalues (the axes of greatest variation) and then find the normal as the cross product of these.

One troublesome issue is that the normal orientation is not necessarily consistent. It may be the case that some normals point towards the interior of the shape and some towards the exterior. Fortunately, for any optical scanning device, we know that all points must be visible from the camera. Assuming the camera is at $\mathbf{o}$ and we arbitrarily desire outward pointing normals, we can simply check whether $\mathbf{n}_i \cdot (\mathbf{o} - \mathbf{p}_i) > 0$ where i is now an index that runs over all points, and $\mathbf{n}_i$ is the normal associated with point i. If $\mathbf{n}_i \cdot (\mathbf{o} - \mathbf{p}_i) < 0$, we invert the normal direction.

In some cases, we have mixed points from several scans, misplaced information about the scanning procedure, or, for some other reason, cannot tell whether the normal points towards the interior or the exterior. In these cases, we can simply enforce a consistent orientation on the points. Hoppe suggested a relatively simple graph algorithm for this purpose in a paper which also discusses the above method for computing normals [3]. However, for some data sets, it might be very difficult to find a consistent normal orientation. For instance, imagine a surface that has interior voids. Alliez et al. propose a method [4], which has the advantage that it obtains the surface normal consistently as a part of the reconstruction algorithm. They also discuss various schemes for finding point normals.

17.2 Poisson Reconstruction

The method presented above in Sect. 17.1 is certainly not the only volumetric method for surface reconstruction. In Sect. 11.3 we discussed another method based on Markov Random Fields [5], and, in fact, a fairly large number of volumetric methods for surface reconstruction have been developed since the early work by Hoppe et al. [3].

Probably, the most well known of these algorithms is *Poisson Reconstruction* due to Kazhdan et al. [6]. Poisson Reconstruction is fairly efficient thanks to an adaptive implementation, it has a simple theoretical explanation, and a publicly available implementation. All of these three virtues are contributory to the success of the method.

The idea behind Poisson Reconstruction is to find an indicator function χ, which is defined as a function which is zero outside the surface and one inside. Unfortunately, this means that the gradient of χ is unbounded on the surface and zero elsewhere. Nevertheless, we wish to obtain a function χ whose gradient field matches a vector field $\mathbf{W}$ given by the (interpolated) normals of the input points, i.e., $\nabla \chi = \mathbf{W}$. Consequently, the authors propose to smooth both sides of the equation, obtaining

$$\nabla(\chi * \psi)(\mathbf{x}) = (\mathbf{W} * \psi)(\mathbf{x}),$$
$$\nabla \Phi(\mathbf{x}) = \mathbf{V}(\mathbf{x}), \tag{17.10}$$

where $\mathbf{V} = \mathbf{W} * \psi$, $\Phi = \chi * \psi$, ψ is a smoothing kernel, and $\mathbf{x}$ is the point at which we evaluate. Thus, we need to invert the gradient operator in order to find Φ. Unfortunately, not all vector fields are gradient fields of a scalar field. Specifically, only curl free vector fields are *integrable*, i.e., the gradient fields of scalar functions. To solve this problem, the authors propose solving the problem in the least squares sense which amount to taking the divergence of both sides of (17.10):

$$\nabla \cdot \nabla \Phi(\mathbf{x}) = \nabla \cdot \mathbf{V}(\mathbf{x}),$$
$$\Delta \Phi(\mathbf{x}) = \nabla \cdot \mathbf{V}(\mathbf{x}). \tag{17.11}$$

In words, we have the Laplacian of Φ on the left hand side and the divergence of a vector field, which is a scalar field, on the right hand side. Such an equation is called a Poisson equation, hence the name of the method.

The actual boundary points where the unknown function χ is discontinuous could in theory be computed by deconvolution of Φ, but in practice the authors propose to take an average of the values of Φ at the input points [6]. This average is then used as the isovalue for isosurface polygonization (cf. Chap. 18).

Solving the Poisson equation on a regular grid would be entirely possible and lead to a method very similar to the one from the previous section. In fact, the differences amount to that

- the right hand side of (17.1) was zero and
- the function, Φ, was forced to interpolate given values at voxels close to the input points.

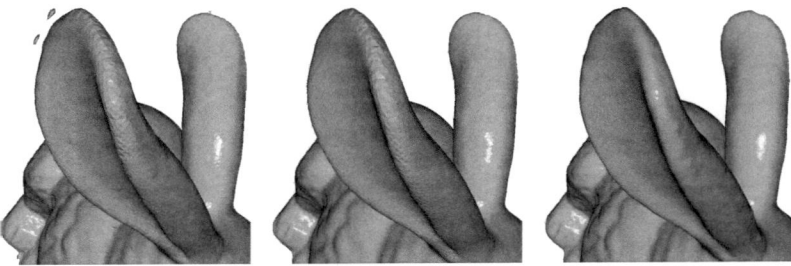

Fig. 17.6 Reconstruction using Laplacian smoothing of the volume and no outlier removal (*left*), the same method with outlier removal (*middle*) and Poisson reconstruction with no outlier removal (*right*). The Poisson method tends to produce a smoother result, but the outliers do not cause spurious components

The Poisson reconstruction method has the advantage that the theory is elegant and the intuition is extremely clear: we are looking for a function whose gradient field matches the point normals. While the isosurface does not precisely correspond to the theoretically correct surface we would get by deconvolution, the method makes up for this potential source of error by representing Φ adaptively using a hierarchical basis of compactly supported functions. This is harder to implement but more space efficient than using a regular voxel grid. It also allows Kazhdan et al. to compute solutions at greater precision. In their paper, the octrees used for the hierarchical representation have depths of up to 10 which corresponds to a voxel grid resolution of 1024^3 [6].

Figure 17.6 compares the method from Sect. 17.1 to Poisson Reconstruction.[2] Outlier removal was deliberately not performed. Consequently, the method based on Laplacian smoothing of the volume contains some spurious surface components due to the outliers. Poisson reconstruction seems to be a bit more robust to outliers. Laplacian reconstruction, however, is simple to implement and works well if outliers are removed.

17.3   The Level Set Method

The reconstruction methods discussed in the first part of this chapter simply produce a volumetric representation of a 3D shape, but once created the shape is static. The Level Set Method (LSM) [7] allows us to deform volumetrically defined shapes. Assume that we are dealing with a surface $S(t) \subset \mathbb{R}^3$ where t is the time parameterization. The surface S is assumed to change according to some *speed function* that pushes S in the normal direction. The speed function may depend on the geometry of S or be completely independent of S. A good example of the latter is a speed function that is always constant causing S to grow by constantly moving in

---

[2]Using MeshLab's implementation: http://meshlab.sourceforge.net/.

the normal direction. A good example of a speed function that does depend on S is one that depends on the curvature of S and pushes the surface in the direction of the curvature center. Such a speed function will smooth the surface and can be very useful.

The Level Set Method tracks the motion of S in the normal direction, and this is expressed by a relationship with an embedding function $\Phi : \mathbb{R}^3 \times \mathbb{R}^+ \to \mathbb{R}$. For all points on S the value of Φ must be zero. This leads to the equation

$$\Phi\big(S(t), t\big) = 0, \tag{17.12}$$

where $S(t)$ denotes a given point on S at time t. Equation (17.12) simply says that $S(t)$ is an isosurface (here called a level set) of $\Phi(\cdot, t)$. Because this holds for any point in time, both S and Φ may evolve but the Level Set Equation continues to hold implying that also

$$d\Phi\big(S(t), t\big)/dt = 0. \tag{17.13}$$

To see how the change of Φ and S are coupled, we take the derivative of (17.12) using the chain rule

$$d\Phi\big(S(t), t\big)/dt = d\Phi\big(Sx(t), Sy(t), Sz(t), t\big)/dt$$
$$= \frac{\partial \Phi}{\partial t} + \nabla \Phi \cdot \frac{dB}{dt}, \tag{17.14}$$

where $\nabla \Phi = \big[\frac{\partial \Phi}{\partial x} \; \frac{\partial \Phi}{\partial y} \; \frac{\partial \Phi}{\partial z}\big]$. Because all motion is in the normal direction, we can write the change of S in terms of a speed function F times the normal, $\frac{\nabla \Phi}{\|\nabla \Phi\|}$,

$$\frac{dS(t)}{dt} = F \frac{\nabla \Phi}{\|\nabla \Phi\|}. \tag{17.15}$$

Plugging this equation back into (17.14), we obtain the Level Set Equation,

$$\frac{\partial \Phi}{\partial t} + F \|\nabla \Phi\| = 0. \tag{17.16}$$

The Level Set Method works on a discrete grid representation of Φ, that is,[3]

$$\Phi^n[i, j, k] = \Phi(i\Delta x, j\Delta y, k\Delta z, n\Delta t).$$

This is a 4D discrete function, but, in general, only one time step is stored. In other words, Φ is really represented as a 3D voxel grid, and the Level Set Method is, essentially, a solver for an initial value problem: given a Φ^0 what is the value at time step n. The initial value, Φ^0 is typically a distance field, but after several iterations of the LSM, Φ drifts from being a distance field and requires reinitialization as we will discuss in Sect. 17.3.2.

---

[3]For simplicity and without loss of generality, we will assume in the following that unit time step is used and that the grid spacing is unit.

17.3.1 Discrete Implementation

An implementation of the LSM on a regular grid requires that de discretize the equations. The first step is to choose discrete operators for the derivatives.

Definition 17.1 (Difference Operators) The forward difference operator is used for approximations of the time derivative:

$$\frac{\partial \Phi}{\partial t} \approx D^{+t}\Phi = \Phi^{n+1}[i, j, k] - \Phi^n[i, j, k]. \tag{17.17}$$

The spatial derivative for x can similarly be approximated using either the forward or the backward difference operator, i.e., either

$$D^{+x} = \Phi^n[i+1, j, k] - \Phi^n[i, j, k] \quad \text{or}$$
$$D^{-x} = \Phi^n[i, j, k] - \Phi^n[i-1, j, k]. \tag{17.18}$$

Of course, the operators for y and z are completely analogous.

Using these definitions, we can discretize (17.16) as follows:

$$\Phi^{n+1}[i, j, k] = \Phi^n[i, j, k] - F\|\nabla\Phi^n\|, \tag{17.19}$$

where the gradient $\|\nabla\Phi^n\|$ must be computed in the *upwind* direction. If $F \geq 0$,

$$\|\nabla\Phi^n\|^2 = \max(D^{-x}, 0)^2 + \min(D^{+x}, 0)^2$$
$$+ \max(D^{-y}, 0)^2 + \min(D^{+y}, 0)^2$$
$$+ \max(D^{-z}, 0)^2 + \min(D^{+z}, 0)^2. \tag{17.20}$$

Conversely, if $F < 0$,

$$\|\nabla\Phi^n\|^2 = \max(D^{+x}, 0)^2 + \min(D^{-x}, 0)^2$$
$$+ \max(D^{+y}, 0)^2 + \min(D^{-y}, 0)^2$$
$$+ \max(D^{+z}, 0)^2 + \min(D^{-z}, 0)^2. \tag{17.21}$$

At first this upwinding seems to be a bit odd; why not simply approximate the gradient with central differences? The answer is that F indicates which way information propagates, and the gradient should be approximated using only voxels that lie in the direction whence information comes. If this principle is not obeyed, the numerical solution can easily become unstable in the presence of discontinuities. A more mathematical explanation is that the upwinding scheme is necessary because Φ might have discontinuities in which case the differential equation does not have a normal solution. However, an integral form of the equation can have a *weak* solution and the upwinding is a part of this weak solution. This is explained in [7] but the discussion of weak solutions is sketchy. To fully appreciate the issues, insight into the field of *conservation laws* [8] is required, but Osher and Fedkiw also provide a more in-depth treatment of the mathematics of the LSM [9].

What time step is appropriate? A condition known as the CFL (Courant Friederichs Lewy) condition asserts that given a first order scheme like the one discussed above, the speed function must obey

$$\max F \leq \frac{\Delta x}{\Delta t}. \tag{17.22}$$

In words, the speed function (in the entire domain) should not exceed the ratio of the spatial grid spacing to the time step. If we consider only grids with unit spacing and unit time step, this reduces to the simple condition that the speed function should not exceed 1. The CFL condition is mentioned by Sethian [7] and explained more deeply by LeVeque [8].

If F depends on the curvature of the evolving surface, discontinuities do not occur because curvature flow in 2D and mean curvature flow in 3D tends to keep things smooth. Sethian suggests using central differences for the first and second order partial derivatives involved in computing the curvature [7].

17.3.2 Maintaining a Distance Field

The Level Set Method does not work well unless we use a method which keeps Φ close to being a distance field. Depending on how we implement the LSM the tendency for Φ to drift away from *distance field'ness* differs, but in all cases, the procedure known as redistancing is generally called for.

Typically, the so-called fast marching method is used for this purpose [10]. However, it is not completely straightforward to implement well as discussed in [11]. It is also not necessarily the fastest option [12]. A very simple alternative is to use the so-called reinitialization equation:

$$\frac{\partial \Phi}{\partial t} + s(\Phi_0)\big(\|\nabla \Phi\| - 1\big) = 0, \tag{17.23}$$

where

$$s(\Phi_0) = \frac{\Phi_0}{\sqrt{\Phi_0{}^2 + \epsilon^2}}, \tag{17.24}$$

where ϵ is a constant often chosen to be about the size of a cell in the grid [13]. The reinitialization equation was introduced by Sussman et al. [14] based on work by Rouy and Tourin [15]. In (17.23), Φ_0 is not quite a distance function but often fairly close. If, for the moment, we ignore s, (17.23) simply links the time derivative of Φ and the length of the spatial gradient of Φ. If the gradient is long, we decrease Φ. If it is short, we increase Φ. After some steps of a discrete implementation, Φ is generally very close to being a distance field.

The sign function $s(\Phi_0)$ is the sign of the original function which must be known for all grid points in advance. Most authors use a sign function which is very small near the interface to avoid moving the interface. Sussman's own choice was (17.24).

Algorithm 17.3 Pseudocode for a simple LSM implementation running simple mean curvature flow

```
float dt = 1.0; // timestep
for(each pixel i,j)
  {
    float d = grid(i,j);
    float l = laplacian(grid, i, j);
    float g = length(grad(grid, i, j));
    tmp_grid(i,j) = d + dt * l * g;
}
grid = tmp_grid;
```

Algorithm 17.4 Pseudocode for the reinitialization equation used to keep Φ close to being a distance field

```
float dt = 0.5; // timestep
for(each pixel i,j)
    sign_grid(i,j) = sign(grid(i,j));

for(each pixel i,j)
  {
    float d = grid(i,j);
    float g = length(grad(grid,i,j));
    float s = sign_grid(i,j);
    tmp_grid(i,j) = d + dt * s * (1.0 - g);
  }
grid = tmp_grid;
```

An important implementation detail is that $\|\nabla\Phi\|$ must be computed in an upwind fashion using the method discussed in the previous section. Pseudocode for an implementation of redistancing using the reinitialization equation is shown in Algorithm 17.4.

17.3.3 Curvature Flow in 2D

To aid in the understanding of the Level Set Method, we provide a simple 2D example. In this example, the speed function is simply the mean curvature. For a 2D distance field, the mean curvature is just the Laplacian. Thus, we are evolving

$$\frac{\partial\Phi}{\partial t} + F\|\nabla\Phi\| = 0 \qquad (17.25)$$

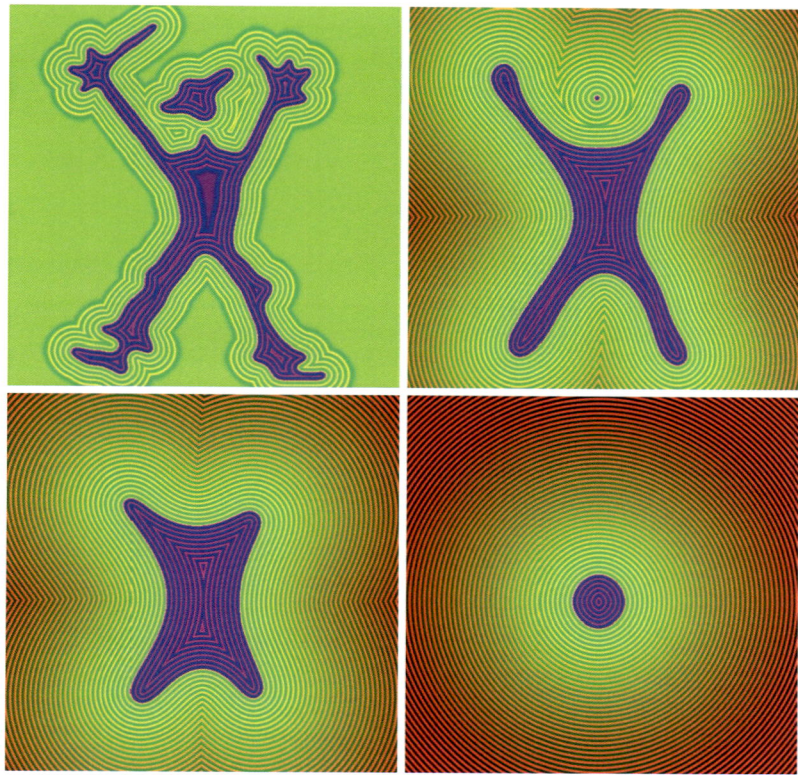

Fig. 17.7 An example of 2D curvature flow. The images show what happens after 30, 660, 1500, and 6030 steps. The *colors* indicate whether we are *inside* or *outside*, and the stripes are added (using a sine function) simply to give an indication of whether the function is indeed a distance function

forward in time where $F = \Delta\Phi$. In pseudocode, one iteration looks as shown in Algorithm 17.3 where grad is implemented in an upwind fashion. The Laplacian is implemented as above. In between every iteration, we run two iterations of reinitialization, which is shown in Algorithm 17.4.

With a few simplifications that do not change the overall picture, this algorithm is precisely what was used to generate the pictures in Fig. 17.7.

17.3.4 3D Examples

While the Level Set Method can be difficult to implement, it is a quite powerful and versatile tool. Some of the possible speed functions are mean curvature which smooths the surface or a constant which expands or contracts the surface. For purposes such as segmentation, we would use a speed function which attracts the surface to certain features in a 2D or 3D image.

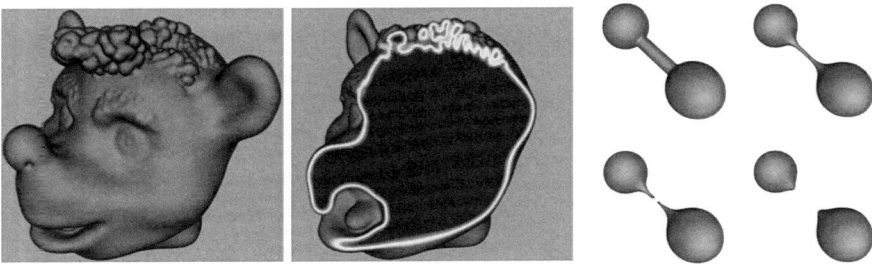

Fig. 17.8 *Left*: a 3D sculpture created using a Level Set Method-based sculpting tool starting from a single cube. *Middle*: the same sculpture but sliced along a plane. The colors indicate how Φ varies. It is clear that voxels farther from the interface than a few voxel grid units are clamped to min and max values. *Right*: Dumbbell deforming under mean curvature flow. Note that it breaks into several pieces

However, the Level Set Method can also be used for 3D sculpting [16]. In this case, we need to be able to make local changes to the model. That is easily done by multiplying the speed function by a spatial window which makes it localized. An example of a model sculpted in this fashion is shown in Fig. 17.8. The main tools used were a constant speed function and mean curvature flow. In both cases, a smooth windowing function was used to restrict the influence, and both positive and negative smoothing was applied: negative mean curvature flow leads to a rather chaotic behavior and was used to sculpt the hair. In the middle image, the model has been sliced with a plane and the value of the Φ function is textured onto the plane. The model was sculpted initially at low volume resolution and then features were added as the resolution was increased.

In Fig. 17.8 (right) we see a dumbbell model deforming under mean curvature flow. The result is that it not only becomes smoother but also breaks into several pieces. This is different from curvature flow in 2D where a shape will never break into several components.

17.4 Converting Triangle Meshes to Distance Fields

The Level Set Method can be initialized such that the level set is a simple geometric shape, say, a sphere. However, in many cases, we would like to convert a triangle mesh or other explicit geometry representation to a signed distance field in order to apply the Level Set Method (LSM). There are also a number of other uses of signed distance fields, cf. [12].

Clearly, computing the distance from a point, $\mathbf{p}$, to a single triangle is a subroutine of any algorithm for computing the distance from $\mathbf{p}$ to a triangle mesh. The distance to a triangle is normally computed by a simple case analysis. First, we compute the closest point, $\mathbf{p}'$, in the plane containing the triangle. If $\mathbf{p}'$ happens to be contained in the triangle, the plane distance is the correct distance. If $\mathbf{p}'$ is outside, the closest point is either an edge or a vertex of the triangle. We can use the same

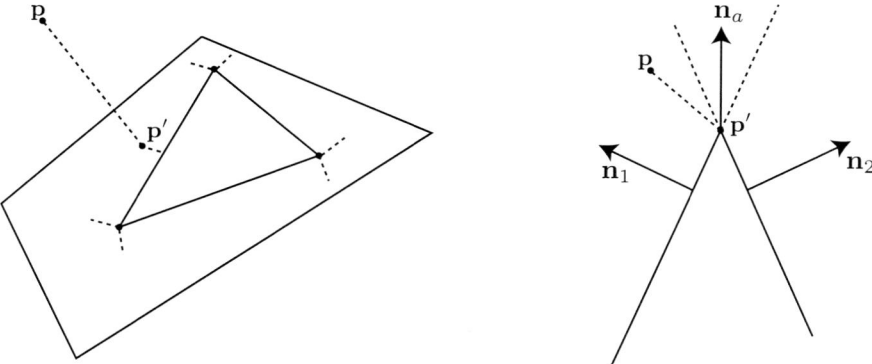

Fig. 17.9 On the *left* **p** has been projected into the plane of a triangle yielding **p**′. A simple case analysis reveals that the closest feature of the triangle is an edge. The *right hand* figure illustrates that the face normals (**n**$_1$ and **n**$_2$) may not give reliable information about whether a point **p** is *inside* or *outside* since the dot products **n**$_{1|2}$ · (**p** − **p**′) have opposite sign. The angle-weighted normal, however, always gives the correct sign. In the figure, **n**$_a$ illustrates the angle-weighted normal

procedure discussed in Delaunay triangulation (see discussion of LeftOf predicate in Sect. 14.3) to test whether the point is inside the triangle.

Otherwise, one or two edges rejected **p**′. Figure 17.9 (left) illustrates the case where a single edge rejected **p**′. We need to project the vertex onto the line containing the rejecting edge to test whether the closest feature is the interior of the edge or one of its vertices. If two edges reject **p**′, the vertex shared by these two edges must be the closest feature. Having found the closest feature and the point on the closest feature, the distance computation amounts simply to computing the distance from the original point **p** to the point on the closest feature.

Of course, not all triangle meshes may be converted to a signed distance field. An important prerequisite is that the triangle mesh must be watertight, i.e., divide space into a part which is outside and a part which is inside. No path may lead from the inside region to the outside without crossing the mesh.

Often, we care only about distances smaller than some threshold. In this case, we say that the distance field is *clamped*. Typically, we clamp to an interval such as [−MAX_DIST, MAX_DIST]. Algorithm 17.5 computes a clamped signed distance field to a triangle mesh. Since the distance is clamped, we only need to visit voxels within a bounding box that contain all points closer to the triangle than MAX_DIST.

The algorithm initially assigns MAX_DIST to every voxel. Next, all triangles are visited, and for each voxel in the bounding box of a triangle, the distance is computed to the triangle. If this distance is smaller than the distance already stored, the new distance is stored.

We compare the absolute value of the distance, since, in the interior, we have negative distances. This leads to the question of how the sign should be computed. The most obvious idea is to simply check on which side of the triangle plane the tested point lies and count the distances negative on one side. This does not work, unfortu-

Algorithm 17.5 Converting a triangle mesh to a clamped, signed distance field

```
for (each voxel i,j,k in grid)
  grid(i,j,k) = MAX_DIST;
for (each triangle T)
{
  for( each voxel i,j,k in bounding_box(T))
  {
    d = T.distance_to(i,j,k);
    if((abs(d) < abs(grid(i,j,k))))
      grid(i,j,k) = d;
  }
}
```

nately. If the closest point is a vertex or an edge, several triangles share that vertex (or edge), and the triangle may lie in front of some and behind others. Figure 17.9 (right) illustrates the issue. Consequently, the normal of the plane containing the triangle is not really useful. Often, the solution has been to use ray casting: we follow a half-line from the voxel at which we desire to compute the sign towards infinity. If we cross the mesh surface just once, we know that we are inside. More generally, an odd number of crossings means that we are inside—and an even number means we are outside. This sounds simple, but in practice it is not trivial to make the method work well. The issue is that the half-line (or ray) may cross the mesh precisely at an edge or a vertex. In these cases, great care must be taken not to count the crossing more than once.

A remedy is the use of the angle-weighted pseudo normal which was discussed in Sect. 8.1. The angle-weighted pseudo normal is defined at the closest feature of the mesh. In other words, if the closest point on the mesh is an edge or a vertex, we do not use a face normal, but instead the angle-weighted pseudo normal at that feature. As discussed, this choice of normal does tell us robustly whether the point is inside or outside the mesh without the need for ray casting.

17.4.1 Alternative Methods

Sean Mauch suggested a somewhat different approach known as *Characteristics Scan Conversion* (CSC) [17] where a cell, denoted a *characteristic*, is associated with each feature of the triangle mesh. These features are similar to Voronoi regions in that they are non-overlapping and enclose the part of space closest to the feature (face, edge, vertex) with which they are associated and the sing of the distance is unambiguously given once we have located the containing characteristic. The CSC method is amenable to GPU implementation [18], however sign issues easily arise and the authors of [18] advocate the use of angle-weighted pseudo normals for correct computation of sign.

If we need an unclamped signed distance field, i.e., the entire voxel grid is contained in the bounding box of every triangle, an altogether different approach is needed to achieve reasonable efficiency. For a given voxel, potentially all triangles could contain the closest distance. To bring that number down, we need a hierarchical representation of the triangle mesh. For more details, the reader is referred to [19] and [20].

17.5 Exercises

Exercise 17.1 (Computing Point Normals) Obtain a point set from the book homepage or an online repository. A triangle mesh where the vertex connectivity is discarded is sufficient. Compute normals for each point using the method described in Sect. 17.1.1 where a given search radius r determines how large a neighborhood should be used. Visualize the normals as short line segments each emanating from their respective points. To avoid dealing with orientation, draw a line segment in both the positive and negative normal direction. Observe how the field of normals becomes smoother as r is increased.

[GEL Users] In the CGAL part of the library, GEL provides functions for computing the exterior product ($\mathbf{ab}^T$) of two vectors as well as a function for computing the eigensolutions to a positive symmetric matrix. GEL also provides appropriate functions for loading meshes in various formats.

Exercise 17.2 (Surface Reconstruction) Implement the diffusion-based reconstruction method from Sect. 17.1 or Poisson Reconstruction on a plain 3D voxel grid using the points and normals from the previous exercise as input. It is advisable to use a fairly low volume resolution (ca. $64 \times 64 \times 64$) since many iterations are needed for convergence on a fine grid.

[GEL Users] In the Geometry part of the library, GEL provides data structures for voxel grids. Also provided are functions for polygonization (cf. Sect. 18).

Exercise 17.3 (2D Level Set Method) Implement the 2D LSM method outlined in Sect. 17.3.3. See if you can smooth a 2D shape represented initially as a 2D black and white image. Use the reinitialization equation to convert the image to a distance field and run iterations of curvature flow as described.

References

1. Davis, J., Marschner, S., Garr, M., Levoy, M.: Filling holes in complex surfaces using volumetric diffusion. In: First International Symposium on 3D Data Processing Visualization and Transmission, 2002. Proceedings, pp. 428–861 (2002)
2. Board, C.E.: CGAL Manuals. http://www.cgal.org/Manual/
3. Hoppe, H., DeRose, T., Duchamp, T., McDonald, J., Stuetzle, W.: Surface reconstruction from unorganized points. Comput. Graph. **26**(2), 71–78 (1992)

4. Alliez, P., Cohen-Steiner, D., Tong, Y., Desbrun, M.: Voronoi-based variational reconstruction of unoriented point sets. In: Proceedings of the Fifth Eurographics Symposium on Geometry Processing, pp. 39–48 (2007). Eurographics Association

5. Paulsen, R.R., Baerentzen, J.A., Larsen, R.: Markov random field surface reconstruction. In: IEEE Transactions on Visualization and Computer Graphics, pp. 636–646 (2009)

6. Kazhdan, M., Bolitho, M., Hoppe, H.: Poisson surface reconstruction. In: Proceedings of the Fourth Eurographics Symposium on Geometry Processing, pp. 61–70 (2006). Eurographics Association

7. Sethian, J.A.: Level Set Methods and Fast Marching Methods, 2nd edn. Cambridge Monographs on Applied and Computational Mathematics. Cambridge University Press, Cambridge (1999)

8. Leveque, R.J.: Numerical Methods for Conservation Laws. Birkhäuser, Boston (1992)

9. Osher, S., Paragios, N. (eds.): Geometric Level Set Methods in Imaging, Vision, and Graphics. Springer, Berlin (2003)

10. Sethian, J.A.: A fast marching level set method for monotonically advancing fronts. Proc. Natl. Acad. Sci. USA **93**(4), 1591–1595 (1996)

11. Bærentzen, J.A.: On the implementation of fast marching methods for 3d lattices. Technical report IMM-REP-2001-13, DTU.IMM (2001). http://www.imm.dtu.dk/~jab/publications.html

12. Jones, M., Bærentzen, A., Sramek, M.: 3D distance fields: a survey of techniques and applications. Trans. Vis. Comput. Graph. **12**(4), 581–599 (2006)

13. Peng, D., Merriman, B., Osher, S., Zhao, H., Kang, M.: A PDE-based fast local level set method. J. Comput. Phys. **155**(2), 410–438 (1999)

14. Sussman, M., Smereka, P., Osher, S.: A level set approach for computing solutions to incompressible two-phase flow. J. Comput. Phys. **114**(1), 146–159 (1994)

15. Rouy, E., Tourin, A.: A viscosity solutions approach to shape-from-shading. SIAM J. Numer. Anal. **29**(3), 867–884 (1992). doi:10.1137/0729053

16. Bærentzen, J.A., Christensen, N.J.: Interactive modelling of shapes using the level-set method. Int. J. Shape Model. **8**(2), 79–97 (2002)

17. Mauch, S.: A fast algorithm for computing the closest point and distance transform. Technical report caltechASCI/2000.077, Applied and Computational Mathematics, California Institute of Technology (2000)

18. Erleben, K., Dohlmann, H.: Signed distance fields using single-pass GPU scan conversion of tetrahedra. In: Nguyen, H. (ed.) GPU Gems 3, pp. 741–763. Addison–Wesley, Upper Saddle River (2008)

19. Bærentzen, J., Aanæs, H.: Signed distance computation using the angle weighted pseudonormal. IEEE Trans. Vis. Comput. Graph. **11**(3), 243–253 (2005)

20. Guéziec, A.: Meshsweeper: Dynamic point-to-polygonal mesh distance and applications. IEEE Trans. Vis. Comput. Graph. **7**(1), 47–60 (2001)

Isosurface Polygonization

Methods for converting point clouds to an implicit representation were discussed extensively in the past two chapters. In Sect. 17.4 we also discussed a method for converting triangle meshes to the implicit representation (specifically, distance fields). In the present chapter, we consider the inverse operation, namely *isosurface polygonization*. Isosurface polygonization is the general term for algorithms that convert implicit surfaces to polygonal (typically triangle) meshes.

One reason why such algorithms are important in the context of geometry processing is that most of the robust algorithms for reconstruction of surfaces from point data go via an implicit representation. From a practical point of view, this means that a point cloud to triangle mesh pipeline is often composed of two steps: the first step converts a point cloud to an implicit surface, and the second step converts the implicit surface to triangle mesh.

There are numerous methods for isosurface polygonization. In this chapter, we restrict ourselves to the cell-based methods, where space is divided into a number of cells (typically cubes or tetrahedra) and the isosurface is then approximated within each cell. In the following section (Sect. 18.1) we introduce the basic idea of cell-based isosurface polygonization. In Sect. 18.2 we go into details with the Marching Cubes algorithm. Finally, in Sect. 18.3 we discuss the merits of the dual contouring type of isosurface polygonization algorithms.

18.1 Cell Based Isosurface Polygonization

To define the problem precisely, we assume that our surface is a level set,

$$S = \Phi^{-1}(\tau) = \{\mathbf{x} \in \mathbb{R}^3 | \Phi(\mathbf{x}) = \tau\}, \tag{18.1}$$

of a function $\Phi : \mathbb{R}^3 \to \mathbb{R}$ where τ is known as the *isovalue*. In some cases, this surface may not exist or it may have singularities. For instance, if $\Phi(\mathbf{x}) = 0$ in all of space, the set S is simply all of space and not a surface. Fortunately, it can be shown that if the gradient ∇f is defined and non-zero at all points where $\Phi(\mathbf{x}) = \tau$ then $\Phi^{-1}(\tau)$ is a closed 2-manifold surface (cf. Sect. 3.10).

J.A. Bærentzen et al., *Guide to Computational Geometry Processing*,
DOI 10.1007/978-1-4471-4075-7_18, © Springer-Verlag London 2012

Fig. 18.1 An illustration of a
voxel grid. We normally
denote the corners voxels,
and this is where we know the
value of the function that
defines our implicit surface.
The cubes are generally
denoted cells

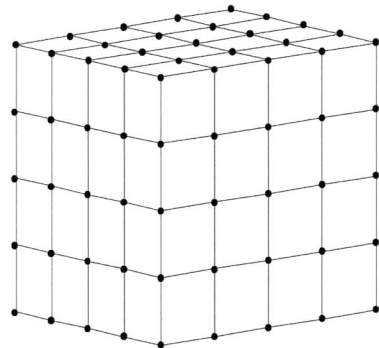

To give a simple example, an ellipsoid may be very compactly represented as
the set of points fulfilling $ax^2 + by^2 + cz^2 - r^2 = 0$. Conversely, approximated by
a triangle mesh, a great number of triangles would be needed to provide a smooth
approximation which would still only be an approximation. However, implicitly represented shapes are sometimes a much less compact representation: medical shape
data often originates from various types of scanning (e.g., MRI or CT scanning)
which produce volume data in the form of voxel grids.

Whatever our implicit representation, we can use the same tools for isosurface
polygonization. If it is necessary, we can obtain a continuous function from a voxel
grid by interpolation. However, that is not necessary in most cases, since the polygonization algorithm may exploit the structure of the voxel grid.

While cell-based approaches (alternatively space decomposition-based approaches) are now common, early approaches for isosurface polygonization looked
at slices of the 3D domain at a time, producing closed curves to approximate the
contour within each slice. Afterwards these contours were stitched together to form
a surface [1]. Another approach which does not use space decomposition is to seed a
triangle somewhere on the implicit surface and then grow the surface by adding adjacent triangles [2]. This can generate a very nice mesh, but special attention needs
to be paid to the situation which arises when two "fronts" meet. Moreover, we do
only get a single surface component per seed point.

Both approaches also tend to be a little more complex than the space decomposition methods on which this chapter will focus. Space decomposition methods divide
space into a number of cells and then polygonize each cell separately. That leaves
open a couple of obvious questions: what cells to use? What kinds of approximation
to use? The most obvious answer to the first question is "cubical cells" since cubes
tile space and it is very easy for a given point to find out which cube contains that
point if we use a regular grid of cubical (or at least rectangular) cells. Moreover, if
we deal with voxel data, we already have a grid of rectangular cells which is clearly
the most obvious choice. A voxel grid is shown in Fig. 18.1.

18.2   Marching Cubes and Variations

Marching Cubes is the name of a well-known algorithm for isosurface polygonization [3] which was preceded slightly by [4]. For a variety of reasons, including that it is a simple, table driven approach, which a competent programmer can quite easily implement, Marching Cubes has become almost synonymous with isosurface polygonization. Like most of the polygonization methods, MC is based on the decomposition of space into a set of cubical cells where the corners of the cells are voxels if our implicit representation is already in the form of a voxel grid. If the implicit surface is given by a some other representation it needs to be sampled at the cell corners.

In either case, to obtain a surface, we approximate the implicit representation with a polygonal surface within each of these cells. Now, given a cubical cell, how do we approximate the surface? To answer that question, we must make some assumptions regarding the surface we polygonize. Most importantly, we assume that we can discover how the surface behaves within the cell simply by looking at the cell corners. For instance, given two cube corners sharing an edge, if $\Phi < \tau$ (the isovalue) at one corner and $\Phi > \tau$ at the other corner, we know that the surface intersects the edge shared by these corners. Thus, to find out whether the isosurface intersects an edge, we just need to classify the corners as above or below the isovalue.

The basic observation in Marching Cubes is that for any two cells, which have the same corner classification, the structure of the surface is the same. Only the precise point location along the edge differs. Thus, we can use a table driven approach where the inside/outside values of each of the eight corners of the cube are used as an index into a polygonization table which tells us what triangles should be generated. As an illustration, in Fig. 18.2 (center) a configuration (shown blue) has been selected based on the corner values. From an implementation point of view, we can observe that a single bit for each of the eight corners can be used to represent the inside–outside state of a corner. Consequently, the table of cell polygonizations should have $256 = 2^8$ elements of "template polygonizations". Due to symmetry, the number of distinct cases is actually much less. However, normally a full table of 256 elements is used since the case analysis is then completely trivial. From an implementation point of view, these templates are usually stored as a collection of triangles. Normally, the triangles form just a single connected component, but in some cases they form two. In principle, we can also store the configuration as a collection of general polygons instead of triangles.

Given a function Φ, we can find out precisely where the surface intersects the edge using a root finding method to find the point where $\Phi = \tau$. Having done this we move the vertices of the template to the intersection points as illustrated in Fig. 18.2 (right).

In the case of voxel data (no known Φ), we simply find the point along the edge where linear interpolation between the values at the corners would yield the isovalue. Let the distance between two voxels be Δ. The linear interpolation, v, of

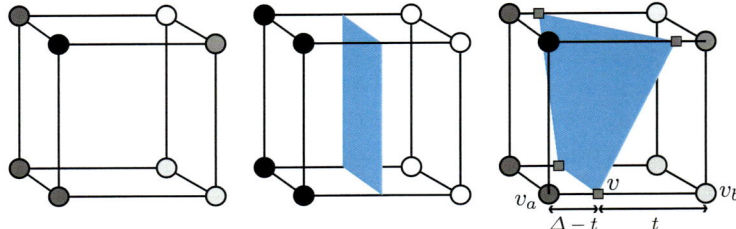

Fig. 18.2 Basic steps of the Marching Cubes algorithm. On the *left* a cell whose eight corners have associated voxel values indicated by *grey* level, and the corners on the *left side* are inside, the ones on the *right* are outside the isosurface. In the *center* image, the appropriate surface which separates the inside corners from the outside corners has been selected (shown *blue*). On the *right*, the vertices of the separating surface have been placed such that the interpolated value at the vertices matches the isovalue (indicated by the *grey color* in the small *squares*)

the values v_a and v_b to a point which is a given by the interpolation parameter t is

$$v = \frac{v_a(\Delta - t) + tv_b}{\Delta}, \tag{18.2}$$

where $t \in [0, \Delta]$ indicates where we wish to interpolate. Now, given an edge connecting voxels with values $v_a < \tau$ and $v_b > \tau$, we can find the precise intersection point by solving for $v = \tau$ in (18.2):

$$t = \Delta \frac{\tau - v_a}{v_b - v_a}. \tag{18.3}$$

This needs to be done for every cell consisting of eight voxels. For an $(L + 1) \times (N + 1) \times (M + 1)$ voxel grid, Marching Cubes proceeds by decomposing space into $L \times N \times M$ cubical cells. For each cell, we classify its corners and use the result to look up the structure of the polygonization for that cell. Once a cell has been polygonized, we proceed to the next cell until all cells have been visited. If we generate a single polygon for each (connected component in a) cell, the result is as shown in Fig. 18.3. However, typically, the result is triangulated as shown in Fig. 18.4.

The use of a table driven approach simplifies the algorithm but it creates the need for a large table whose generation is likely to be error prone. A different approach was proposed by Bloomenthal [5]. The essence of his algorithm is to start from an intersection on an edge and then systematically find the other intersection points while staying on one side of the isosurface. From an algorithmic point of view, this is more complicated, but it does remove the need for a table.

18.2.1 Ambiguity Resolution

While cubical cells have a number of advantages, the choice of cubes entails some ambiguities.

Fig. 18.3 The output of
Marching Cubes if we
produce a single polygon for
each (component in a) cell.
Normally, we produce
triangles, but if we produce
just a single polygon for each
cell, the structure of the
polygonization is more clear:
we note that the edges of the
polygons form three families
of 2D contours

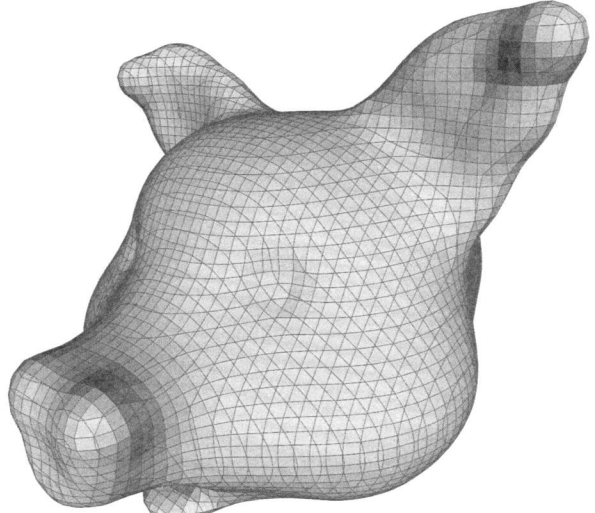

Fig. 18.4 The typical
triangulated Marching Cubes
output

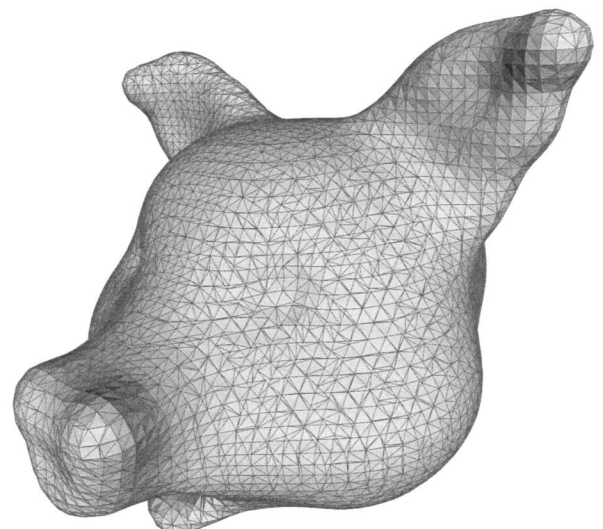

Assume that you are given a cube and the two diagonally opposite corners are
outside the isosurface while the remaining corners are inside. This configuration has
two interpretations. Either the surface has two different components—one for each
exterior corner—or the surface has a tubular hole and the two corners are in the hole.
The 2D analogue is a bit easier to grasp and shown in Fig. 18.5.

The 2D analogue is a problem in 3D triangulation, since it corresponds precisely
to the situation on a cube face which is also open to two interpretations. When we
use an algorithm such as Marching Cubes, we need to consistently choose the same

Fig. 18.5 Ambiguous face
configuration

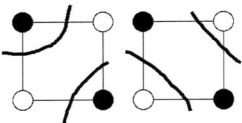

configuration for both polygonization cells sharing a given face. If we always connect or always divide surface components then we obtain a consistent result, but if we do not choose the same face configuration when polygonizing adjacent cells using, e.g., Marching Cubes, we do not get a watertight surface. Nielson and Hamann observed that if we use bilinear interpolation across the face, the intersection curves where the implicit surface cuts the face form two hyperbolic arcs which each join a pair of intersection points [6]. They then find the value of the function where the asymptotes of the two hyperbola branches intersect and decide based on the value of that point whether to separate or connect the surface components. Aptly, this method is known as the *asymptotic decider*.

Apart from consistently choosing a configuration or choosing a configuration based on the asymptotic decider there is one more solution: choose a different type of cell. A tetrahedron is the simplest possible polyhedron, and we can easily divide our cubical cells into tetrahedra. The advantage of using a tetrahedral decomposition is that we can always separate the interior corners from the exterior corners by slicing the tetrahedron with a single plane. The intersection of a tetrahedron and a plane is also relatively simple; it is either a triangle or a quadrilateral, whereas the intersection of a cube and a plane can be a more complex polygon.

The next question is how we obtain these tetrahedra. Perhaps the most frequently used method is to insert a corner at the center of the cube dividing it into six pyramids. Each pyramid can then be divided into a tetrahedron by slicing it along the diagonal of the bottom. That leads to a total of twelve tetrahedra.

However, a cube can also be divided into just five tetrahedra: one tetrahedron is formed in the center of the cube by connecting a diagonal in the bottom face of the cube with the perpendicular diagonal in the top face. The remaining four corners of the cube then form tetrahedra together with the faces of the central tetrahedron.

Unfortunately, using tetrahedra rather than cubes leads to more triangles—and not necessarily better shaped triangles. Hence, since the ambiguities inherent when using cubical cells can be resolved, it is not clear that it is better to use tetrahedra.

18.3   Dual Contouring

The main issue with the contouring methods described so far is that they always place vertices on the edges of the polygonization cells and the isosurface may intersect an edge very close to a corner. If that happens, very small and often poorly shaped triangles are the result. Such triangles are abundant in the example in Fig. 18.4. A solution to this problem is called dual contouring. Probably, the idea was first envisioned by Frisken who called the method surface nets and used it on binary voxel data [7].

Fig. 18.6 Dual contouring
before vertices have been
pushed onto the surface

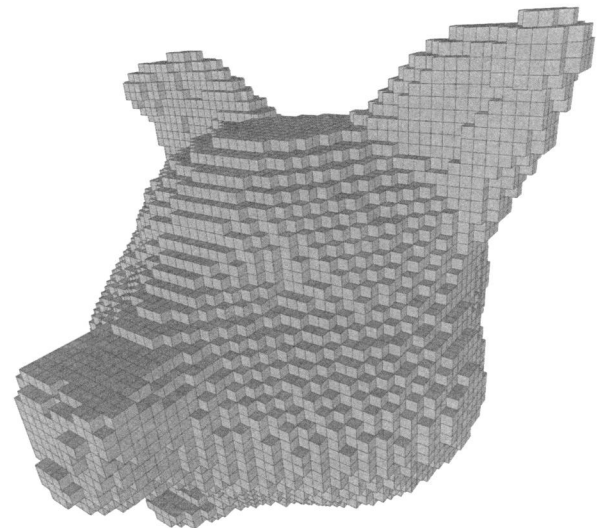

Fig. 18.7 A 2D illustration
of two cubes around voxels at
positions $[i, j, k]$ and
$[i + 1, j, k]$ which are on
either side of the isosurface.
The *bold line* is the interface
between the two cubes which
the polygonizer will output

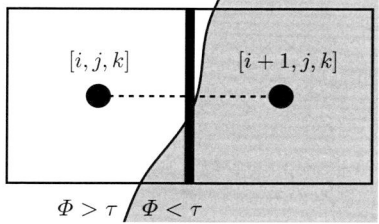

The idea is to place a cube around each voxel in a regular 3D voxel grid. The cube should have a side length equal to the distance between two voxels and be centered exactly on the location of the voxel. These cubes are dual to the cubes used as polygonization cells since in a polygonization cell, the voxel are at the corners rather than centers. To avoid confusion in the following, we will refer to the cubes centered on voxels simply as *cubes*. The cubes are dual to the polygonization cells (whose voxels are corners), which we will denote *cells*.

Now, if a cube belongs to a voxel which is inside the isosurface and it shares a face with a cube that is outside, that shared face belongs to our polygonization. Thus, we proceed by visiting all cube faces and simply output the faces that are shared by cubes on opposite sides of the surface. The result is a sugar cubes model as shown in Fig. 18.6.

To give a very simple example, assume that a voxel at position i, j, k in the voxel grid is inside the isosurface, say $\Phi(i, j, k) > \tau$, where τ is the isovalue, and that its neighboring voxel along the x axis is outside, i.e., $\Phi(i + 1, j, k) < \tau$. This situation is illustrated in Fig. 18.7.

Then a dual contouring polygonizer would output a quadrilateral face with the four corners

$$[i + 0.5, j, k] + [0, -0.5, -0.5],$$

$$[i + 0.5, j, k] + [0, 0.5, -0.5],$$

$$[i + 0.5, j, k] + [0, 0.5, 0.5],$$

$$[i + 0.5, j, k] + [0, -0.5, 0.5]$$

Thus for each voxel where $\Phi > \tau$ we check all six neighboring voxels in the positive and negative X, Y, and Z directions. For each neighbor where $\Phi < \tau$ we output a quadrilateral.

The resulting set of quadrilaterals does not form a connected mesh. If we output the coordinates of each corner for each quad, we will simply have a "soup" of unconnected quads. To get a connected mesh, the procedure is to identify quad corners that lie at the same point with a single vertex. Then we can store our quads as indexed faces: i.e., we have a list of vertices and each quad is represented by four vertex indices. This also means that we can do smooth shading since it is possible to compute a normal for each face and then compute a normal for each vertex as the average of the incident face's normals.

However, if we would like a manifold mesh, the problem arises that dual contouring does not remove the consistency issues encountered by MC and related methods. Cases arise where two cubes are connected by a single edge or a vertex. Such configurations are clearly non-manifold (cf. Sect. 5.2). We can solve most of the problems by being consistent, but non-manifold situations will still arise: in particular, a pair of vertices that occur twice in each others 1-ring is a situation that could arise. Instead of simply identifying quad corners at the same geometric position with a single vertex, we can stitch the quads together. From the initial set of quads, we can create an edge-based representation (e.g. halfedge based) where each face is a manifold onto itself. Then, we stitch the edges the edges together. For each boundary edge, we locate another boundary edge with end points at the same geometric location and in opposite order. This pair is then welded together which, in itself, is not a trivial algorithm to implement, because there is a number of special cases to consider. On the other hand, one should but bear in mind that it is necessary only if we need a manifold representation of the mesh, and, for a number of purposes, such as visualization, it is not necessary that the mesh is manifold.

18.3.1 Placing Vertices

So far, we have said nothing about how the cube vertices are placed. By default they are at the centers of the cells, but in general, we just constrain them to remain inside the cells. This means that they are constrained to a 3D region and not to an edge. Consequently, they can be placed much more judiciously leading to a nicer triangulation.

Fig. 18.8 Dual contouring
after vertices have been
pushed onto the surface

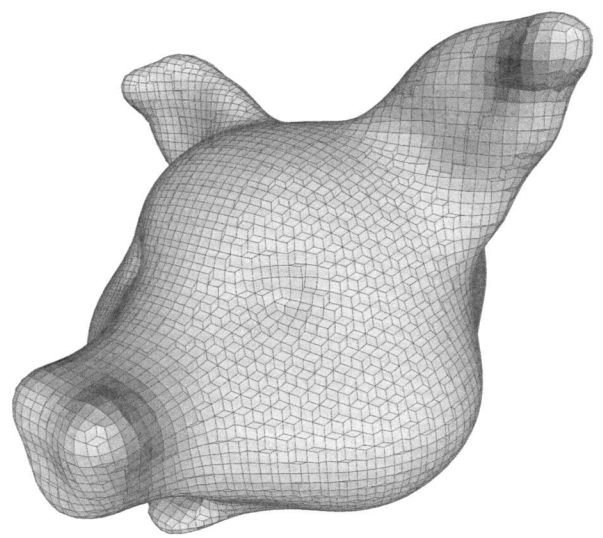

Frisken simply relaxed the mesh by considering all edges to be springs of zero rest length. The vertices then move to tauten the mesh but subject to the constraint that they have to remain inside cells. This approach is particularly useful if we have no real isosurface because we only have a voxel classification, i.e., we know whether a voxel is inside or outside but not its precise value. That is the case, if we are dealing with segmented volume data.

Provided we have a smooth function Φ (e.g., defined by interpolation) we can project the vertices of the mesh onto the isosurface using the method mentioned in Sect. 11.3, which we will describe in more details here. It is simply Newton's method in 3D. Given a function Φ, an estimate of the distance to the τ isosurface at a point $\mathbf{p}$ is

$$\frac{\Phi(\mathbf{p}) - \tau}{\|\nabla\Phi(\mathbf{p})\|}. \tag{18.4}$$

For a linear or an affine function, the above equation is exact. In general, it is simply an approximation. In order to get to the isosurface, we need to multiply by the normalized gradient $\frac{\nabla\Phi}{\|\nabla\Phi\|}$ and subtract from the original point. Plugging in, we obtain

$$\mathbf{x} \leftarrow \mathbf{x} - \big(\Phi(\mathbf{x}) - \tau\big)\frac{\nabla\Phi(\mathbf{x})}{\|\nabla\Phi(\mathbf{x})\|^2}, \tag{18.5}$$

where $\mathbf{x}$ is initialized to be the original cube corner (vertex). Since we start quite close to the isosurface, a few iterations should suffice to place $\mathbf{x}$ on a nearby point on the isosurface. That procedure was used in Fig. 18.8.

Fig. 18.9 Dual contouring
after vertices have been
pushed onto the surface and
the faces have been
triangulated

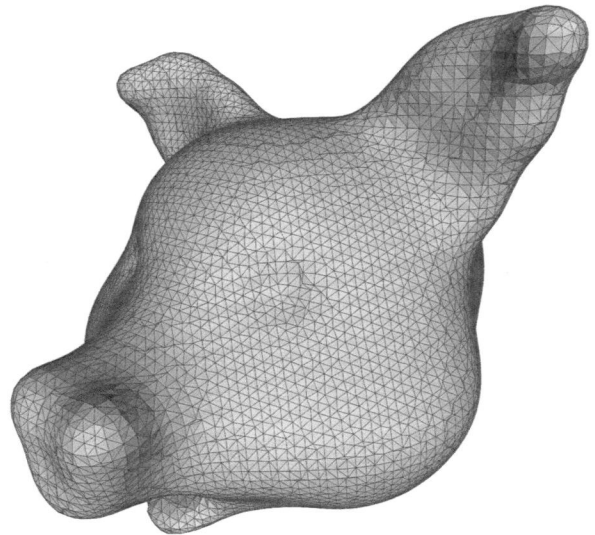

Dual contouring also has the property that it produces quadrilateral faces rather
than triangles. If triangles are desired, it is easy to divide the quadrilaterals along a
diagonal. A sound strategy is to divide along the shortest diagonal. For the example
above, this leads to the result in Fig. 18.9.

The name dual contouring was actually coined in [8] by Tao Ju et al. The method
was a good deal more complex than described above due to the fact that they used
an octree rather than a regular grid of cubes. They also desired to reconstruct surfaces with sharp edges. That is not normally possible with isosurface polygonization
methods because the sharp features do not normally coincide with the vertices generated. However, in dual contouring we can place the vertex anywhere within a cell.
Assuming Φ contains a sharp feature which intersects the cell, we can place our
vertex on that feature. Of course, this requires us to have precise knowledge of the
feature, but that is often the case. For instance, the implicit representation Φ may
represent the intersection of two implicitly defined solids, and in this case we can
detect nearby points on the intersection curve of their surfaces.

18.4   Discussion

A topic not discussed above that often causes some difficulty is the choice of coordinate system. It is natural to assign integer coordinates to voxels, but the function we
polygonize generally does not exist in a coordinate system that precisely matches
these voxel coordinates. Consequently, we often need a function that maps between
voxel coordinates and the coordinates in which our function is defined.

There are a number of well-known strategies for accelerating polygonization
methods. The fact that we need to visit all cells is a little wasteful since typically
only on the order of N^2 cells contain parts of the surface, while we visit N^3 cells.

One strategy for improving on this aspect is to track surface components. Say, we are visiting a polygonization cell which happens to intersect the surface. We then visit only those neighboring cells that share a face which is intersected by the surface. By recursively visiting cells next to cells which contain the surface, we can find all of the cells intersected by the surface, but only for one connected component. Thus, if we know the isosurface has only one component we can get a speedup from visiting far fewer cells (on the order of N^2 rather than N^3 where N is the side length in voxels of the volume).

18.5 Exercises

Exercise 18.1 Implement dual contouring isosurface polygonization. Start by defining a very simple implicit shape such as a sphere and sample the function defining the sphere i.e., $f(x, y, z) = (x - x_c)^2 + (y - y_c)^2 + (z - z_c)^2 - r^2$, where $[x_c, y_c, z_c]^T$ is the center and r is the radius on a regular grid in $\mathbb{R}^3$ (illustrated in Fig. 18.1). Choose center and radius such that the sphere lies in the center of the voxel grid. In this exercise, only the basic algorithm should be implemented and the result will be a sugar cube polygonization like the one in Fig. 18.6. Render the resulting quads.

[GEL Users] A voxel grid data structure and several tools are provided in GEL. A small example program on the book homepage should serve as a starting point for the exercise.

Exercise 18.2 Continuing the exercise above, use a small, fixed number of iterations to project vertices onto the isosurface using (18.5). You should now see a sphere.

Exercise 18.3 Continuing the exercise above, stitch the quads together. This should be done before the vertices are projected onto the isosurface since we then have to do less work. A simple form of stitching is to simply merge vertices which share the same geometric point.

[GEL Users] GEL users can take advantage of the `stitch_mesh` function to do this exercise with a single function call.

References

1. Fuchs, H., Kedem, Z.M., Uselton, S.P.: Optimal surface reconstruction from planar contours. Commun. ACM **20**(10), 693–702 (1977)
2. Akkouche, S., Galin, E.: Adaptive implicit surface polygonization using marching triangles. Comput. Graph. Forum **20**(2), 67–80 (2001)
3. Lorensen, W.E., Cline, H.E.: Marching cubes: a high resolution 3D surface construction algorithm. In: ACM Computer Graphics (1987)
4. Wyvill, B., McPheeters, C., Wyvill, G.: Data structure for soft objects. Vis. Comput. **2**(4), 227–234 (1986)

5. Bloomenthal, J.: Polygonization of implicit surfaces. Comput. Aided Geom. Des. **5**(4), 341–355 (1988)
6. Nielson, G.M., Hamann, B.: The asymptotic decider: Resolving the ambiguity in marching cubes. In: Nielson, G.M., Rosenblum, L.J. (eds.) IEEE Visualization '91, pp. 83–91. IEEE Comput. Soc., Los Alamitos (1991)
7. Gibson, S.F.F.: Constrained elastic surface nets: generating smooth surfaces from binary segmented data. In: First International Conference. Medical Image Computing and Computer-Assisted Intervention—MICCAI'98. Proceedings, pp. 888–898 (1998)
8. Ju, T., Losasso, F., Schaefer, S., Warren, J.: Dual contouring of hermite data. ACM Trans. Graph. **21**(3), 339–346 (2002)

Index

Symbols
1-ring, 84

A
Adaptive generalized kD tree, 218
Adaptive kD tree, 217
Affine combination, 34
Aligning point 3D sets, 268
Angle, 26
Angle weighted pseudo normal, 145
Annealing schedule, 204
Area, 46
Asymptotic directions, 52

B
B-spline, 102
    derivative, 104
Backward difference, 66
Barycentric coordinates, 35, 36, 78
Basis, 17
Basis function, 100
Bernstein polynomials, 18, 110, 111, 116
Bijective, 19
Bilinear interpolation, 78
Boundary conditions, 68, 71, 72
    Dirichlet, 72
    Neumann, 72
Boundary value problem, 65
BSP tree, 219

C
C^n function, 14
C^n map, 15
Caps, 87
Catenoid, 53, 63
Cauchy–Schwartz inequality, 26
Central differences, 66
Characteristic map, 129
Characteristic polynomial, 29, 60
Circle
    empty, 242

Circulators, 92
Circumcircle of edge, 243
Circumcircle of triangle, 243
Closed set, 40
Closure, 40
Compact, 41
Conformal, 181
Conformal map, 48
Consistency, 70
Consistent, 75
Continuous, 41
Continuous map, 41
Control point, 100, 103
Convergence, 70
Convergent sequence, 40
Converges, 75
Convex, 37
Convex combination, 35
Convex combination mappings, 183
Convex hull, 37, 227
Convex set, 227
Convexity, 227
Convolution, 160
Coordinate system, 34
Costa's minimal surface, 53
Courant–Friederichs–Lewy Condition, 76, 300
Covariance matrix, 295
Curvature vector, 50

D
Darboux frame, 50
De Boor's algorithm, 104
Delaunay
    divide and conquer algorithm, 254
    flip algorithm, 246, 253
    geometric primitives, 251
    lifted circle, 249
Delaunay edge, 243
Delaunay Lemma, 245
Delaunay triangle, 243
Delaunay triangulation, 227, 241, 243

J.A. Bærentzen et al., *Guide to Computational Geometry Processing*,
DOI 10.1007/978-1-4471-4075-7, © Springer-Verlag London 2012

FSC
www.fsc.org
MIX
Papier | Fördert
gute Waldnutzung
FSC® C083411

Zeitfracht Medien GmbH
Ferdinand-Jühlke-Straße 7
99095 Erfurt, Deutschland
produktsicherheit@kolibri360.de